On different concepts for the linearization of matrix polynomials and canonical decompositions of structured matrices with respect to indefinite sesquilinear forms

Von der
Carl-Friedrich-Gauß-Fakultät
der Technischen Universität Carolo-Wilhelmina zu Braunschweig

zur Erlangung des Grades eines
Doktors der Naturwissenschaften (Dr. rer. nat.)
genehmigte

Dissertation

von
Philip Martin Saltenberger
geboren am 15.11.1988
in Bad Homburg v. d. H.

Eingereicht am: 29.01.2019
Disputation am: 23.04.2019

1. Referentin: Prof. Dr. Heike Faßbender
2. Referent: Prof. Dr. Volker Mehrmann
3. Referent: Prof. Dr. Paul Van Dooren

2019

Bibliografische Information der Deutschen Nationalbibliothek

Die Deutsche Nationalbibliothek verzeichnet diese Publikation in der
Deutschen Nationalbibliografie; detaillierte bibliografische Daten sind
im Internet über http://dnb.d-nb.de abrufbar.

ISBN 978-3-8325-4914-5

Logos Verlag Berlin GmbH
Comeniushof, Gubener Str. 47,
10243 Berlin
Tel.: +49 (0)30 42 85 10 90
Fax: +49 (0)30 42 85 10 92
INTERNET: https://www.logos-verlag.de

Zusammenfassung

In dieser Arbeit werden Fragestellungen aus der angewandten und numerischen linearen Algebra untersucht.

Reelle *polynomielle Eigenwertprobleme* (PEPs) treten in einer Vielzahl von Anwendungen in Erscheinung [11, 95]. Die Beschreibung derartiger Probleme erfolgt meist über die Angabe eines *Matrixpolynoms* $A(\lambda)$, d. h. eines Polynoms

$$A(\lambda) = \sum_{j=0}^{k} A_j \lambda^j = A_k \lambda^k + A_{k-1} \lambda^{k-1} + \cdots + A_1 \lambda + A_0, \qquad (1)$$

dessen Koeffizienten $A_0, \ldots, A_k$ reelle $m \times n$ Matrizen sind [56]. Die Zusammenfassung der Terme in (1) gestattet es, das Matrixpolynom $A(\lambda)$ als *polynomielle Matrix* $\tilde{A}(\lambda) = [a_{ij}(\lambda)]_{ij}$ mit Einträgen $a_{ij}(\lambda) \in \mathbb{R}[\lambda]$ aufzufassen. Als Grad von $A(\lambda)$ wird der höchste Grad unter den Polynomen $a_{ij}(\lambda)$ verstanden. Da $\mathbb{R}[\lambda]$ einen Unterring des rationalen Funktionenkörpers $\mathbb{R}(\lambda)$ darstellt, lassen sich Matrixpolynome alternativ als Matrizen über $\mathbb{R}(\lambda)$ charakterisieren, wodurch sich viele matrixanalytische Konzepte der linearen Algebra auf Matrixpolynome übertragen lassen [16].

Ist $m \neq n$ oder $\det(\tilde{A}(\lambda)) = 0$, so heißt $A(\lambda)$ singulär, andernfalls regulär. Als (endliche) *Eigenwerte* von $A(\lambda)$ bezeichnet man all jene Skalare $\lambda_0 \in \mathbb{C}$ mit der Eigenschaft, dass der Rang von $A(\lambda_0)$ (aufgefasst als komplexe Matrix) geringer ist als der Rang von $A(\lambda)$ (aufgefasst als Matrix über dem Körper $\mathbb{R}(\lambda)$). Ist $\lambda_0 \in \mathbb{C}$ ein Eigenwert von $A(\lambda)$, so wird der Nullraum von $A(\lambda_0)$ als zugehöriger *Eigenraum* interpretiert. Ist ferner $A(\lambda) = \lambda I_n - B$ für eine $n \times n$ Matrix B, so decken sich diese Definitionen mit denen der Eigenwerte und Eigenräume von B. Für singuläre Matrixpolynome $A(\lambda)$ betrachtet man neben den Eigenräumen oft zusätzlich die (linksseitigen und rechtsseitigen) *Nullräume* von $A(\lambda)$, d.h. Vektoren $x(\lambda) \in \mathbb{R}[\lambda]^n$ mit $A(\lambda)x(\lambda) = 0$ bzw. Vektoren $x(\lambda) \in \mathbb{R}[\lambda]^m$ mit $x(\lambda)^T A(\lambda) = 0$ (oder beides). Ist $x_1(\lambda), \ldots, x_p(\lambda) \in \mathbb{R}[\lambda]^n$ eine Basis des rechtsseitigen Nullraums von $A(\lambda)$, so bezeichnet man die einzelnen Grade der Vektoren $x_1(\lambda), \ldots, x_p(\lambda)$ (aufsteigend geordnet) als rechtsseitige *Minimalindizes* von $A(\lambda)$, sofern keine Basis $\tilde{x}_1(\lambda), \ldots, \tilde{x}_p(\lambda) \in \mathbb{R}[\lambda]^n$ desselben Raumes existiert, deren Gesamtgradsumme ihrer Basisvektoren kleiner ist als die Gesamtgradsumme der Vektoren $x_1(\lambda), \ldots, x_p(\lambda)$ [32]. Linksseitige Minimalindizes zu $A(\lambda)$ sind analog definiert.

Im Kern kann die Lösung eines polynomiellen Eigenwertproblems als Bestimmung der Eigenwerte und Eigenräume eines gegebenen Matrixpolynoms verstanden werden. Gelegentlich wird darunter zusätzlich die Bestimmung der Minimalindizes von $A(\lambda)$ verstanden [31]. Ein populärer Ansatz zur Lösung dieser Probleme ist die *Linearisierung* [32]. Dabei wird (1) in ein lineares Matrixpolynom größerer Dimension $\mathcal{L}(\lambda) = \lambda B_1 + B_0$ mit denselben Eigenwerten überführt. Das zugehörige verallgemeinerte Eigenwertproblem $\lambda B_1 x = -B_0 x$ kann dann beispielsweise mit dem QZ-Algorithmus gelöst werden [71, 99]. Eine zusätzliche wünschenswerte Eigenschaft einer Linearisierung $\mathcal{L}(\lambda)$ von $A(\lambda)$ ist, dass sich die Eigenräume und Minimalindizes von $A(\lambda)$ aus denen von $\mathcal{L}(\lambda)$ auf einfache Weise zurückgewinnen lassen. Zudem ist man im Fall eines strukturierten Matrixpolynoms (z. B. falls $m = n$ gilt und es sich um ein symmetrisches Matrixpolynom $A(\lambda) = A(\lambda)^T$ handelt) an *strukturerhaltenden Linearisierungen* interessiert, die diese Struktur gleichermaßen besitzen. In [89] werden die Vektorräume $\mathbb{L}_1(A), \mathbb{L}_2(A)$ und $\mathbb{DL}(A) = \mathbb{L}_1(A) \cap \mathbb{L}_2(A)$ zu einem gegebenen (quadratischen) Matrixpolynom $A(\lambda)$ definiert. Diese sogenannten *Ansatzräume* ermöglichen die systematische Konstruktion einer Vielzahl von Linearisierungen $\mathcal{L}(\lambda)$ für $A(\lambda)$ mit vielfältigen zusätzlichen Eigenschaften wie z. B. der Strukturerhaltung verschiedenster Matrixpolynomstrukturen. Diese Ansatzräume wurden in einer Reihe wissenschaftlicher Publikationen untersucht und weiterentwickelt, z. B. [65, 87, 101] u. v. m. Der erste Teil dieser Arbeit schließt nach einer Einführung in die Theorie der Matrixpolynome in Kapitel 2 an diese Entwicklung an.

In Kapitel 3 dieser Arbeit wird die Familie der *Block-Kronecker-Ansatzräume* $\mathbb{G}_{\eta+1}(A), 0 \leq \eta \leq k - 1$, für $A(\lambda)$ in (1) eingeführt und untersucht. Wie bei $\mathbb{L}_1(A), \mathbb{L}_2(A)$ und $\mathbb{DL}(A)$ handelt es sich dabei um Vektorräume linearer Matrixpolynome, die als Lösungsmengen linearer Gleichungssysteme (der *Ansatzgleichungen*) auftreten. Die Räume $\mathbb{G}_{\eta+1}(A)$ werden analysiert und Linearisierungen für $A(\lambda)$ darin umfassend charakterisiert. Auf weitere elementare Eigenschaften wie die Rückgewinnung der Eigenräume und Minimalindizes von $A(\lambda)$ unter Kenntnis der Eigenräume und Minimalindizes einer Linearisierung wird gleichermaßen eingegangen. Aufbauend auf den Ergebissen aus [21] und [35] wird auch der Zusammenhang zu Fiedler-Linearisierungen aufgezeigt. Da zu jedem Matrixpolynom $A(\lambda)$ der Form (1) genau k strukturell verschiedene Block-Kronecker-Ansatzräume existieren, können (im Fall $m = n$) deren Schnittmengen untersucht werden. Die *doppelten Block-Kronecker-Ansatzräume* $\mathbb{DG}_{\eta+1}(A)$ bilden erneut Vektorräume und sind definiert als

$$\mathbb{DG}_{\eta+1}(A) := \mathbb{G}_{\eta+1}(A) \cap \mathbb{G}_{k-\eta}(A), \qquad k = 0, \ldots, \lceil k/2 \rceil.$$

Eine genauere Betrachtung dieser Ansatzräume erfolgt in Kapitel 4. Hier wird eine exakte Darstellung der Elemente aus $\mathbb{DG}_{\eta+1}(A)$ präsentiert und die Struktur der Ansatzräume $\mathbb{DG}_{\eta+1}(A)$ beschrieben. Es wird gezeigt, dass die Räume $\mathbb{DG}_{j+1}(A)$ für aufsteigende j eine Folge nichtleerer, ineinander enthaltener Vektorräume bilden. Ferner wird die Konstruktion (block-)symmetrischer Elemente in $\mathbb{DG}_{\eta+1}(A)$ im Zuge der Strukturerhaltung diskutiert. Die Kapitel 3 und 4 basieren in weiten Teilen auf der Publikation [42], in der die Block-Kronecker-Ansatzräume erstmalig definiert wurden.

Kapitel 5 ist der praktischen Lösung verallgemeinerter Eigenwertprobleme unter Berücksichtigung der Ergebnisse aus den Kapiteln 3 und 4 gewidmet. Es wird gezeigt, wie der Einsatz spezieller Linearisierungen aus $\mathbb{G}_{\eta+1}(A)$ zur Eigenwertberechnung in

Krylov-Unterraumverfahren die allgemeine Verfahrenskomplexität reduzieren kann. Detailliert werden diese Ideen für den EVEN-IRA Algorithmus [94] beschrieben, der ein Krylov-Unterraumverfahren zur Lösung speziell strukturierter verallgemeinerter Eigenwertprobleme darstellt. Die praktische Anwendbarkeit dieses Ansatzes wird an numerischen Experimenten belegt. In diesem Zusammenhang wird auch auf weitere Vereinfachungen im Kontext symmetrischer Linearisierungen aus $\mathbb{DG}_{\eta+1}(A)$ für symmetrische Matrixpolynome explizit eingegangen. Dieses Kapitel baut auf den Ergebnissen aus [43] auf. Die Verknüfung der Ergebnisse aus den Kapiteln 3 und 4 mit den Ansatzräumen aus [89] ist Gegenstand von Kapitel 6. Hierbei wird eine alternative Charakterisierung von $\mathbb{L}_1(A)$ und $\mathbb{L}_2(A)$ hergeleitet und auf dieser Grundlage die Isomorphie zwischen speziellen Unterräumen von $\mathbb{L}_1(A)$ und den Block-Kronecker-Ansatzräumen gezeigt. Die neu entwickelte Betrachtungsweise von $\mathbb{L}_1(A)$ und $\mathbb{L}_2(A)$ erweist sich insbesondere bei der Betrachtung von Matrixpolynomen $A(\lambda)$ in anderen Polynombasen als vorteilhaft. Konkret ist die Konstruktion der Ansatzräume $\mathbb{L}_1(A), \mathbb{L}_2(A)$ und $\mathbb{DL}(A)$ speziell für quadratische Matrixpolynome $A(\lambda)$ der Form (1) vorgesehen. Ist $A(\lambda)$ in der Form

$$A(\lambda) = \sum_{j=0}^{k} A_j \phi_j(\lambda) = A_k \phi_k(\lambda) + \cdots + A_1 \phi_1(\lambda) + A_0 \phi_0(\lambda) \tag{2}$$

für eine beliebige Polynombasis $\phi_0(\lambda), \phi_1(\lambda), \ldots$ gegeben, so ergibt eine geeignete Anpassung der Definition von $\mathbb{L}_1(A), \mathbb{L}_2(A)$ und $\mathbb{DL}(A)$ die sogenannten *verallgemeinerten Ansatzräume* $\mathbb{M}_1(A), \mathbb{M}_2(A)$ und $\mathbb{DM}(A) = \mathbb{M}_1(A) \cap \mathbb{M}_2(A)$ zur gegebenen Polynombasis. Diese Räume wurden bereits in [88, 101] definiert und werden für orthogonale Polynombasen in Kapitel 7 untersucht. Insbesondere wird dabei die strukturelle Ähnlichkeit zur Charakterisierung von $\mathbb{L}_1(A)$ und $\mathbb{L}_2(A)$ aus Kapitel 6 herausgestellt. Ferner werden die Rückgewinnung von Eigenräumen und Minimalindizes, die Linearisierung singulärer Matrixpolynome und die Konstruktion strukturerhaltender Elemente diskutiert. Detailliert wird zudem auf die Charakterisierung der Linearisierungen in $\mathbb{DM}(A)$ eingegangen. Die Basis dieses Kapitels ist die Publikation [45].

Ist ein strukturiertes Matrixpolynom $A(\lambda)$ wie in (1) (mit $m = n$) geeignet und strukturerhaltend linearisiert worden ($\mathcal{L}(\lambda) = \lambda B_1 + B_0$), so lässt sich das entstandene verallgemeinerte Eigenwertproblem $\lambda B_1 x = -B_0 x$ oft als Standardeigenwertproblem $-B_1^{-1} B_0 x = \lambda x$ umformulieren. Diese Beobachtung bildet den Ausgangspunkt für die Kapitel 8 bis 10, den zweiten Teil dieser Arbeit, auch wenn die dort vorgestellten Resultate *per se ipsum* unabhängig von der zuvor diskutierten Thematik verstanden werden können. Matrizen der Form $B_1^{-1} B_0$ für Hermitesche bzw. schief-Hermitesche Matrizen B_0, B_1 treten im Zusammenhang mit Sesquilinearformen auf.

Eine Sesquilinearform auf $\mathbb{C}^n \times \mathbb{C}^n$ ist eine Abbildung $[\cdot, \cdot]$ nach $\mathbb{C}$, sodass

$$[u, \lambda v + \mu w] = \lambda[u, v] + \mu[u, w] \quad \text{und} \quad [\lambda u + \mu v, w] = \overline{\lambda}[u, w] + \overline{\mu}[v, w]$$

für alle $u, v, w \in \mathbb{C}^n$ und alle $\lambda, \mu \in \mathbb{C}$ gelten. Jede Sesquilineaform über $\mathbb{C}^n \times \mathbb{C}^n$ kann dargestellt werden als $[u, v] = u^H B v$ für eine feste Matrix $B \in \mathrm{M}_{n \times n}(\mathbb{C})$, wobei B hier stets als nichtsingulär angenommen wird. Ist $B = I_n$ so ergibt sich das Euklidische Skalarprodukt. Zu jeder Matrix $A \in \mathrm{M}_{n \times n}(\mathbb{C})$ ist ihre adjungierte Matrix $A^\star$ eindeutig durch die Gültigkeit der Gleichung $[Au, v] = [u, A^\star v]$ für alle $u, v \in \mathbb{C}^n$ festgelegt. Eine Matrix $A \in \mathrm{M}_{n \times n}(\mathbb{C})$ heißt B-normal, falls $AA^\star = A^\star A$ gilt. Innerhalb der Klasse der B-normalen Matrizen (wie auch in dieser Arbeit) nehmen die selbstadjungierten Matrizen $(A = A^\star)$ und die schiefadjungierten Matrizen $(A = -A^\star)$ eine wichtige Rolle ein. Insbesondere lassen sich Matrizen der Form $B_1^{-1} B_0$ für Hermitesche bzw. schief-Hermitesche Matrizen B_0, B_1 als selbstadjungiert bzw. schiefadjungiert bzgl. der Sesquilinearform $[u, v] = u^H B_1 v$ auffassen. Eine vierte wichtige Matrizenklasse bilden die Automorphismen $Q \in \mathrm{M}_{n \times n}(\mathbb{C})$ bezüglich einer Sesquilinearform, welche den Wert $[u, v]$ beim Übergang zu $[Qu, Qv]$ für alle $u, v \in \mathbb{C}^n$ unverändert lassen. Im Falle des Euklidischen Skalarprodukt sind die selbstadjungierten und schiefadjungierten Matrizen gerade durch die Hermiteschen bzw. schief-Hermiteschen Matrizen gegeben. Automorphismen bezüglich dieser Sesquilinearform nennt man unitär.

In Kapitel 8 werden grundlegende Ergebnisse zur Theorie der Sesquilinearformen bereitgestellt. Dabei werden insbesondere Sesquilinearformen $[u, v] = u^H Bv$ über $\mathbb{C}^n \times \mathbb{C}^n$ betrachtet, die durch eine (komplexe) verallgemeinerte Permutationsmatrix $B \in \mathrm{M}_{n \times n}(\mathbb{C})$ induziert sind. Verschiedene Eigenschaften dieser Klasse von Sesquilinearformen werden dabei speziell entwickelt. Ferner werden elementare Eigenschaften der durch eine Sesquilinearform $[v, w] = v^H Bw$ induzierten B-normalen, selbstadjungierten, schiefadjungierten und automorphen Matrizen präsentiert. B-normale Matrizen bezüglich einer Sesquilinearform werden im nachfolgenden Kapitel 9 unter dem Gesichtspunkt der automorphen Diagonalisierung analysiert. Konkret wird der Frage nachgegangen, wann eine B-normale Matrix $A \in \mathrm{M}_{n \times n}(\mathbb{C})$ mittels eines Automorphismus $Q \in \mathrm{M}_{n \times n}(\mathbb{C})$ durch Ähnlichkeitstransformation $Q^{-1} AQ$ auf Diagonalform gebracht werden kann. Explizite Bedingungen zur automorphen Diagonalisierbarkeit werden hergeleitet, wobei zunächst verschiedene Zusammenhänge zwischen den Eigenräumen einer B-normalen Matrix A und ihrer Adjungierten $A^\star$ aufgezeigt werden. Bilden die Spalten von $U \in \mathrm{M}_{n \times n}(\mathbb{C})$ eine gemeinsame Eigenbasis von A und $A^\star$ so wird die Form der Matrix $U^H BU$ eingehend analysiert und zur Beantwortung der Existenzfrage einer automorphen Diagonalisierung von A herangezogen. Zum Abschluss des Kapitels werden die gefundenen Ergebnisse auf spezielle Sesquilinearformen von praktischer Relevanz angewandt.

Aufbauend auf den Erkenntnissen des neunten Kapitels werden in Kapitel 10 selbstadjungierte und schiefadjungierte Matrizen bezüglich konkreter Sesquilinearformen betrachtet, die zugleich normal im Sinne des Euklidischen Skalarprodukts sind (d. h. es gilt die zusätzliche Eigenschaft $A^H A = AA^H$). Für solche Matrizen existiert stets eine gemeinsame orthogonale Eigenbasis. Es werden kanonische (multiplikative) Faktorisierungen derartiger Matrizen konstruiert und darauf aufbauend kanonische additive Zerlegungen hergeleitet. Es wird u. a. gezeigt wie sich auf der Grundlage dieser Zerlegungen Aussagen über die Existenz einer automorphen Diagonalisierung herleiten lassen.

Contents

List of Symbols

$\mathbb{R}, \mathbb{C}$	Field of real/complex numbers
$\mathbb{R}[\lambda]$	Algebra of polynomials over $\mathbb{R}$
$\mathbb{R}[\lambda]^m$	m-dimensional vector space of vector polynomials
$M_{m \times n}(\mathbb{R}), M_{m \times n}(\mathbb{C})$	Vector Space of real/complex $m \times n$ matrices
$GL_n(\mathbb{R}), GL_n(\mathbb{C})$	General linear group of $n \times n$ matrices over $\mathbb{R}, \mathbb{C}$
$M_{m \times n}(\mathbb{R}[\lambda]), M_{m \times n}(\mathbb{R})[\lambda]$	Vector Space of real $m \times n$ matrix polynomials
$\deg(\cdot)$	Degree of a (vector/matrix) polynomial
$\det(\cdot)$	Determinant of a matrix (polynomial)
$\mathrm{span}(\cdot)$	Linear hull/span of a subset from $\mathbb{R}^n$ or $\mathbb{C}^n$
$\mathcal{S}_n$	Symmetric group on $\{1, \ldots, n\}$
$\mathrm{sgn}(\cdot)$	Parity of a permutation from $\mathcal{S}_n$
I_n	$n \times n$ identity matrix
$0_{m \times n}$	$m \times n$ zero matrix
$\mathrm{rev}_d, \mathrm{rev}$	Reversal of a matrix polynomial
$A(\lambda)^{\mathcal{B}}$	Block-transpose of the matrix polynomial $A(\lambda)$
$\mathrm{rank}(\cdot)$	Rank of a real or complex matrix
$\mathrm{nrank}(\cdot)$	Normal rank of a matrix polynomial
$\lvert \cdot \rvert$	Cardinality of a set, absolute value of a number
$\emptyset$	Empty set
$\sigma_f(A), \sigma(A)$	(finite) spectrum of the matrix polynomial $A(\lambda)$
$\mathrm{geomult}(A, \lambda_0)$	Geometric multiplicity of the eigenvalue λ_0 for the matrix polynomial $A(\lambda)$
$\mathrm{algmult}(A, \lambda_0)$	Algebraic multiplicity of the eigenvalue λ_0 for the matrix polynomial $A(\lambda)$
$\mathrm{Eig}_r(A, \lambda_0), \mathrm{Eig}_\ell(A, \lambda_0)$	Right (left) eigenspace of the matrix polynomial $A(\lambda)$ for the eigenvalue λ_0
$\mathrm{null}(\cdot)$	Kernel/Nullspace of a real/complex matrix or a function
$\dim(\cdot)$	Dimension of a subspace of either $\mathbb{R}^n, \mathbb{C}^n$ or $\mathbb{R}(\lambda)^n$
$\mathcal{N}_r(A), \mathcal{N}_\ell(A)$	Right and left nullspaces of a (singular) matrix polynomial $A(\lambda)$
$\mathrm{ord}(\cdot)$	Order of a vector polynomial basis of some subspace of $\mathbb{R}[\lambda]^n$
SCF	Abbreviation for Smith canonical form
$\mathrm{Frob}_1(\lambda), \mathrm{Frob}_2(\lambda)$	First and second Frobenius companion form (for a matrix polynomial $A(\lambda)$)

GEP	Abbreviation for generalized eigenvalue problem
$\otimes$	Kronecker product of matrices/matrix polynomials
$\mathbb{L}_1(A), \mathbb{L}_2(A), \mathbb{DL}(A)$	Classical Ansatz Spaces of the matrix polynomial $A(\lambda)$
$\mathbb{H}(A), \mathbb{S}(A)$	Hermitian and symmetric ansatz spaces for the matrix polynomial $A(\lambda)$
$\mathbb{EDL}(A)$	Extended Ansatz Space for the matrix polynomial $A(\lambda)$
$\mathbb{B}_1(A), \mathbb{B}_2(A), \mathbb{DB}(A)$	Ansatz Spaces for the matrix polynomial $A(\lambda)$ expressed in the Bernstein basis
$\mathbb{T}_1(A)$	Ansatz Spaces for the matrix polynomial $A(\lambda)$ expressed in the Chebyshev basis
$\mathbb{G}_{\eta+1}(A)$	Block Kronecker Ansatz Space for the matrix polynomial $A(\lambda)$
$\Theta(\mathcal{L}(\lambda))$	Short-hand-notation for $(\Lambda_\eta(\lambda) \otimes I_m)\mathcal{L}(\lambda)(\Lambda_\epsilon(\lambda)^T \otimes I_n)$
$(\mathcal{F}_{\alpha,\eta,A}(\lambda))_{\alpha\in\mathbb{R}}$	Family of anchor pencils/block Kronecker pencils used for the construction of $\mathbb{G}_{\eta+1}(A)$
range$(\cdot)$	Range/Image of a matrix or a function
im$(\cdot)$	Image of a matrix
SBMB pencil	Abbreviation for strong block minimal bases pencil
$\mathbb{DG}_{\eta+1}(A)$	Double Block Kronecker Ansatz Space
$\mathcal{C}(\mathcal{L})$	Core part of a pencil $\mathcal{L}(\lambda)$ in a double block Kronecker Ansatz Space
$\mathbb{BG}_{\eta+1}(A)$	Block-symmetric block Kronecker Ansatz Space
$\mathbb{SG}_{\eta+1}(A)$	Symmetric block Kronecker Ansatz Space
$\Xi(\cdot)$	Fundamental isomorphism between $\mathbb{G}_{\eta+1}(A)$ and $\mathbb{L}_1(A)|_{\text{span}(e_\ell)}$
$\mathbb{M}_1(A), \mathbb{M}_2(A), \mathbb{DM}(A)$	Generalized Ansatz Spaces for the matrix polynomial $A(\lambda)$ expressed in some orthogonal basis
$\Phi = \{\phi_j(\lambda)\}_{j=0}^\infty$	Orthogonal polynomial basis of $\mathbb{R}[\lambda]$
$F_\Phi^A(\lambda)$	Anchor pencil used for the construction of $\mathbb{M}_1(A)$ for the matrix polynomial $A(\lambda)$ expressed in the basis Φ
$[\cdot,\cdot]$	Notation for a arbitrary sesquilinear form
$E(\mu, A)$	Abbreviation for $\text{Eig}_r(\mathcal{L}, \mu)$ with $\mathcal{L}(\lambda) = \lambda I_n - A$ and $A \in \text{M}_{n\times n}(\mathbb{C})$
T/S	Notation for the quotient vector space of T modulo S
$\sim$	Notation for an equivalence relation
$\simeq$	Notation for an isomorphism (in particular for isomorphic vector spaces)
$S^{[\perp]}$	Orthogonal companion of a subspace $S \subseteq \mathbb{C}^n$ with respect to some sesquilinear form $[\cdot,\cdot]$
$\oplus$	Denotes either the direct sum of matrices or of subspaces
$A^\star$	Adjoint of the matrix $A \in \text{M}_{n\times n}(\mathbb{C})$ with respect to some sesquilinear form $[\cdot,\cdot]$
diag$[A]$	Diagonal matrix with the same diagonal as the matrix A

$\mathbb{G}(B)$	Automorphism group with respect to some sesquilinear form $[x, y] = x^H B y$
$\mathbb{J}(B)$	Jordan algebra of selfadjoint matrices with respect to some sesquilinear form $[x, y] = x^H B y$
$\mathbb{L}(B)$	Lie algebra of skewadjoint matrices with respect to some sesquilinear form $[x, y] = x^H B y$
$\mathbb{D}_n(\mathbb{C})$	Set of all $n \times n$ complex diagonal matrices
$\mathcal{N}(S)$	Centralizer of a set $S \subset \mathrm{M}_{n \times n}(\mathbb{C})$
$\widehat{\mathbb{D}}$	Subset of $\mathbb{D}_n(\mathbb{C})$ consisting of diagonal matrices with pairwise distinct entries
$\mathbb{GP}_n(\mathbb{C})$	Set of complex generalized permutation matrices
$i(M)$	Number of nonzero diagonal entries of $M \in \mathbb{GP}_n(\mathbb{C})$
$t(M)$	Abbreviation for $\frac{1}{2}(n - i(M))$ for a Hermitian or skew-Hermitian matrix $M \in \mathbb{GP}_n(\mathbb{C})$
$\mathcal{UP}(2n)$	Set of unitary-perplectic complex $2n \times 2n$ matrices
$\mathcal{US}(2n)$	Set of unitary-symplectic complex $2n \times 2n$ matrices

Dedicated to my family.

Chapter 1

Introduction

Polynomial eigenvalue problems (PEPs) arise in a variety of applications: the solution of systems of differential equations, cf. [56, Chap. 8], the vibration analysis of buildings and vehicles, cf. [76, 125], the theory of damped oscillatory systems, [56, Chap. 13], the dynamical analysis of structures or the study of corner singularities in anisotropic elastic materials, [95]. Due to their diversified applications and practical importance, polynomial eigenvalue problems have received ongrowing attention in the scientific community on numerical linear algebra over the last decades. This resulted in the development of new and efficient numerical solution techniques for PEPs and was also accompanied by extensive research on their theoretical foundation. From this point of view, matrix polynomials and their linearizations lie at the heart of almost any numerical or theoretical analysis of these problems. A cornerstone in the theoretical investigation of matrix polynomials themselves is certainly the publication of the book *Matrix Polynomials* by Israel Gohberg, Peter Lancaster and Leiba Rodman in 1982. From this time on, the consideration of polynomial and generalized eigenvalue problems was extended by the systematic study of matrix polynomials and linearizations in general. These are the concepts the first part of this work is primarily concerned with.

In fact, working with a linearization of the matrix polynomial corresponding to a given PEP is a standard technique for the solution of these problems. Hereby, the linearization process turns the polynomial eigenvalue problem into a generalized eigenvalue problem which is comparatively easier to solve. Furthermore, a well developed theory for generalized eigenvalue problems along with several efficient solution algorithms exist (see [57] and the references therein). Linearizations of matrix polynomials leading to equivalent generalized eigenvalue problems can be constructed in many ways and it is not a priori clear, which one is best in view of numerical issues such as rounding errors or backward error stability etc. Therefore, starting with [89], the systematic construction of linearizations for matrix polynomials with specific properties became an active field of research. In the first part of this work we present a new approach for the construction of linearizations for matrix polynomials. Our linearizations are located in special vector spaces and share a variety of important properties with the linearizations from the classical "ansatz spaces" in [89]. According to the introduction of these ansatz spaces and the recent work [34] (which is an extended version of [35]), we named them "Block Kronecker Ansatz

Spaces". This new family of vector spaces will be comprehensively and completely analyzed in the first part of this thesis. This will shed new light on well-known results for the classical ansatz spaces and provide new insights on features that are special to the block Kronecker ansatz spaces. In particular, incorporating the work from [34] and [21] we will be able to relate our ansatz space framework to the linearization techniques for matrix polynomials based on Fiedler-pencils initiated in [3]. From a numerical point of view we will show, that linearizations from block Kronecker ansatz spaces are particularly valuable for eigenvalue computations based on Krylov subspace methods. In this case, a few eigenvalues of a matrix polynomial can be computed via its linearization and an appropriate Krylov subspace method incorporating a suitable restart strategy at low computational costs. Beside our theoretical investigations, this confers practical importance to the linearizations from block Kronecker ansatz spaces as well. Finally, we will use the results on the block Kronecker ansatz spaces and revisit the concepts from [89] in an extended framework. We consider and comprehensively analyze the classical ansatz spaces for matrix polynomials expressed in nonstandard polynomial bases. This situation has already been analyzed in, e.g., [101]. However, our approach is totally different to the one from [101] and leads to a very simple intrinsic characterization of these spaces. Moreover, it provides a series of simple proofs to various old and new results.

In the second part of this work indefinite sesquilinear forms over the complex numbers are considered. These occur in the study of algebraic Riccati equations [75], time-invariant differential equations, optimal control problems or polynomial eigenvalue problems [55, Sec. 12, 13]. They also arise frequently in infinite-dimensional spaces, cf. [13]. In particular, structured matrices with respect to some indefinite form (such as, for instance, Hamiltonian or symplectic matrices) are omnipresent in mathematics and establish important concepts in applied and theoretical investigations [40, 96]. Under some circumstances, these matrices also occur from the transformation of specially-structured generalized eigenvalue problems into standard eigenproblems. In contrast to the Euclidean case, structured linear transformations in finite-dimensional indefinite inner product spaces are in many regards not as well-behaved as their companions in Hilbert spaces. We will consider linear transformations that are normal with respect to some indefinite sesquilinear form (e.g. matrices coming from its Lie or Jordan algebra) and analyze their properties concerning structure-preserving diagonalization (via similarity transformations) with matrices from the corresponding Lie group. For the special case of Hamiltonian matrices and bilinear forms this problem has been considered in [30]. In particular, the fact that Hermitian and skew-Hermitian matrices are always unitarily diagonalizable (which is a well-known result in the context of the standard Euclidean scalar product) will turn out to be false as soon as selfadjoint and skewadjoint matrices for indefinite sesquilinear forms are considered. We will derive necessary and sufficient conditions on when such a diagonalization is possible. Our conditions show how the structural properties of the indefinite inner product and those of the eigenvalues and eigenspaces of the matrix at hand have to be related for a structure-preserving diagonalization to exist. Moreover, we show how our results apply to certain indefinite forms occurring frequently in practice [83]. Finally, based on these results, we investigate the situation for matrices that are structured with respect to some indefinite form and, additionally, normal with respect to the standard Euclidean scalar product. For this type of matrices we derive new canonical factorizations and, in particular, a new additive decomposition represented by a sum of a normal matrix, its adjoint and a special doubly-structured normal matrix. These results are closely related to the existence of a structure-preserving diagonalization and the question whether the matrix at hand has an invariant Lagrangian subspace. This, in fact, is a question of practical importance in the theory of matrix equations such as algebraic Riccati equations [75].

1.1 Publications

The work presented in this thesis has already partially been published. In particular, the content of Chapters 3, 4, 6 and 7 is mainly based upon the two articles

[42] H. Faßbender and P. Saltenberger, *Block Kronecker Ansatz Spaces for Matrix Polynomials*, Linear Algebra and its Applications, 542 (2018), pp. 118–148.

[45] H. Faßbender and P. Saltenberger, *On Vector Spaces of Linearizations for Matrix Polynomials in orthogonal Bases*, Linear Algebra and its Applications, 525 (2017), pp. 59–83.

although these chapters have been extended by several new results and some of the proofs have been reworked. The new results have been individually listed in Section 2.5 (and within each chapter). The work from Sections 5.2 and 5.3.1 has already been published in

[43] H. Faßbender and P. Saltenberger, *On a modification of the EVEN-IRA algorithm for the solution of T-even polynomial eigenvalue problems*, Proceedings in Applied Mathematics and Mechanics (PAMM), (2019),

while the remaining sections of Chapter 5 extend the basic ideas from [43] to make them applicable in a more general setting. The work from Chapters 8, 9 and 10 has not been published in any scientific journal at the time of this publication.

1.2 Acknowledgement

My first debt of gratitude goes to Heike Faßbender for supervising this thesis. Without her unstinting support since I joined her research group at TU Braunschweig this work would certainly not exist. Furthermore, special thanks are due to all of my colleagues in the numerics group, Matthias Bollhöfer, Anna and Christian Bertram, Lena Vestweber and Tanja Schenk. It's hard to imagine a finer group of people to spend my days with at the office and doing mathematics. Beyond them I am very grateful to Nikta Shayanfar for encouraging and supporting me in so many ways.

I would like to thank Volker Mehrmann and Paul van Dooren for reporting on this thesis. Finally, I am grateful to Peter Benner, Froilán Dopico, Steve Mackey, Javier Pérez Alvaro and Erna Bégovic-Kovac for many inspiring mathematical discussions and common work.

Chapter 2

A primer on matrix polynomials

> *Well, I don't believe in magic, but the closest thing to it, in my opinion, is matrix multiplication.*

Linear Algebra Professor[1]

In this chapter, we present the mathematical background on matrix polynomials and linearizations required for the discussion in Sections 3 to 7. The material presented in Sections 2.1 to 2.3 is standard and can be found in many articles or books on the topic, e.g. [56, 80]. We also refer the reader to [16] which presents the theory of matrices over arbitrary commutative rings certainly applying here as well.

Section 2.1 introduces the basic notion of matrix polynomials along with their elementary properties. The standard terminology on eigenvalues and eigenvectors for matrix polynomials is presented in Section 2.2. Together with two important examples, the Frobenius companion forms, the concept of linearization is motivated and introduced in Section 2.3. An overview of the two standard approaches to linearize matrix polynomials - the ansatz space approach and the Fiedler linearizations - is presented in Section 2.4. Finally, Section 2.5 provides an outlook on the content of the upcoming sections.

2.1 Matrices, matrix polynomials and polynomial matrices

Although we assume that the reader is familiar with the standard concepts in linear algebra, we briefly review the basic tools and notions required in the sequel. The fields of real and complex numbers are denoted by $\mathbb{R}$ and $\mathbb{C}$, respectively. The set of natural numbers (including zero) is denoted by $\mathbb{N}_0$. The notation $\bar{a}$ for some complex number $a = b + ic \in \mathbb{C}$ refers to its complex conjugate $b - ic$. The vector space of all $m \times n$ matrices over $\mathbb{F}$ (with $\mathbb{F}$ being either $\mathbb{R}$ or $\mathbb{C}$) is denoted by $\mathrm{M}_{m \times n}(\mathbb{F})$. For any $A \in \mathrm{M}_{m \times n}(\mathbb{F})$, null($A$) refers to the kernel of A, i.e. the set of all $x \in \mathbb{F}^n$

---

[1]http://mathprofessorquotes.com/tagged/linear-algebra (04-01-2019)

with $Ax = 0_{m\times 1}$. Here and in the following, $0_{r\times s}$ will always denote the $r \times s$ zero vector/matrix whenever the dimensions are necessary to be specified. Otherwise, any zero matrix or vector is simply denoted by 0. The image of A is denoted by $\mathrm{im}(A)$ and consist of all vectors $y \in \mathbb{F}^m$ such that there exists some $x \in \mathbb{F}^n$ with $y = Ax$. For any $A \in \mathrm{M}_{m\times n}(\mathbb{F})$, $\mathrm{im}(A)$ and $\mathrm{null}(A)$ are vector subspaces of $\mathbb{F}^m$ and $\mathbb{F}^n$, respectively. The dimension of $\mathrm{im}(A)$ is called the rank of A and is equal to the number of linear independent columns vectors of A. The multiplicative group of nonsingular matrices in $\mathrm{M}_{n\times n}(\mathbb{F})$ is throughout denoted by $\mathrm{GL}_n(\mathbb{F})$ and is called the general linear group. The notation A^T and A^H is used to denote the transpose and the Hermitian transpose, respectively, of a vector or a matrix. Moreover, the symbols $\oplus$ and $\otimes$ are used to denote the direct sum of matrices and the Kronecker product of vectors/matrices.

Let $A \in \mathrm{M}_{n\times n}(\mathbb{F})$. Any vector $v \in \mathbb{C}^n$ for which there exists some scalar $\mu \in \mathbb{C}$ such that $Av = \mu v$ is called an eigenvector of A corresponding to the eigenvalue μ (of A). The set of all eigenvectors v corresponding to a single eigenvalue μ of A is a subspace of $\mathbb{C}^n$. It is called the eigenspace for μ and is denoted $E(A, \mu)$. The dimension of $E(A, \mu)$ is called the geometric multiplicity of μ (as an eigenvalue of A). All eigenvalues of $A \in \mathrm{M}_{n,n}(\mathbb{F})$ can be obtained as the zeros of its characteristic polynomial $\det(\lambda I_n - A) \in \mathbb{F}[\lambda]$, where $\mathbb{F}[\lambda]$ denotes the algebra of polynomials over $\mathbb{F}$ and I_n is the $n \times n$ identity matrix. The multiplicity of any eigenvalue $\mu \in \mathbb{C}$ of A as a zero of $\det(\lambda I_n - A)$ is called its algebraic multiplicity.

An element $a(\lambda) \in \mathbb{F}[\lambda]$ is called a (scalar) polynomial over $\mathbb{F}$. It is said to have degree $k \in \mathbb{N}_0$ if $a(\lambda)$ can be expressed as $\sum_{j=0}^k a_j \lambda^j$ for some coefficients $a_0, \ldots, a_k \in \mathbb{F}$ with $a_k \neq 0$. If $a(\lambda)$ has degree $k \geq 0$ we use the notation $\deg(a(\lambda)) = k$. The scalar a_k is then called the leading coefficient of $a(\lambda)$. Moreover, the $(k+1)$-dimensional subspace of $\mathbb{F}[\lambda]$ consisting only of polynomials of degree less than or equal to k is now and then used and denoted by $\mathbb{F}_k[\lambda]$. In particular, $\mathbb{F}_0[\lambda] = \mathbb{F}$.

A real *matrix polynomial* $A(\lambda)$ is an element from $\mathrm{M}_{m\times n}(\mathbb{R})[\lambda]$, the $\mathbb{R}$-vector space of polynomials over $\mathrm{M}_{m\times n}(\mathbb{R})$. That is, for any matrix polynomial $A(\lambda) \in \mathrm{M}_{m\times n}(\mathbb{R})[\lambda]$ there exist matrices $A_0, A_1, \ldots, A_k \in \mathrm{M}_{m\times n}(\mathbb{R})$ so that $A(\lambda)$ has the formal expression

$$A(\lambda) = \sum_{j=0}^k A_j \lambda^j = A_k \lambda^k + A_{k-1}\lambda^{k-1} + \cdots + A_1 \lambda + A_0 \tag{2.1}$$

for some $k \in \mathbb{N}_0$. The variable λ in (2.1) is usually referred to as the indeterminate whereas $A_0, \ldots, A_k$ are called matrix coefficients for $A(\lambda)$. If $m = n$ the matrix polynomial $A(\lambda)$ is called square whereas it is called rectangular otherwise. In the former case, $\mathrm{M}_{n\times n}(\mathbb{R})[\lambda]$ has the structure of an $\mathbb{R}$-algebra with its standard (scalar) multiplication and addition. If $m = 1$ or $n = 1$, any $A(\lambda) \in \mathrm{M}_{m\times n}(\mathbb{R})[\lambda]$ is referred to as a (row or column) vector polynomial. For m-dimensional column vector polynomials we use the notation $\mathbb{R}[\lambda]^m$ instead of $\mathrm{M}_{m\times 1}(\mathbb{R}[\lambda])$. Certainly $\mathrm{M}_{1\times 1}(\mathbb{R})[\lambda] = \mathbb{R}[\lambda]$.

Now let $A(\lambda) \in \mathrm{M}_{m\times n}(\mathbb{R})[\lambda]$ be given as in (2.1). The degree $\deg(A(\lambda))$ of $A(\lambda)$ is defined to be the highest index $j \in \mathbb{N}_0$ such that λ^j has a nonzero matrix coefficient

A_j in the expression (2.1). The row and column degrees of $A(\lambda)$ are the degrees of its row and column vector polynomials, i.e.

$$a^j(\lambda) := \sum_{s=0}^{k} \left(e_j^T A_s\right)\lambda^s \in \mathrm{M}_{1 \times n}(\mathbb{R})[\lambda] \quad \text{and} \quad a_t(\lambda) := \sum_{\ell=0}^{k} \left(A_\ell e_t\right)\lambda^\ell \in \mathbb{R}[\lambda]^m$$

for $j = 1, \ldots, m$ and $t = 1, \ldots, n$. Here and subsequently, $e_1, e_2, \ldots, e_\ell$ will always denote standard unit vectors with $e_j \in \mathbb{R}^p, p \geq 1$, having 1 at its j-th position and zeros elsewhere. The transpose $A(\lambda)^T \in \mathrm{M}_{n \times m}(\mathbb{R})[\lambda]$ of $A(\lambda)$ denotes the matrix polynomial obtained from $A(\lambda)$ by the transposition of its matrix coefficients.

Notice that if $\deg(A(\lambda)) = 0$, then $A(\lambda) = A \in \mathrm{M}_{m \times n}(\mathbb{R})$ is independent of λ. Moreover, if $\deg(A(\lambda)) \leq 1$ then $A(\lambda)$ is a constant or linear matrix polynomial which will be called matrix pencil[2]. For linear matrix polynomials $\lambda A_1 + A_0$ the nomenclature matrix pencil and its origin are explained in [106, p. 339] as follows: "The rather strange use of the word pencil comes from optics and geometry: an aggregate of (light) rays converging to a point does suggest the sharp end of a pencil and, by a natural extension, the term came to be used for any *one parameter family* of curves, spaces, matrices or other mathematical objects." If $A(\lambda)$ has degree $k \geq 0$ we define the $m \cdot n$ scalar polynomials

$$a_{ij}(\lambda) := e_i^T A(\lambda)e_j = \sum_{s=0}^{k} \left(e_i^T A_s e_j\right)\lambda^s \in \mathbb{R}[\lambda] \quad i = 1, \ldots, m; \; j = 1, \ldots, n.$$

Now, any matrix polynomial $A(\lambda) \in \mathrm{M}_{m \times n}(\mathbb{R})[\lambda]$ of the form (2.1) can be uniquely identified with the polynomial matrix $\tilde{A}(\lambda) := [a_{ij}(\lambda)]_{ij}$. Here and in the following, $[a_{ij}(\lambda)]_{ij}$ denotes a matrix polynomial having the entry $a_{ij}(\lambda)$ in its (i,j)-th position. Obviously, $\tilde{A}(\lambda)$ belongs to the $\mathbb{R}$-vector space of $m \times n$ matrices with entries from $\mathbb{R}[\lambda]$ which we denote by $\mathrm{M}_{m \times n}(\mathbb{R}[\lambda])$. We have the following result:

Proposition 2.1 ([16, Lem. 7.18]). *For any two numbers $m, n \in \mathbb{N}$, the vector spaces $\mathrm{M}_{m \times n}(\mathbb{R}[\lambda])$ and $\mathrm{M}_{m \times n}(\mathbb{R})[\lambda]$ over $\mathbb{R}$ are isomorphic. A canonical isomorphism is given by mapping $A(\lambda) \in \mathrm{M}_{m \times n}(\mathbb{R})[\lambda]$ to $\tilde{A}(\lambda) = [e_i^T A(\lambda)e_j]_{ij} \in \mathrm{M}_{m \times n}(\mathbb{R}[\lambda])$ and vice versa. Moreover, if $m = n$, both $\mathrm{M}_{m \times n}(\mathbb{R}[\lambda])$ and $\mathrm{M}_{m \times n}(\mathbb{R})[\lambda]$ are $\mathbb{R}$-algebras and isomorphic with respect to their standard addition and (scalar) multiplication.*

In accordance with Proposition 2.1, the terms $A(\lambda) = \sum_{j=0}^{k} A_j \lambda^j \in \mathrm{M}_{m \times n}(\mathbb{R})[\lambda]$ and $A(\lambda) = [a_{ij}(\lambda)]_{ij} \in \mathrm{M}_{m \times n}(\mathbb{R}[\lambda])$ will be used interchangeably. Moreover, we will also refer to elements from $\mathrm{M}_{m \times n}(\mathbb{R}[\lambda])$ as matrix polynomials instead of polynomial matrices or λ-matrices [76]. For any matrix polynomial $A(\lambda) \in \mathrm{M}_{m \times n}(\mathbb{R}[\lambda])$ as in (2.1) the mapping

$$\mathrm{M}_{m \times n}(\mathbb{R}[\lambda]) \ni A(\lambda) \mapsto A(\lambda_0) = \sum_{i=0}^{k} A_i \lambda_0^i \in \mathrm{M}_{m \times n}(\mathbb{C}) \tag{2.2}$$

---

[2]Calling a matrix polynomial $A(\lambda)$ of degree zero a matrix pencil is not standard terminology but serves our purposes well.

is well-defined and produces a complex matrix $A(\lambda_0) \in \mathrm{M}_{m \times n}(\mathbb{C})$ for every $\lambda_0 \in \mathbb{C}$. We say that the matrix polynomial $A(\lambda)$ is *evaluated* at λ_0.

The determinant of matrix polynomials is an important concept defined by the Leibniz formula analogously to real or complex matrices for square matrix polynomials below.

Definition 2.1 (Determinant, [16, Def. 2.16]). *Let $A(\lambda) = [a_{ij}(\lambda)]_{ij} \in \mathrm{M}_{n \times n}(\mathbb{R}[\lambda])$. We define the determinant of $A(\lambda)$ as*

$$\det(A(\lambda)) := \sum_{\sigma \in \mathcal{S}_n} \mathrm{sgn}(\sigma) a_{1\sigma(1)}(\lambda) a_{2\sigma(2)}(\lambda) \cdots a_{n\sigma(n)}(\lambda). \qquad (2.3)$$

In Definition 2.1, $\mathcal{S}_n$ denotes the set of all bijections on $\{1, 2, \ldots, n\}^3$ while $\mathrm{sgn}(\sigma)$ denotes the parity of $\sigma \in \mathcal{S}_n$ [16]. Certainly, $\det(A(\lambda)) \in \mathbb{R}[\lambda]$ as is obvious from (2.3). Moreover, it is immediate that $\deg(\det(A(\lambda))) \leq n \cdot \deg(A(\lambda))$ holds. As usual for matrices over $\mathbb{R}$ or $\mathbb{C}$ we have $\det(A(\lambda)B(\lambda)) = \det(A(\lambda))\det(B(\lambda))$ for two arbitrary square matrix polynomials of the same size [16, p. 14]. Applying Definition 2.1 enables us now to distinguish between certain classes of matrix polynomials:

- A matrix polynomial $A(\lambda) \in \mathrm{M}_{m \times n}(\mathbb{R}[\lambda])$ is called *unimodular* (that is, invertible) if $m = n$ and there exists a matrix polynomial $A(\lambda)^{-1} \in \mathrm{M}_{n \times n}(\mathbb{R}[\lambda])$ such that $A(\lambda)A(\lambda)^{-1} = A(\lambda)^{-1}A(\lambda) = I_n$. It can be shown that $A(\lambda)$ is unimodular iff $\det(A(\lambda)) \in \mathbb{R} \setminus \{0\}$ [16, Cor. 2.21].

- A matrix polynomial $A(\lambda) \in \mathrm{M}_{m \times n}(\mathbb{R}[\lambda])$ is called *regular* if $m = n$ and $\det(A(\lambda)) \in \mathbb{R}[\lambda] \setminus \{0\}$. Although regular matrix polynomials are not necessarily unimodular, they are still invertible considered as matrices over $\mathbb{R}(\lambda)$, the field of rational functions. Thus, the regular matrix polynomials form a multiplicative group in $\mathrm{M}_{n \times n}(\mathbb{R}(\lambda))$. Moreover, it is easily seen that unimodular matrix polynomials form a subgroup among the regular matrix polynomials.

- A matrix polynomial $A(\lambda) \in \mathrm{M}_{m \times n}(\mathbb{R}[\lambda])$ which is not regular is called *singular* [32, Sec. 2]. The singularity of $A(\lambda)$ is thus equivalent to $\det(A(\lambda)) = 0$ if $A(\lambda)$ is square.

As matrices over $\mathbb{R}$ and $\mathbb{C}$, matrix polynomials can be structured. An overview of structures that are frequently arising in practical applications is provided in Table 2.1. Moreover, a nice collection of examples of structured matrix polynomials along with their applications can be found in [11]. Some particular structures require the important notion of the reversal matrix polynomial.

Definition 2.2 (Reversal, [32, Def. 2.12]). *Let $A(\lambda) \in \mathrm{M}_{m \times n}(\mathbb{R}[\lambda])$. We define for any $d \geq \deg(A(\lambda))$*

$$\mathrm{rev}_d A(\lambda) := \lambda^d A(\lambda^{-1}).$$

---

[3] The set $\mathcal{S}_n$ is called *symmetric group* or *permutation group* on $\{1, 2, \ldots, n\}$.

Then $\text{rev}_d A(\lambda)$ *is again a matrix polynomial, i.e.* $\text{rev}_d A(\lambda) \in \mathrm{M}_{m \times n}(\mathbb{R}[\lambda])$. *It is called the d-reversal corresponding to* $A(\lambda)$. *In case* $d = \deg(A(\lambda))$ *we call* $\text{rev} A(\lambda) := \text{rev}_d A(\lambda)$ *the reversal of* $A(\lambda)$.

Considering a matrix polynomial $A(\lambda) \in \mathrm{M}_{m \times n}(\mathbb{R})[\lambda]$ as in (2.1) the reversal $\text{rev} A(\lambda)$ is easily determined as

$$\text{rev} A(\lambda) = \sum_{j=0}^{k} A_{k-j} \lambda^j = A_0 \lambda^k + A_1 \lambda^{k-1} + \cdots + A_{k-1} \lambda + A_k.$$

Structure	Property of $A(\lambda) \in \mathrm{M}_{n \times n}(\mathbb{R}[\lambda])$
symmetric	$A(\lambda)^T = A(\lambda)$
skew-symmetric	$A(\lambda)^T = -A(\lambda)$
palindromic	$A(\lambda)^T = \text{rev} A(\lambda)$
anti-palindromic	$A(\lambda)^T = -\text{rev} A(\lambda)$
T-even	$A(\lambda)^T = A(-\lambda)$
T-odd	$A(\lambda)^T = -A(-\lambda)$

TABLE 2.1: Structures frequently considered for matrix polynomials.

One particular structure of interest later (that is not appearing in Table 2.1) is block-symmetry. It requires the notion of the block-transpose of a particular square matrix polynomial as defined below. Here and in the following, the symbol $\otimes$ always denotes the standard Kronecker product of either matrices or matrix polynomials.

Definition 2.3 (Block-Transpose, [65, Def. 2.1]). *Let* $A(\lambda) \in \mathrm{M}_{kn \times \ell n}(\mathbb{R}[\lambda])$ *for some* $k, \ell, n \in \mathbb{N}$. *Whenever* $A(\lambda)$ *may be expressed as*

$$A(\lambda) = \sum_{i=1}^{k} \sum_{j=1}^{\ell} e_i e_j^T \otimes A_{ij}(\lambda) \tag{2.4}$$

for certain $n \times n$ *matrix polynomials* $A_{ij}(\lambda)$, *we call* $A(\lambda)^{\mathcal{B}} = \sum_{i=1}^{k} \sum_{j=1}^{\ell} e_j e_i^T \otimes A_{ij}(\lambda)$ *the* (k, n)-*block-transpose of* $A(\lambda)$. *We have* $A(\lambda)^{\mathcal{B}} \in \mathrm{M}_{\ell n \times kn}(\mathbb{R}[\lambda])$.

Whenever block-transposition is used in the sequel, the parameters k, ℓ and n will be clear from the context. For this reason, we simply call $A(\lambda)^{\mathcal{B}}$ the block-transpose of $A(\lambda)$. Any $A(\lambda) \in \mathrm{M}_{kn \times kn}(\mathbb{R}[\lambda])$ with $A(\lambda) = A(\lambda)^{\mathcal{B}}$ according to Definition 2.3 is called block-symmetric; whenever $A(\lambda) = -A(\lambda)^{\mathcal{B}}$ holds, $A(\lambda)$ is called block-skew-symmetric.

2.2   Eigenvalues and eigenvectors of matrix polynomials

In this section we introduce the main notions of eigenvalues and eigenvectors for matrix polynomials. Our main reference for this section is [32], although we sometimes confine ourselves to slightly less general definitions that better fit our purposes. The basic tool for most upcoming concepts is introduced in Theorem 2.1 below.

Theorem 2.1 (Smith Canonical Form, [56, Thm. S1.1],[51]). *Let $A(\lambda) \in M_{m \times n}(\mathbb{R}[\lambda])$. Then there exist two unimodular matrix polynomials $W_1(\lambda) \in M_{m \times m}(\mathbb{R}[\lambda])$ and $W_2(\lambda) \in M_{n \times n}(\mathbb{R}[\lambda])$ such that*

$$
W_1(\lambda)A(\lambda)W_2(\lambda) =
\begin{bmatrix}
\begin{matrix}
d_1(\lambda) & 0 & \cdots & 0 \\
0 & d_2(\lambda) & & \vdots \\
\vdots & & \ddots & \vdots \\
0 & \cdots & 0 & d_r(\lambda)
\end{matrix}
& 0_{r \times (n-r)} \\
0_{(m-r) \times r} & 0_{(m-r) \times (n-r)}
\end{bmatrix}
=: D(\lambda)
$$

(2.5)

with scalar polynomials $d_1(\lambda), \ldots, d_r(\lambda) \in \mathbb{R}[\lambda]$ that have leading coefficient equal to one and form a divisibility chain, i.e., for any $d_i(\lambda), d_{i+1}(\lambda)$ there exists some $g_i(\lambda) \in \mathbb{R}[\lambda]$ such that $d_{i+1}(\lambda) = g_i(\lambda)d_i(\lambda)$ holds for $i = 1, \ldots, r - 1$.

For every $A(\lambda) \in M_{m \times n}(\mathbb{R}[\lambda])$ the matrix $D(\lambda)$ in (2.5) is unique [56]. It is named the *Smith canonical form* (SCF) of $A(\lambda)$. In addition, the scalar polynomials $d_1(\lambda), \ldots, d_r(\lambda)$ are called the invariant factors for $A(\lambda)$ [122, Def. 2.3].

The number $r \in \mathbb{N}_0$ in (2.5) is called the normal rank of $A(\lambda)$. It is denoted nrank(A) and can be understood as the rank of $A(\lambda) \in M_{m \times n}(\mathbb{R}[\lambda])$ considered as a matrix over the field of rational functions $\mathbb{R}(\lambda)$ [34]. If nrank($A(\lambda)$) = min$\{m, n\}$ the matrix polynomial $A(\lambda)$ is said to have full (normal) rank. It is immediate from (2.5) that a square matrix polynomial $A(\lambda)$ is regular (singular) iff it has full (deficient) normal rank. The following definition introduces eigenvalues for matrix polynomials by means of the (loss of) normal rank.

Definition 2.4 (Finite Eigenvalue, [32, Def. 2.8]). *Let $A(\lambda) \in M_{m \times n}(\mathbb{R}[\lambda])$. Any scalar $\lambda_0 \in \mathbb{C}$ so that*

$$\text{rank}(A(\lambda_0)) < \text{nrank}(A(\lambda)) \tag{2.6}$$

is called a finite eigenvalue of $A(\lambda)$. The set of all finite eigenvalues of $A(\lambda)$ is subsequently denoted $\sigma_f(A)$ and is called the finite spectrum of $A(\lambda)$.

Certainly $|\sigma_f(A)|$ is always finite, where $|\cdot|$ denotes the cardinality of the set in consideration. In particular, if $A(\lambda) \in M_{n \times n}(\mathbb{R}[\lambda])$ is regular, $\lambda_0 \in \sigma_f(A)$ holds iff rank$(A(\lambda_0)) < n$ (since nrank$(A(\lambda)) = n$), i.e. det$(A(\lambda_0)) = 0$. As a consequence, in the regular case, the eigenvalues $\lambda_0 \in \sigma_f(A)$ depend continuously on the matrix coefficients of $A(\lambda)$. Moreover, in view of (2.2), they depend continuously on λ [76]. This continuous dependence does not hold for singular matrix polynomials in general, see [119]. Do not overlook that, according to Definition 2.4, if $Q(\lambda) \in M_{n \times n}(\mathbb{R}[\lambda])$ is unimodular we have $\sigma_f(Q) = \emptyset$.

Definition 2.5 (Geometric Multiplicity, [32]). *Let* $A(\lambda) \in M_{m \times n}(\mathbb{R}[\lambda])$. *The geometric multiplicity of a finite eigenvalue* $\lambda_0 \in \mathbb{C}$ *of* $A(\lambda)$ *is defined to be*

$$\text{geomult}(A, \lambda_0) := \text{nrank}(A(\lambda)) - \text{rank}(A(\lambda_0)). \tag{2.7}$$

Moreover, we define $\text{Eig}_r(A, \lambda_0) := \text{null}(A(\lambda_0))$ *and* $\text{Eig}_\ell(A, \lambda_0) = \text{null}(A(\lambda_0)^T)$ *to be the corresponding right and left eigenspaces for* λ_0.

The elements of $\text{Eig}_r(A, \lambda_0)$ and $\text{Eig}_\ell(A, \lambda_0)$ are called right (left, respectively) eigenvectors of $A(\lambda)$ for λ_0. Whenever $\lambda_0 \in \sigma_f(A)$ and $A(\lambda)$ is square, notice that $\dim(\text{Eig}_r(A, \lambda_0)) = \dim(\text{Eig}_\ell(A, \lambda_0))$ always holds but

$$\dim(\text{Eig}_r(A, \lambda_0)) = \dim(\text{Eig}_\ell(A, \lambda_0)) \geq \text{geomult}(A, \lambda_0) \tag{2.8}$$

might hold for singular matrix polynomials with strict inequality [32, Def. 2.8]. For instance, consider

$$A(\lambda) = \begin{bmatrix} \lambda(\lambda - 1) & 0 \\ 0 & 0 \end{bmatrix} \in M_{2 \times 2}(\mathbb{R}[\lambda])$$

where geomult$(A, 0) = $ geomult$(A, 1) = 1$ holds but $\text{Eig}_r(A, 0) = \text{Eig}_\ell(A, 0) = \text{Eig}_r(A, 1) = \text{Eig}_\ell(A, 1) = \mathbb{C}^2$. Moreover, although $e_2 \in \text{null}(A(\mu))$ for any $\mu \in \mathbb{C}$, $\lambda_0 = 0$ and $\lambda_1 = 1$ are the only finite eigenvalues of $A(\lambda)$ according to Definition 2.4. Notice further that eigenspaces for different finite eigenvalues might intersect.

Remark 2.1 (Partial Multiplicities, [32, Def. 2.6, Rem. 2.7]). *Let* $A(\lambda) \in M_{m \times n}(\mathbb{R}[\lambda])$ *be of degree* $k \geq 1$ *and let* $D(\lambda) \in M_{m \times n}(\mathbb{R}[\lambda])$ *as in* (2.5) *be the Smith canonical form of* $A(\lambda)$ *so that* $r = \text{nrank}(A(\lambda))$.

(i) *Every* $d_j(\lambda) \in \mathbb{R}[\lambda]$ *with* $1 \leq j \leq r$ *can be factored uniquely as*

$$d_j(\lambda) = (\lambda - \lambda_0)^{\alpha_j} q_j(\lambda), \quad \alpha_j \in \mathbb{N}_0, \ q_j(\lambda_0) \neq 0,$$

for any $\lambda_0 \in \mathbb{C}$. *The sequence* $(\alpha_1, \ldots, \alpha_r) \in \mathbb{N}_0^r$ *is called the partial multiplicity sequence of* $A(\lambda)$ *at* λ_0.

(ii) *If* $(\alpha_1, \ldots, \alpha_r) \in \mathbb{N}_0^r$ *is the partial multiplicity sequence of* $A(\lambda)$ *at* λ_0, *its nonzero exponents* $\alpha_j, 1 \leq j \leq r$, *are called structural indices (or partial multiplicities) of* $A(\lambda)$ *at* λ_0. *Moreover, the factors* $(\lambda - \lambda_0)^{\alpha_j}, \alpha_j \neq 0$, *are referred to as the elementary divisors of* $A(\lambda)$ *for* λ_0.

As a consequence of Definition 2.4 and Theorem 2.1, $\lambda_0 \in \sigma_f(A)$ iff the partial multiplicity sequence $(\alpha_1, \ldots, \alpha_r)$ of $A(\lambda)$ at λ_0 is not identically zero. In particular, $\alpha_i > 0$ implies $\alpha_j > \alpha_i$ for any $i \leq j \leq r$ due to the divisibility condition (see Theorem 2.1). Moreover, in view of Remark 2.1 (i) and (2.7), the geometric multiplicity of some $\lambda_0 \in \sigma_f(A)$ coincides with the number of its structural indices.

Definition 2.6 (Algebraic Multiplicity, [32, Def. 2.8]). *Let* $A(\lambda) \in M_{m \times n}(\mathbb{R}[\lambda])$. *For any finite eigenvalue* $\lambda_0 \in \sigma_f(A)$ *with partial multiplicity sequence* $(\alpha_1, \ldots, \alpha_r) \in \mathbb{N}_0^r$ *the algebraic multiplicity of* λ_0 *is defined as* $\mathrm{algmult}(A, \lambda_0) := \alpha_1 + \cdots + \alpha_r$.

Beside finite eigenvalues, a matrix polynomial $A(\lambda) = \sum_{j=0}^{k} A_j \lambda^j$ of degree $k \geq 1$ has eigenvalues at infinity iff $\mathrm{rank}(A_k) < \mathrm{nrank}(A(\lambda))$ [32, Rem. 2.14]. Their definition requires the notion of the reversal (recall Definition 2.2). In fact, for any $A(\lambda) \in M_{m \times n}(\mathbb{R}[\lambda])$ it is easy to show that $\lambda_0 \in \mathbb{C}, \lambda_0 \neq 0$, is an eigenvalue of $A(\lambda)$ iff $1/\lambda_0$ is an eigenvalue of $\mathrm{rev}A(\lambda)$ and vice versa. This motivates the concept of infinite eigenvalues as introduced below.

Definition 2.7 (Infinite Eigenvalue, [32, Def. 2.13]). *Let* $A(\lambda) \in M_{m \times n}(\mathbb{R}[\lambda])$. *Then* $\lambda_0 = \infty$ *is called an (infinite) eigenvalue of* $A(\lambda)$ *if*

$$\mathrm{rank}(\mathrm{rev}A(0)) < \mathrm{nrank}(A(\lambda)) = \mathrm{nrank}(\mathrm{rev}A(\lambda)). \tag{2.9}$$

That is, $\lambda_0 = \infty$ *is an eigenvalue of* $A(\lambda)$ *iff zero is an eigenvalue of* $\mathrm{rev}A(\lambda)$.

If $\lambda_0 = \infty$ is an eigenvalue of $A(\lambda)$, the *spectrum* $\sigma_f(A) \cup \{\infty\}$ of $A(\lambda)$ is denoted $\sigma(A)$. The partial multiplicity sequence of $\lambda_0 = \infty$ is defined via $\mathrm{rev}A(\lambda)$ according to Remark 2.1 as are the (algebraic and geometric) multiplicities and the corresponding eigenspaces [32, Def. 2.15]. A proof of $\mathrm{nrank}(A) = \mathrm{nrank}(\mathrm{rev}A)$ from (2.9) can be found in [86].

Beside its eigenspaces for its eigenvalues, if $A(\lambda) \in M_{m \times n}(\mathbb{R}[\lambda])$ is singular, it has nontrivial right and left nullspaces for itself:

Definition 2.8 (Nullspace, [32, Sec. 2.2]). *For any* $A(\lambda) \in M_{m \times n}(\mathbb{R}[\lambda])$ *we define*

$$\mathcal{N}_r(A(\lambda)) := \{x(\lambda) \in \mathbb{R}(\lambda)^n \mid A(\lambda)x(\lambda) = 0\} \quad \text{and}$$
$$\mathcal{N}_\ell(A(\lambda)) := \{x(\lambda) \in \mathbb{R}(\lambda)^m \mid x(\lambda)^T A(\lambda) = 0\}.$$

Here, as before, $\mathbb{R}(\lambda)^p$ denotes the p-dimensional vector space over the field of rational functions $\mathbb{R}(\lambda)$. It is easily seen that $A(\lambda) \in M_{m \times n}(\mathbb{R}[\lambda])$ is singular iff $\mathcal{N}_r(A(\lambda))$ or $\mathcal{N}_\ell(A(\lambda))$ (or both) are nontrivial (this follows directly from Theorem 2.1, see also [32]). Certainly $\mathcal{N}_r(A(\lambda))$ and $\mathcal{N}_\ell(A(\lambda))$ are (by definition) subspaces of $\mathbb{R}(\lambda)^n$ and $\mathbb{R}(\lambda)^m$, respectively. If $A(\lambda)$ is regular, we have $\mathcal{N}_r(A(\lambda)) = \mathcal{N}_\ell(A(\lambda)) = \{0\}$. Moreover, if $A(\lambda)$ is square (i.e. $m = n$), then it holds that

$$\mathrm{nrank}(A(\lambda)) = n - \dim(\mathcal{N}_r(A(\lambda))) = n - \dim(\mathcal{N}_\ell(A(\lambda))).$$

In particular $\dim(\mathcal{N}_r(A(\lambda))) = \dim(\mathcal{N}_\ell(A(\lambda)))$ [31, (2.2)].

It is easily seen that any subspace $\mathcal{V} \subseteq \mathbb{R}(\lambda)^k$ has bases consisting entirely of vector polynomials. In fact, if $y_1(\lambda), \ldots, y_p(\lambda) \subseteq \mathbb{R}(\lambda)^k$ is a basis of $\mathcal{V}$, multiplying with the least common multiple of all denominators in elements from $y_1(\lambda), \ldots, y_p(\lambda)$ turns each $y_t(\lambda)$ into a vector polynomial $x_t(\lambda) \in \mathbb{R}[\lambda]^k$. In particular, both $y_1(\lambda), \ldots, y_p(\lambda)$ and $x_1(\lambda), \ldots, x_p(\lambda)$ are a basis of $\mathcal{V}$. We define the order of the vector polynomial basis $x_1(\lambda), \ldots, x_p(\lambda) \subset \mathbb{R}[\lambda]^k$ of $\mathcal{V} \subseteq \mathbb{R}(\lambda)^k$ as

$$\mathrm{ord}(x_1, \ldots, x_p) := \deg(x_1(\lambda)) + \deg(x_2(\lambda)) + \cdots + \deg(x_p(\lambda)). \tag{2.10}$$

This motivates the important concept of a minimal basis for $\mathcal{V} \subseteq \mathbb{R}(\lambda)^k$ as follows:

A basis $x_1(\lambda), \ldots, x_p(\lambda) \subseteq \mathbb{R}[\lambda]^k$ of $\mathcal{V} \subseteq \mathbb{R}(\lambda)^k$ with least order among all bases of $\mathcal{V}$ consisting of vector polynomials is called a minimal basis of $\mathcal{V}$ [32, Def. 2.19]. For convenience, we will also refer to a matrix polynomial $N(\lambda) \in \mathrm{M}_{m \times n}(\mathbb{R}[\lambda])$ as a minimal basis if the columns of $N(\lambda)^T$ are a minimal basis of the m-dimensional subspace $\mathcal{V} \subseteq \mathbb{R}(\lambda)^n$ they span. A simple criterion to identify such minimal bases $N(\lambda)$ was given in [34, Thm. 2.2] but will not be of further importance for our discussion.

Related to $\mathcal{N}_r(A(\lambda))$ and $\mathcal{N}_\ell(A(\lambda))$, this definition of minimal bases leads to the notion of minimal indices for singular matrix polynomials:

Definition 2.9 (Minimal Indices, [32, Def. 2.20]). *Let $A(\lambda) \in \mathrm{M}_{m \times n}(\mathbb{R}[\lambda])$ be a singular matrix polynomial. Moreover, let $x_1(\lambda), \ldots, x_p(\lambda) \in \mathbb{R}[\lambda]^n$ be a minimal basis of $\mathcal{N}_r(A(\lambda))$ and $y_1(\lambda), \ldots, y_s(\lambda) \in \mathbb{R}[\lambda]^m$ be a minimal basis of $\mathcal{N}_\ell(A(\lambda))$.*

1. *If $x_1(\lambda), \ldots, x_p(\lambda)$ are ordered so that $0 \le \deg(x_1(\lambda)) \le \cdots \le \deg(x_p(\lambda))$, then the scalars $\eta_j := \deg(x_j(\lambda)), j = 1, \ldots, p$, are referred to as the right minimal indices of $A(\lambda)$.*

2. *If $y_1(\lambda), \ldots, y_s(\lambda)$ are ordered so that $0 \le \deg(y_1(\lambda)) \le \cdots \le \deg(y_s(\lambda))$, then the scalars $\epsilon_j := \deg(y_j(\lambda)), j = 1, \ldots, s$, are referred to as the left minimal indices of $A(\lambda)$.*

Regardless of the minimal basis considered, the (ordered sequence of) minimal indices of any singular $A(\lambda) \in \mathrm{M}_{m \times n}(\mathbb{R}[\lambda])$ is always the same (as is its order) [32]. Therefore, Definition 2.9 is indeed consistent and minimal indices are well-defined. Moreover, the concept of dual minimal bases is important in the sequel:

Definition 2.10 (Dual Minimal Bases, [34, Def. 2.5]). *Two matrix polynomials $M(\lambda) \in \mathrm{M}_{r \times n}(\mathbb{R}[\lambda])$ and $N(\lambda) \in \mathrm{M}_{s \times n}(\mathbb{R}[\lambda])$ are called (a pair of) dual minimal bases, if both $M(\lambda)$ and $N(\lambda)$ are minimal bases with $r + s = n$ and they satisfy $M(\lambda)N(\lambda)^T = 0_{r \times s}$.*

The pair of dual minimal bases that is most important in our context is given by

$$L_k(\lambda) := \begin{bmatrix} -1 & \lambda & & & \\ & -1 & \lambda & & \\ & & \ddots & \ddots & \\ & & & -1 & \lambda \end{bmatrix} \in \mathrm{M}_{k \times (k+1)}(\mathbb{R}[\lambda])$$

and $\Lambda_k(\lambda) := [\lambda^k \; \lambda^{k-1} \; \cdots \; \lambda \; 1] \in M_{1 \times (k+1)}(\mathbb{R}[\lambda])$. The dual minimal basis property of $L_k(\lambda)$ and $\Lambda_k(\lambda)$ was proven in, e.g., [34, (2.3), (2.4)].

In particular, it is seen directly that $L_k(\lambda)\Lambda_k(\lambda)^T = 0$ holds. Regarding the previous discussion it is easily confirmed that the columns of $L_k(\lambda)^T$ form a basis (over $\mathbb{R}(\lambda)$) of $\mathcal{N}_r(\Lambda_k(\lambda))$. In particular, as $L_k(\lambda)$ and $\Lambda_k(\lambda)$ are dual minimal bases, the right minimal indices of $\Lambda_k(\lambda)$ are all equal to one. Moreover, since $L_k(\lambda)^T$ has full normal rank, we have $\mathcal{N}_r(L_k(\lambda)^T) = \{0\}$ and there are no right minimal indices of $L_k(\lambda)^T$. For $L_k(\lambda)$ it holds that $\dim(\mathcal{N}_r(L_k(\lambda))) = 1$, i.e. $\mathcal{N}_r(L_k(\lambda)) = \mathbb{R}(\lambda) \cdot \Lambda_k(\lambda)^T$. In particular, $L_k(\lambda)$ has one right minimal index equal to k. Notice that $L_k(\lambda) \otimes I_n$ and $\Lambda_k(\lambda) \otimes I_n$ are also dual minimal bases [34].

We introduce in Definition 2.11 below what we will understand to be the complete eigenstructure of a matrix polynomial.

Definition 2.11 ([31, Def. 2.6]). *The complete eigenstructure of any matrix polynomial $A(\lambda) \in M_{m \times n}(\mathbb{R}[\lambda])$ is determined by the quantities stated in (i) and (ii) below.*

(i) *The finite and infinite elementary divisors of $A(\lambda)$, that is its (finite and infinite) partial multiplicities of all eigenvalues, in particular their algebraic and geometric multiplicities.*

(ii) *The left and right minimal indices in case $A(\lambda)$ is singular.*

Before we pass on to the next section introducing linearizations of matrix polynomials, we state the *Index Sum Theorem* which displays the connection between the structural indices, the minimal indices, the degree and the normal rank of a matrix polynomial. The proof can be found in [32, Sec. 6] or [108].

Theorem 2.2 (Index Sum Theorem, [32, Thm. 6.5, Lem. 6.1]). *Let $A(\lambda) \in M_{m \times n}(\mathbb{R}[\lambda])$ be arbitrary of degree $k \in \mathbb{N}$. Then it holds that*

$$\sum_{\lambda_0 \in \sigma(A)} \mathrm{algmult}(A, \lambda_0) + \sum_{i=1}^{p} \eta_i + \sum_{i=1}^{s} \epsilon_i = k \cdot \mathrm{nrank}(A)$$

where, as in Definition 2.9, $\eta_1 \leq \cdots \leq \eta_p$ and $\epsilon_1 \leq \cdots \leq \epsilon_s$ denote the right and left minimal indices of $A(\lambda)$. In particular, $\sum_{\lambda_0 \in \sigma(A)} \mathrm{algmult}(A, \lambda_0) = kn$ if $m = n$ and $A(\lambda)$ is regular.

2.3 Linearizations of matrix polynomials

The concept of linearization lies at the heart of the first part of this work. It is introduced in Definition 2.13 below and heavily relies on the notions of equivalence and strict equivalence of matrix polynomials. In fact, we have used the equivalence relation between matrix polynomials already in Theorem 2.1 and the subsequent discussion. Nevertheless, we begin this subsection with a rigorous formal definition.

Definition 2.12 (Equivalence, [34, Def. 3.1]). *Let $A(\lambda), B(\lambda) \in M_{m \times n}(\mathbb{R}[\lambda])$.*

(i) *The matrix polynomials $A(\lambda)$ and $B(\lambda)$ are called (unimodular) equivalent if there exists two unimodular matrix polynomials $U(\lambda) \in M_{m \times m}(\mathbb{R}[\lambda])$ and $V(\lambda) \in M_{n \times n}(\mathbb{R}[\lambda])$ so that $B(\lambda) = U(\lambda)A(\lambda)V(\lambda)$.*

(ii) *The matrix polynomials $A(\lambda)$ and $B(\lambda)$ are called strict equivalent if there exists two nonsingular matrices $U \in GL_m(\mathbb{R})$ and $V \in GL_n(\mathbb{R})$ so that $B(\lambda) = UA(\lambda)V$.*

Certainly both concepts from Definition 2.12 establish equivalence relations on $M_{m \times n}(\mathbb{R}[\lambda])$. It is easily seen that two strict equivalent matrix polynomials have the same degree while this is in general not the case for (unimodular) equivalence. The typical canonical form considered under unimodular equivalence of matrix polynomials is the Smith canonical form, cf. Theorem 2.1. In fact, two matrix polynomials $A(\lambda), B(\lambda) \in M_{m \times n}(\mathbb{R}[\lambda])$ are unimodular equivalent iff they have the same Smith canonical form. The following definition introduces linearizations, a concept based on unimodular equivalence.

Definition 2.13 (Linearization, [32, Def. 3.3]). *Let $A(\lambda) \in M_{m \times n}(\mathbb{R}[\lambda])$.*

(i) *Any matrix pencil $\mathcal{L}(\lambda) = \lambda X + Y$ that can be expressed as*

$$\mathcal{L}(\lambda) = U(\lambda) \begin{bmatrix} I_s & 0 \\ 0 & A(\lambda) \end{bmatrix} V(\lambda) \in M_{(m+s) \times (n+s)}(\mathbb{R}[\lambda]) \qquad (2.11)$$

for some $s \in \mathbb{N}_0$ and unimodular matrix polynomials $V(\lambda) \in M_{(n+s) \times (n+s)}(\mathbb{R}[\lambda])$ and $U(\lambda) \in M_{(m+s) \times (m+s)}(\mathbb{R}[\lambda])$ is called a linearization for $A(\lambda)$.

(ii) *Assume $\deg(A(\lambda)) = k$. A linearization $\mathcal{L}(\lambda)$ for $A(\lambda)$ as in (2.11) is called strong (linearization) whenever $\mathrm{rev}_1\mathcal{L}(\lambda)$ is a linearization for $\mathrm{rev}_k A(\lambda) = \mathrm{rev}A(\lambda)$, too.*

A even more general definition of equivalence and (strong) linearizations for matrix polynomials can be found in [32, Def. 3.3 (a)]. However, the above definition serves our purposes. Notice that, whenever $D(\lambda)$ as in (2.5) is the SCF of $A(\lambda) \in M_{m \times n}(\mathbb{R}[\lambda])$ and $\mathcal{L}(\lambda)$ is a linearization for $A(\lambda)$ satisfying (2.11), then

$$\widetilde{D}(\lambda) := \begin{bmatrix} I_s & 0 \\ 0 & D(\lambda) \end{bmatrix} = \begin{bmatrix} I_s & 0 \\ 0 & W_1(\lambda) \end{bmatrix} U(\lambda)^{-1}\mathcal{L}(\lambda)V(\lambda)^{-1} \begin{bmatrix} I_s & 0 \\ 0 & W_2(\lambda) \end{bmatrix} \qquad (2.12)$$

$$=: \widetilde{U}(\lambda)\mathcal{L}(\lambda)\widetilde{V}(\lambda)$$

is the SCF of $\mathcal{L}(\lambda)$ (since $\widetilde{U}(\lambda)$ and $\widetilde{V}(\lambda)$ are unimodular and the SCF is unique due to Theorem 2.1). Thus any linearization $\mathcal{L}(\lambda)$ as in (2.11) of $A(\lambda)$ has the same number of minimal indices, the same elementary divisors and structural indices as $A(\lambda)$ [32, Thm. 4.1 (1)]. Furthermore, if $\mathcal{L}(\lambda)$ is strong, its infinite elementary divisors and structural indices coincide with those of $A(\lambda)$, too [32, Thm. 4.1 (2)]. It was shown in [32, Thm. 4.1] that the reverse implications hold as well. Beside these theoretical aspects, the following corollary highlights the most important practical consequences of (2.12) and Definition 2.13:

Corollary 2.1. *Let $A(\lambda) \in M_{m \times n}(\mathbb{R}[\lambda])$ be of degree $k \geq 1$.*

(i) *If $\mathcal{L}(\lambda)$ is a linearization of $A(\lambda)$ then $\sigma_f(\mathcal{L}) = \sigma_f(A)$. Regarding $\mathcal{L}(\lambda)$ and $A(\lambda)$, the algebraic and geometric multiplicities of all finite eigenvalues are the same.*

(ii) *If $\mathcal{L}(\lambda)$ is a strong linearization of $A(\lambda)$ then $\sigma(\mathcal{L}) = \sigma(A)$ and the algebraic and geometric multiplicities of all eigenvalues coincide.*

The most frequently used linearizations for matrix polynomials are the Frobenius companion forms. These are introduced in the following example.

Example 2.1. *Let $A(\lambda) = \sum_{j=0}^{k} A_j \lambda^j \in M_{m \times n}(\mathbb{R})[\lambda]$ be of degree $k \geq 1$. The first Frobenius companion form $\mathrm{Frob}_1(\lambda) \in M_{km \times kn}(\mathbb{R}[\lambda])$ for $A(\lambda)$ is defined as*

$$
\mathrm{Frob}_1(\lambda) := \begin{bmatrix} A_k & 0 & \cdots & 0 \\ 0 & I_n & & \vdots \\ \vdots & & \ddots & 0 \\ 0 & \cdots & 0 & I_n \end{bmatrix} \lambda + \begin{bmatrix} A_{k-1} & A_{k-2} & \cdots & A_0 \\ -I_n & 0 & \cdots & 0 \\ & \ddots & \ddots & \vdots \\ 0 & & -I_n & 0 \end{bmatrix}. \tag{2.13}
$$

As a consequence of [34, Thm 5.2], $\mathrm{Frob}_1(\lambda)$ is always a strong linearization for $A(\lambda)$. Analogously, the second Frobenius companion form

$$
\mathrm{Frob}_2(\lambda) := \begin{bmatrix} A_k & 0 & \cdots & 0 \\ 0 & I_m & & \vdots \\ \vdots & & \ddots & 0 \\ 0 & \cdots & 0 & I_m \end{bmatrix} \lambda + \begin{bmatrix} A_{k-1} & -I_m & & 0 \\ A_{k-2} & 0 & & \ddots \\ \vdots & \vdots & \ddots & -I_m \\ A_0 & 0 & \cdots & 0 \end{bmatrix} \tag{2.14}
$$

is always a strong linearization for $A(\lambda)$, too. For square (and regular) matrix polyomials, $\mathrm{Frob}_1(\lambda)$ and $\mathrm{Frob}_2(\lambda)$ can be found in, e.g., [56] or [89].

Notice that, if $m \neq n$, then $\mathrm{Frob}_1(\lambda)$ and $\mathrm{Frob}_2(\lambda)$ as in Example 2.1 have different sizes. This shows that strong linearizations for singular matrix polynomials need not all have the same dimension [32, Sec. 5.1]. However, if $A(\lambda) \in M_{n \times n}(\mathbb{R}[\lambda])$ is regular and of degree $k \geq 1$, any strong linearization for $A(\lambda)$ must have size $kn \times kn$ as $\mathrm{Frob}_1(\lambda)$ and $\mathrm{Frob}_2(\lambda)$ in this case [32, Rem. 4.12].

Given the task of finding the eigenvalues of some $A(\lambda) \in M_{m \times n}(\mathbb{R}[\lambda])$, Corollary 2.1 states that this problem may equivalently be solved by considering any (strong) linearization $\mathcal{L}(\lambda)$ instead of $A(\lambda)$. Finding the eigenvalues of a matrix pencil $\mathcal{L}(\lambda) = \lambda X + Y$ is known as the *generalized eigenvalue problem* (GEP).

2.4 Linearizations from vector spaces and Fiedler-pencils

In this section, we begin considering the systematic construction of (strong) linearizations $\mathcal{L}(\lambda)$ for matrix polynomials $A(\lambda) \in M_{m \times n}(\mathbb{R}[\lambda])$. This problem is both of theoretical and numerical interest. In particular, from a numerical point of view, due to conditioning, structure preservation or sparsity issues it might not always be appropriate to linearize any $A(\lambda)$ via its Frobenius companion form (see Example 2.1) for solving the corresponding GEP. So it seems desirable, to have a "toolbox" at hand for the construction of strong linearizations with specified properties. From a more theoretical viewpoint, linearizations can play a key role to reduce an arbitrary matrix polynomial to a simpler form such as block-diagonal or upper-Hessenberg [102].

It is immediate from Definition 2.12 that any matrix pencil $\mathcal{L}'(\lambda) = U(\lambda)\mathcal{L}(\lambda)V(\lambda)$ being unimodularly equivalent to $\mathcal{L}(\lambda) = \lambda X + Y$ is a linearization for $A(\lambda) \in M_{m \times n}(\mathbb{R}[\lambda])$ whenever $\mathcal{L}(\lambda)$ is a linearization for $A(\lambda)$. The same statement obviously holds for strict equivalence and strong linearizations. Thus, a first (unsystematic) approach to generate numerous strong linearizations is given by applying Definition 2.12 (ii) and Corollary 2.1 to $\mathrm{Frob}_j(A), j = 1, 2$, giving new strong linearizations of the form $\mathcal{L}(\lambda) := U\mathrm{Frob}_j(\lambda)V$ if U and V are both nonsingular.

A primer approach towards the development of linearizations is due to Lancaster, see [74, 76]. However, the first systematic approach in that regard is probably [89]. The main insight that led to the theory established there is the following observation: notice that for a matrix polynomial $\Lambda(\lambda) \in M_{n \times n}(\mathbb{R}[\lambda])$ of degree $k \geq 1$, the Frobenius companion forms introduced in Example 2.1 satisfy the equalities

$$\mathrm{Frob}_1(\lambda)\big(\Lambda_{k-1}(\lambda)^T \otimes I_n\big) = e_1 \otimes A(\lambda) \quad \text{and} \quad \big(\Lambda_{k-1}(\lambda) \otimes I_n\big)\mathrm{Frob}_2(\lambda) = e_1^T \otimes A(\lambda)$$

with $e_1 = [1 \ 0 \ \cdots \ 0]^T \in \mathbb{R}^k$. These equations have been generalized by the introduction of the ansatz spaces $\mathbb{L}_1(A), \mathbb{L}_2(A)$ and $\mathbb{DL}(A)$.

Definition 2.14 (Ansatz Spaces $\mathbb{L}_1(A), \mathbb{L}_2(A)$ and $\mathbb{DL}(A)$, [80, 89]). *Let a matrix polynomial $A(\lambda) \in M_{n \times n}(\mathbb{R}[\lambda])$ of degree $k \geq 1$ be given.*

(i) *The ansatz space $\mathbb{L}_1(A)$ for $A(\lambda)$ is defined to be the set of all $kn \times kn$ matrix pencils $\mathcal{L}(\lambda) \in M_{kn \times kn}(\mathbb{R}[\lambda])$ that satisfy $\mathcal{L}(\lambda)(\Lambda_{k-1}(\lambda)^T \otimes I_n) = v \otimes A(\lambda)$ for some ansatz vector $v \in \mathbb{R}^k$. This equation is called the ansatz equation for $\mathbb{L}_1(A)$.*

(ii) *The ansatz space $\mathbb{L}_2(A)$ for $A(\lambda)$ is defined to be the set of all $kn \times kn$ matrix pencils $\mathcal{L}(\lambda) \in M_{kn \times kn}(\mathbb{R}[\lambda])$ that satisfy $(\Lambda_{k-1}(\lambda) \otimes I_n)\mathcal{L}(\lambda) = v^T \otimes A(\lambda)$ for some ansatz vector $v \in \mathbb{R}^k$. This equation is called the ansatz equation for $\mathbb{L}_2(A)$.*

(iii) *The double ansatz space $\mathbb{DL}(A)$ is defined as $\mathbb{DL}(A) := \mathbb{L}_1(A) \cap \mathbb{L}_2(A)$.*

A short survey on the (classical) ansatz spaces $\mathbb{L}_1(A), \mathbb{L}_2(A)$ and $\mathbb{DL}(A)$ and their various derivatives that evolved subsequently is given in the next section.

2.4.1   The classical ansatz spaces and their derivatives

If $A(\lambda) \in M_{n \times n}(\mathbb{R}[\lambda])$ is a matrix polynomial of degree $k \geq 1$ it was shown in [89] that $\mathbb{L}_1(A) \cong \mathbb{L}_2(A)$ are vector spaces over $\mathbb{R}$ of dimension $k(k-1)n^2 + k$ whereas their intersection space $\mathbb{DL}(A)$ has dimension k [80, Thm. 3.4.2]. By far the most important property of those vector spaces of matrix pencils is that, beside $\text{Frob}_1(\lambda)$ and $\text{Frob}_2(\lambda)$ being elements of these spaces, almost every matrix pencil $\mathcal{L}(\lambda)$ in $\mathbb{L}_1(A)$ or $\mathbb{L}_2(A)$ is a strong linearization for $A(\lambda)$ regardless whether $A(\lambda)$ is regular or singular, cf. [31, Thm. 4.4]. This genericity result also holds for $\mathbb{DL}(A)$ if $A(\lambda)$ is regular [89, Thm. 6.8]. However, as shown in [32, Thm. 6.1], the double ansatz space $\mathbb{DL}(A)$ lacks of strong linearizations if $A(\lambda)$ is singular and of degree $k \geq 2$.

The characterization of all pencils in $\mathbb{L}_1(A), \mathbb{L}_2(A)$ and $\mathbb{DL}(A)$ was given in [89] along with simple criteria to identify strong linearizations. In particular, strong linearizations can be determined by means of two important theorems, the *Strong Linearization Theorem* and the *Eigenvalue Exclusion Theorem* see [89, Thm. 4.3, 6.7]. The former theorem states that, if $A(\lambda)$ is regular, any pencil $\mathcal{L}(\lambda)$ in $\mathbb{L}_1(A), \mathbb{L}_2(A)$ or $\mathbb{DL}(A)$ is automatically a strong linearization for $A(\lambda)$ if $\mathcal{L}(\lambda)$ is regular. In addition, these ansatz spaces admit simple recovery rules for eigenvectors and minimal indices, i.e. the eigenvectors and minimal indices of $A(\lambda)$ can be determined by those of any strong linearization $\mathcal{L}(\lambda)$ from $\mathbb{L}_1(A), \mathbb{L}_2(A)$ or $\mathbb{DL}(A)$ via simple formulae [89, Thm. 3.8, Thm. 3.14], [31, Sec. 5].

The ansatz space $\mathbb{DL}(A)$ received special research attention. It has been investigated with respect to all matrix polynomial structures presented in Table 2.1 to construct structure preserving linearizations [80]. If $A(\lambda)$ is structured, these are linearizations $\mathcal{L}(\lambda) \in \mathbb{DL}(A)$ (or linearizations $\widetilde{\mathcal{L}}(\lambda) \notin \mathbb{DL}(A)$ that can easily be obtained from those in $\mathbb{DL}(A)$) belonging to the same structure class as $A(\lambda)$.

For instance, the symmetric/Hermitian case was considered deeply in [65], where the subspaces $\mathbb{H}(A) \subset \mathbb{L}_1(A)$ and $\mathbb{S}(A) \subset \mathbb{L}_1(A)$ of Hermitian and symmetric pencils have been introduced. Considering particularly the symmetric case, it should also be mentioned that in [22] the space $\mathbb{DL}(A)$ and its standard basis (see [80, Thm. 3.3.2]) were used to construct a family of new ansatz spaces $\mathbb{EDL}(A)$ (called *extended* ansatz spaces). These vector spaces were shown to contain a great many of block-symmetric strong linearizations for $A(\lambda)$ [22, Thm. 2.2]. Palindromic and alternating structures have been handled in [80, 87]. In addition, the conditioning of linearizations from $\mathbb{DL}(A)$ has been thoroughly analyzed, see [62, 64, 80]. Extensions for the ansatz equations satisfied by pencils from $\mathbb{L}_1(A)$ and $\mathbb{L}_2(A)$ (see Definition 2.14) were presented in [84, 88, 101] considering

$$\mathcal{L}(\lambda)\Big(\mathcal{T}_{k-1}(\lambda)^T \otimes I_n\Big) = v \otimes A(\lambda) \quad \text{and} \quad \Big(\mathcal{T}_{k-1}(\lambda) \otimes I_n\Big)\mathcal{L}(\lambda) = v^T \otimes A(\lambda) \quad (2.15)$$

instead of $\mathcal{L}(\lambda)(\Lambda_{k-1}(\lambda)^T \otimes I_n) = v \otimes A(\lambda)$ and $(\Lambda_{k-1}(\lambda) \otimes I_n)\mathcal{L}(\lambda) = v^T \otimes A(\lambda)$. Hereby $\mathcal{T}_{k-1}(\lambda)^T = [\, \tau_{k-1}(\lambda) \; \cdots \; \tau_1(\lambda) \; \tau_0(\lambda)\,]^T \in \mathbb{R}[\lambda]^k$ is a vector polynomial consisting of any polynomial basis $\tau_0(\lambda), \ldots, \tau_{k-1}(\lambda)$ of $\mathbb{R}_{k-1}[\lambda]$. These ansatz spaces are particularly suitable if the matrix polynomial $A(\lambda)$ given to be linearized is already available as $A(\lambda) = \sum_{j=0}^{k} A_j \tau_j(\lambda)$. Moreover, it was shown in [72] that considering matrix polynomials of large degree not in the monomial basis might be advantageous in view of numerical stability.

The ansatz space framework for matrix polynomials in nonstandard polynomial bases has already been analyzed for special bases. For instance, the corresponding vector spaces of matrix pencils for the Bernstein basis $\mathbb{B}_1(A), \mathbb{B}_2(A)$ and $\mathbb{DB}(A)$ were introduced in [84], the ansatz space $\mathbb{T}_1(A)$ for the Chebyshev basis was considered in [46] and the ansatz spaces $\mathcal{N}_1(A), \mathcal{N}_2(A)$ and $\mathcal{DN}(A)$ for the Newton basis in [85]. In conclusion, the ansatz spaces $\mathbb{L}_1(A), \mathbb{L}_2(A)$ and $\mathbb{DL}(A)$ and their derivatives provide a convenient setting to systematically construct and analyze (strong) linearizations for (structured and unstructured) matrix polynomials $A(\lambda)$ expressed in standard and nonstandard polynomial bases. There is a large amount of literature on $\mathbb{L}_1(A), \mathbb{L}_2(A)$ and $\mathbb{DL}(A)$ and the ansatz space setting in general, see also [31, 63, 87, 95] and the references therein.

2.4.2 Fiedler-pencils and Fiedler-like linearization families

Beside the construction of (strong) linearizations for $A(\lambda) \in M_{n \times n}(\mathbb{R}[\lambda])$ via $\mathbb{L}_1(A)$, $\mathbb{L}_2(A)$, $\mathbb{DL}(A)$ or their derivatives mentioned in Section 2.4.1, there exist strong linearizations for $A(\lambda)$ that are also build systematically, but totally different. The following Definition 2.15 introducing Fiedler-pencils is taken from [34, Sec. 4.1], although the concept traces back to [3]. The name *Fiedler-pencil* was chosen in honor of [49] where the concept was introduced by M. Fiedler for the construction of companion forms for monic scalar polynomials.

Definition 2.15. *Let $A(\lambda) = \sum_{j=0}^{k} A_j \lambda^j \in M_{n \times n}(\mathbb{R})[\lambda]$ be of degree $k \geq 1$. We define the matrices $P_0, P_1, \ldots, P_k \in M_{kn \times kn}(\mathbb{R})$ as follows*

$$P_k := \begin{bmatrix} A_k & \\ & I_{(k-1)n} \end{bmatrix}, \qquad P_0 := \begin{bmatrix} I_{(k-1)n} & \\ & -A_0 \end{bmatrix}$$

and

$$P_i := \begin{bmatrix} I_{(k-i-1)n} & & & \\ & -A_i & I_n & \\ & I_n & 0_n & \\ & & & I_{(i-1)n} \end{bmatrix}, \qquad i = 1, 2, \ldots, k-1.$$

Let $\sigma : \{0, \ldots, k-1\} \rightarrow \{1, \ldots, k\}$ be any bijection. Then, the Fieder-pencil for $A(\lambda)$ associated with σ is defined to be

$$F_\sigma(\lambda) = \lambda P_k - P_\sigma = \lambda P_k - P_{\sigma^{-1}(1)} P_{\sigma^{-1}(2)} \cdots P_{\sigma^{-1}(k)}. \qquad (2.16)$$

Notice that different bijections might yield the same Fielder-pencil (due to commutativity relations among $P_0, \ldots, P_d$), [34, Sec. 4.1]. For square $A(\lambda) \in M_{n \times n}(\mathbb{R}[\lambda])$,

a premier indicator for the usefulness of Fiedler-pencils is that both Frobenius companions forms (2.13) and (2.14) can be expressed as in Definition 2.15 as particular
Fiedler-pencils, that is

$$\text{Frob}_1(\lambda) = \lambda P_k - P_{k-1}P_{k-2}\cdots P_0 \quad \text{and} \quad \text{Frob}_2(\lambda) = \lambda P_k - P_0 P_1 \cdots P_{k-1},$$

see also [3, Lem. 2.1]. Beyond that, the intensive study of Fiedler-pencils can be
justified by the fact that any Fiedler-pencil $F_\sigma(\lambda)$ as in (2.16) associated with any
bijection σ is a strong linearization for $A(\lambda)$ [3, Thm. 2.3]. Beside their strong
linearization property, Fiedler-pencils are obviously quite easy to construct. Moreover,
they might share structural properties with the matrix polynomial $A(\lambda)$ [34].

Beyond Definition 2.15, the idea of Fiedler-pencils was extended in many ways. For
instance, Fiedler-linearizations for rectangular matrix polynomials were introduced in
[121], an extension that is not available for the ansatz spaces $\mathbb{L}_1(A), \mathbb{L}_2(A)$ and $\mathbb{DL}(A)$
so far. Therefore, the in-depth study of Fiedler-pencils is in no way inferior to that of
the ansatz spaces. For instance, the recovery of eigenvectors and minimal indices for
Fiedler-pencils (and their rectangular extensions) was considered in, e.g., [20, 120].
The concept of generalized Fiedler-pencils traces back to [127] and was reconsidered
and extended to generalized Fiedler-pencils with repetition in [22]. In addition,
structured linearizations for Fiedler-pencils with repetitions have been studied in
[17–19] and extensions related to other polynomial bases were treated in [104]. All
these related but different types of Fiedler-pencils are often summarized under the
notion of *Fiedler-like linearizations*. The scientific work on Fiedler-like linearizations
is extensive, see [17–21, 34, 120, 121] and the references therein. However, we opt
out from a deeper investigation of Fiedler-pencils since our focus is primarily on the
ansatz space framework.

2.4.3 From Fielder-linearizations to block Kronecker matrix pencils

Beside their useful properties, the analysis of (generalized) Fiedler-pencils (with
repetitions) is highly nontrivial, for example with respect to conditioning and backward
error analysis. A simplification was achieved in [34] where it was shown that all
(classical) Fiedler-pencils according to Definition 2.15 can be permuted to a particular
block-pencil form. Certainly any permutation can be realized as a strong equivalence
(in the sense of Definition 2.12). The linearizations obtained this way were called
block Kronecker linearizations [34, Sec. 5]. A general definition is given below.

Definition 2.16 (Block Kronecker Pencil, [34, Def. 5.1])**.** *Let $k \in \mathbb{N}$ and $\epsilon, \eta \in \mathbb{N}_0$ be
arbitrary such that $k = \epsilon + \eta + 1$. Then any $((\eta + 1)m + \epsilon n) \times ((\epsilon + 1)n \times \eta m)$ matrix
pencil $\mathcal{L}(\lambda)$ of the form*

$$\mathcal{L}(\lambda) = \begin{bmatrix} \lambda M_1 + M_0 & L_\eta(\lambda)^T \otimes I_m \\ L_\epsilon(\lambda) \otimes I_n & 0 \end{bmatrix} \tag{2.17}$$

is called a (real) (ϵ, n, η, m)-block Kronecker pencil.

The most striking property of block Kronecker pencils can be stated as follows: according to [34, Thm. 5.2], any block Kronecker pencil $\mathcal{L}(\lambda)$ of the form (2.17) is a strong linearization of

$$A(\lambda) := \Big(\Lambda_\eta(\lambda) \otimes I_m\Big)\Big(\lambda M_1 + M_0\Big)\Big(\Lambda_\epsilon(\lambda)^T \otimes I_n\Big) \in M_{m \times n}(\mathbb{R}[\lambda]) \qquad (2.18)$$

whenever $A(\lambda)$ has degree $k = \epsilon + \eta + 1$[4]. Given any matrix polynomial $A(\lambda) \in M_{m \times n}(\mathbb{R}[\lambda])$ of degree $k = \epsilon + \eta + 1$ this implies that, whenever a matrix pencil $\lambda M_1 + M_0$ satisfying (2.18) can be found, the matrix pencil $\mathcal{L}(\lambda)$ in (2.17) will be a strong linearization for $A(\lambda)$. A rigorous backward error analysis of block Kronecker pencils was presented in [34]. Moreover, in [21] the authors showed that many more types of Fiedler-like linearizations as surveyed in Section 2.4.2 (beside the classical Fiedler-pencils) admit a strict equivalence transformation to a block Kronecker-like form similar to (2.17). This relation will be discussed in Section 3.4.

2.5 Outlook on Chapters 3 to 7

The ansatz space concept and the approach via Fiedler pencils are the two *standard techniques* to linearize matrix polynomials. However, due to their diversity, both approaches do not have much in common[5] and both theories have been developed almost parallel to each other over the last decades. Although the ansatz space concept is well understood and was investigated and further developed in a considerable amount of scientific work, cf. Section 2.4.1, there are still a few fundamental questions open:

For example, the ansatz space concept from $\mathbb{L}_1(A), \mathbb{L}_2(A)$ and $\mathbb{DL}(A)$ was introduced only for square matrix polynomial and the question arises, whether it admits similar extensions to the rectangular case as Fiedler pencils. In fact, the ansatz space approach was considered to be "valid only for square polynomials" in [34, Sec. 1]. A recent development in this direction is [28]. Furthermore, it is an open question whether Fiedler-pencils might also belong to some vector space (i.e. ansatz space) not recognized so far having an ansatz equation which has not been discovered yet. In addition to that, the analysis of the ansatz equations given in (2.15) for nonmonomial bases did yet not receive the attention it deserves. For instance, a simple and intrinsic characterization and analysis of the pencils satisfying (2.15) beside the bivariate approach from [101] is not available so far. This first part of this work is mainly concerned with these questions.

---

[4]Actually, [34, Thm. 5.2] is more general since it uses the concept of grade instead of degree. This allows to consider any $A(\lambda) \in M_{m \times n}(\mathbb{R}[\lambda])$ of degree k as a matrix polynomial of grade $d > k$ by adding a zero term of the form $\lambda^d A_d + \cdots + A_{k+1} \lambda^{k+1}$ to $A(\lambda)$ with zero matrix coefficients $A_d, \ldots, A_{k+1}$. In this regard [34, Thm. 5.2] states that any block Kronecker pencil $\mathcal{L}(\lambda)$ as in (2.17) is always a strong linearization for $A(\lambda)$ in (2.18) considered as a matrix polynomial of grade $\epsilon + \eta + 1$. We will not discuss this linearization concept here.

[5]Nevertheless, there are a few common aspect to both approaches. For example, the standard basis of $\mathbb{DL}(A)$ (see [80, Thm. 3.3.2]) can be expressed via Fiedler pencils [22, Thm. 2.2] and the ansatz spaces $\mathbb{EDL}(A)$ are related to Fiedler pencils as well.

In particular, the main contribution of Chapter 3 is that both fundamental questions concerning ansatz spaces and Fiedler-pencils raised above can essentially be answered positively. The main observation in this regard is that for any choice of ϵ and η in Definition 2.16 any (ϵ, n, η, m)-block Kronecker pencil can naturally be embedded into a new sort of ansatz space. We present and characterize this new family of vector spaces we named *block Kronecker ansatz spaces* in Section 3.1 and 3.2 and show that it serves as an abundant source of strong linearizations. Here we give a simple proof that the *Strong Linearization Theorem* from [89] extends to block Kronecker ansatz spaces. In Section 3.3 we show how eigenvectors for the linearized matrix polynomial can be easily recovered from those of its linearizations in block Kronecker ansatz spaces. Finally, Section 3.4 points out the relationship between various Fiedler-like linearization families and this new ansatz space concept.

Inspired by the analysis of the double ansatz space $\mathbb{DL}(A)$ in [80, 89], Chapter 4 is dedicated to the investigation of pencils that are contained in two or more block Kronecker ansatz spaces simultaneously. A comprehensive characterization of these matrix pencils is presented in Section 4.1 followed by a deeper structural analysis in Section 4.2. The whole flexibility of this concept shows up in Section 4.3 where we prove the *superpartition property* of pencils belonging to two block Kronecker ansatz spaces. Section 4.5 is concerned with the construction of block-symmetric pencils from block Kronecker ansatz spaces while Section 4.6 discusses structure-preserving symmetric linearizations for symmetric matrix polynomials. This is followed by a short summary of the results from Chapter 3 and 4 in Section 4.7.

In Chapter 5 our focus turns back to the utilization of linearizations from block Kronecker ansatz spaces for the solution of polynomial eigenvalue problems. In particular, Section 5.1 presents a way to reformulate a generalized eigenvalue problem as a standard eigenvalue problem. Section 5.1 presents a few facts about Krylov subspace methods for eigenvalue problems and shows that Krylov-based eigenvalue solvers are in general a valuable tool to solve problems as those of Section 5.1. To make this appraoch efficient, Section 5.2 shows how linear systems with linearizations from the block Kronecker ansatz spaces can be solved at low computational costs. We show how this approach can effectively be applied for T-even matrix polynomials in Section 5.3. In particular, we consider the EVEN-IRA algorithm from [94] and apply our results from Section 5.2 to reduce the computational costs of this algorithm. This is followed by some numerical examples. The special case of symmetric matrix polynomials and their eigenvalue computations via symmetric linearizations is briefly discussed in Section 5.4.

In Chapter 6 we turn our attention to the classical ansatz spaces $\mathbb{L}_1(A), \mathbb{L}_2(A)$ and $\mathbb{DL}(A)$ from [89]. In particular, Section 6.1 presents a new and simple characterization of $\mathbb{L}_1(A)$ and $\mathbb{L}_2(A)$ based on the results from Chapter 3. In Section 6.2 the *Fundamental Isomorphism Theorem* relating the block Kronecker ansatz spaces to the classical ansatz spaces is derived followed by some conclusions in Section 6.3. Chapter 7 is dedicated to the study of the ansatz spaces defined by (2.15) for nonmonomial, i.e. orthogonal polynomial bases. The basic framework and a simple characterization of the pencils satisfying (2.15) are presented in Section 7.1 beside a criterion for the identification of strong linearizations. How eigenvectors may be recovered from those linearizations is shown in Section 7.2. In Section 7.3 the focus lies on singular matrix polynomials. Here we derive new linearization conditions based on certain relationships for left eigenvectors, ansatz vectors and minimal indices. The investigation of pencils satisfying both ansatz equations from (2.15) (similarly to $\mathbb{DL}(A)$) is the content of Section 7.4 where we also derive a recursion-based algorithm for their construction. In Section 7.5 the *eigenvector exclusion theorem* is proven which yields a new insight on the *eigenvalue exclusion theorem* from [89] in the nonmonomial case. The results from Chapter 7 are summarized in Section 7.6.

The main results from Chapter 3 and 4 (despite Section 4.6 on symmetric linearization) have been published in [42]. Furthermore, this thesis contains results that cannot be found in [42]. In particular, the effect of block-transposition on pencils from the block Kronecker ansatz spaces has been emphasized directly in Proposition 3.3 and the recovery of nullspaces and minimal indices for singular matrix polynomials has been addressed (see Theorem 3.5 and Proposition 3.4). The connection to Fiedler-like linearizations has been laid out in more detail in Section 3.4. Moreover, in order to achieve a better comprehensibility, the proofs of Theorem 4.1 and Theorem 4.3 have been reworked (the first resulted in the new Corollary 4.2). In Chapter 5 the results from Section 5.3.1 and (partially) Section 5.3.2 can be found in [43]. The content of all other sections of Chapter 5 has not been published and extends the main ideas underlying Section 5.3.1. Regarding Chapter 6, some findings from Section 6.1 arise as special cases from the results of Chapter 7 (in particular Theorem 6.1, Corollary 6.1 and Corollary 6.2) whereas the results from Section 6.2 have not been published yet. In turn, Chapter 7 is mainly based on the publication [45]. Nevertheless, some new results are included. In particular, Theorem 7.3 has been extended while Theorem 7.8 and Corollary 7.6 are new. The findings from Proposition 7.5 and Theorem 7.14 also appear here for the first time. Throughout Chapters 3 to 7 results from any of the previously mentioned publications are referenced.

Chapter 3

Block Kronecker ansatz spaces

> *If you really want to impress your friends and confound your enemies, you can invoke tensor products... People run in terror from the $\otimes$ symbol.*

Professor at Stanford University[1]

In this chapter we introduce the family of *block Kronecker ansatz spaces* $\mathbb{G}_{\eta+1}(A)$ for arbitrary matrix polynomials $A(\lambda) \in M_{m \times n}(\mathbb{R}[\lambda])$. These large-dimensional vector spaces of matrix pencils provide - beside the classical ansatz spaces $\mathbb{L}_1(A), \mathbb{L}_2(A)$ and $\mathbb{DL}(A)$ - a new ansatz space framework for the construction of strong linearizations for $A(\lambda)$. In this and the subsequent chapter this new concept will be comprehensively analyzed and investigated. Moreover, based on the results from [34] and [21] we will show that the theoretical gap between the Fiedler-like linearization families and the conceptual ansatz space framework initiated in [89] is closed by the block Kronecker ansatz spaces. In addition, the block Kronecker ansatz spaces $\mathbb{G}_{\eta+1}(A)$ can be seen as a natural vector-space-setting covering all the block Kronecker linearizations for $A(\lambda)$ introduced in [34, Sec. 5]

In Section 3.1 we begin with a formal definition and a comprehensive characterization of these new ansatz spaces. Thereby we give a simple condition to identify strong linearizations $\mathcal{L}(\lambda)$ for $A(\lambda)$ in $\mathbb{G}_{\eta+1}(A)$. An advanced structural analysis of $\mathbb{G}_{\eta+1}(A)$ follows in Section 3.2 where we will show that almost every pencil $\mathcal{L}(\lambda) \in \mathbb{G}_{\eta+1}(A)$ is a strong linearization for $A(\lambda)$ and prove an important equivalence theorem (the *strong linearization theorem*) in case $A(\lambda)$ is regular. This theorem was proven in [89] for the ansatz space $\mathbb{L}_1(A)$ and extends unchanged to the block Kronecker ansatz spaces. How eigenvectors of $A(\lambda)$ can easily be recovered from those of any strong linearization $\mathcal{L}(\lambda) \in \mathbb{G}_{\eta+1}(A)$ will be shown in Section 3.3. Finally, Section 3.4 provides the connection between the block Kronecker ansatz spaces and the various families of Fiedler-like linearizations.

It is appropriate mentioning that some ideas and results presented in this and the subsequent chapter have also been observed or developed independently in [21].

---

[1]https://jeremykun.com/2014/01/17/how-to-conquer-tensorphobia/ (04-01-2019)

Those similarities are emphasized in [42] and [21], so detailed reference within the upcoming discussion is not given.

3.1 Introduction of $\mathbb{G}_{\eta+1}(A)$ and basic properties

The following definition is central for the theory developed subsequently. Here and from now on, we always assume ϵ and η to be nonnegative integers, i.e. $\epsilon, \eta \in \mathbb{N}_0$.

Definition 3.1 (Block Kronecker Ansatz Equation, [42, Def. 1]). *For any matrix polynomial $A(\lambda) \in M_{m \times n}(\mathbb{R}[\lambda])$ of degree $k = \epsilon + \eta + 1$ we define $\mathbb{G}_{\eta+1}(A)$ to be the set of all matrix pencils $\mathcal{L}(\lambda)$ satisfying*

$$\left[\begin{array}{c|c} \Lambda_\eta(\lambda) \otimes I_m & 0 \\ \hline 0 & I_{\epsilon n} \end{array}\right] \mathcal{L}(\lambda) \left[\begin{array}{c|c} \Lambda_\epsilon(\lambda)^T \otimes I_n & 0 \\ \hline 0 & I_{\eta m} \end{array}\right] = \left[\begin{array}{c|c} \alpha A(\lambda) & 0 \\ \hline 0 & 0_{\epsilon n \times \eta m} \end{array}\right] \quad (3.1)$$

for some $\alpha \in \mathbb{R}$. Equation (3.1) is called block Kronecker ansatz equation for $A(\lambda)$.

Any pencil $\mathcal{L}(\lambda)$ satisfying (3.1) has size $((\eta+1)m + \epsilon n) \times ((\epsilon+1)n + \eta m)$. Thus, $\mathcal{L}(\lambda) \in \mathbb{G}_{\eta+1}(A)$ is square iff $A(\lambda)$ is, i.e. $m = n$. Since the relation $\deg(A(\lambda)) = k = \epsilon + \eta + 1$ has to hold, any choice of $\eta = 0, 1, \ldots, k-1$ uniquely determines ϵ and thus yields a structurally different ansatz equation for $\mathbb{G}_{\eta+1}(A)$ in (3.1). In particular, the notion $\mathbb{G}_{\eta+1}(A)$ is unambiguous although ϵ does not explicitly appear.

Notice that (3.1) can be written in short hand notation as

$$\left((\Lambda_\eta(\lambda) \otimes I_m) \oplus I_{\epsilon n}\right)\mathcal{L}(\lambda)\left((\Lambda_\epsilon(\lambda)^T \otimes I_n) \oplus I_{\eta m}\right) = \alpha A(\lambda) \oplus I_{\epsilon n \times \eta m}. \quad (3.2)$$

Before we analyze the solutions $\mathcal{L}(\lambda)$ of (3.1), we make the following observation:

Proposition 3.1 ([42, Lem. 1]). *Let $A(\lambda) \in M_{m \times n}(\mathbb{R}[\lambda])$ be of degree $k = \epsilon + \eta + 1$. Then, for any choice of $\eta, 0 \leq \eta \leq k-1$, the set $\mathbb{G}_{\eta+1}(A)$ is a vector space over $\mathbb{R}$.*

Since Proposition 3.1 is quite obvious, the proof is omitted. The vector spaces $\mathbb{G}_{\eta+1}(A), \eta = 0, 1, \ldots, k-1$, obtained via (3.1) for any $A(\lambda) \in M_{m \times n}(\mathbb{R}[\lambda])$ of degree $k \geq 1$ are called *block Kronecker ansatz spaces* for $A(\lambda)$. This name was chosen in compliment of the ansatz spaces $\mathbb{L}_1(A), \mathbb{L}_2(A)$ and $\mathbb{DL}(A)$ (see Definition 2.14) established in [89] and the block Kronecker pencils (cf. Definition 2.15) introduced in [34, Sec. 5]. How the main ideas of both concepts may be unified using the concept of block Kronecker ansatz spaces is one primary concern of this work. In fact, various relationships will be developed throughout the subsequent sections.

Our first goal is to comprehensively characterize the spaces $\mathbb{G}_{\eta+1}(A), 0 \leq \eta \leq k-1$. To this end, first notice that equation (3.1) may be restated as

$$\left[\begin{array}{c|c} \Lambda_\eta(\lambda) \otimes I_m & 0 \\ \hline 0 & I_{\epsilon n} \end{array}\right] \left[\begin{array}{c|c} \mathcal{L}_{11}(\lambda) & \mathcal{L}_{12}(\lambda) \\ \hline \mathcal{L}_{21}(\lambda) & \mathcal{L}_{22}(\lambda) \end{array}\right] \left[\begin{array}{c|c} \Lambda_\epsilon(\lambda)^T \otimes I_n & 0 \\ \hline 0 & I_{\eta m} \end{array}\right] = \left[\begin{array}{c|c} \alpha A(\lambda) & 0 \\ \hline 0 & 0 \end{array}\right]$$

$$(3.3)$$

where we have expressed $\mathcal{L}(\lambda) \in \mathbb{G}_{\eta+1}(A)$ as a 2×2 block-pencil with $\mathcal{L}_{11}(\lambda)$ having dimension $(\eta + 1)m \times (\epsilon + 1)n$. Following [34, Def. 5.1], this structured 2×2 block-notation of $\mathcal{L}(\lambda)$ is called its natural partition. Equation (3.3), in particular its left-hand-side, now explicitly takes the form

$$\left[\begin{array}{c|c} (\Lambda_\eta(\lambda) \otimes I_m)\mathcal{L}_{11}(\lambda)(\Lambda_\epsilon^T \otimes I_n) & (\Lambda_\eta \otimes I_m)\mathcal{L}_{12}(\lambda) \\ \hline \mathcal{L}_{21}(\lambda)(\Lambda_\epsilon(\lambda)^T \otimes I_n) & \mathcal{L}_{22}(\lambda) \end{array}\right] = \left[\begin{array}{c|c} \alpha A(\lambda) & 0 \\ \hline 0 & 0_{\epsilon n \times \eta m} \end{array}\right]. \quad (3.4)$$

For $(\Lambda_\eta(\lambda) \otimes I_m)\mathcal{L}_{11}(\lambda)(\Lambda_\epsilon(\lambda)^T \otimes I_n)$ in (3.4) we will steadily be using the short-hand-notation $\Theta(\mathcal{L}_{11}(\lambda))$ assuming the parameters ϵ, η, m and n involved are clear from the context. For instance, (3.4) implies $\Theta(\mathcal{L}_{11}(\lambda)) = \alpha A(\lambda)$.

We will now characterize the solutions $\mathcal{L}_{ij}(\lambda), i, j = 1, 2$, of the four equations in (3.4) independently. Clearly, $\mathcal{L}_{22}(\lambda)$ has to be the $\epsilon n \times \eta m$ zero-matrix. To determine the possible forms of $\mathcal{L}_{21}(\lambda)$ and $\mathcal{L}_{12}(\lambda)$ we need the following proposition.

Proposition 3.2 ([42, Lem. 2]). *Let* $\mathcal{K}(\lambda)$ *be an* $\mu m \times (\nu + 1)n$ *matrix pencil and assume*

$$\mathcal{K}(\lambda)\left(\Lambda_\nu(\lambda)^T \otimes I_n\right) = 0_{\mu m \times n}. \quad (3.5)$$

Then $\mathcal{K}(\lambda) = C(L_\nu(\lambda) \otimes I_n)$ *for some matrix* $C \in \mathrm{M}_{\mu m \times \nu n}(\mathbb{R})$.

Proof. Assume $\mathcal{K}(\lambda) = [\, k_1 \mid K_1\,]\lambda + K_0$ with $k_1 \in \mathrm{M}_{\mu m \times n}(\mathbb{R})$ satisfies (3.5). Then

$$\Delta\mathcal{K}(\lambda) = \mathcal{K}(\lambda) - K_1\left(L_\nu(\lambda) \otimes I_n\right) =: \left[\, d_1(\lambda) \mid D_1 \,\right]$$

is independent of λ in all but its first block-column $d_1(\lambda) \in \mathrm{M}_{\mu m \times n}(\mathbb{R}_1[\lambda])$. However, for $\Delta\mathcal{K}(\lambda)(\Lambda_\nu(\lambda)^T \otimes I_n)$ we now obtain

$$\mathcal{K}(\lambda)\left(\Lambda_\nu(\lambda)^T \otimes I_n\right) - K_1\left(L_\nu(\lambda) \otimes I_n\right)\left(\Lambda_\nu(\lambda)^T \otimes I_n\right) = 0_{\mu m \times n},$$

so $\Delta\mathcal{K}(\lambda)$ also satisfies (3.5). Notice that $\Delta\mathcal{K}(\lambda)(\Lambda_\nu(\lambda)^T \otimes I_n)$ has dimension $\mu m \times n$ and that every $m \times n$ block is a matrix polynomial in the variables $1, \lambda, \lambda^2, \ldots, \lambda^{\nu+1}$. Due to the basis property of the monomials this implies $\Delta\mathcal{K}(\lambda) \equiv 0$. Therefore we obtain $\mathcal{K}(\lambda) = K_1(L_\nu(\lambda) \otimes I_n)$ which proves the statement. $\square$

As a direct consequence of Proposition 3.2, any $\epsilon n \times (\epsilon + 1)n$ pencil $\mathcal{L}_{21}(\lambda)$ satisfying $\mathcal{L}_{21}(\lambda)(\Lambda_\epsilon(\lambda)^T \otimes I_n) = 0$ has the form $C(L_\epsilon(\lambda) \otimes I_n)$ for some matrix $C \in \mathrm{M}_{\epsilon n \times \epsilon n}(\mathbb{R})$. Furthermore, it can be seen from Proposition 3.2 by block-transposition (recall Definition 2.3) that any arbitrary $(\mu + 1)m \times \nu n$ pencil $\mathcal{K}(\lambda)$ satisfying $(\Lambda_\mu(\lambda) \otimes I_m)\mathcal{K}(\lambda) = 0_{m \times \nu n}$ admits an expression as $\mathcal{K}(\lambda) = (L_\mu(\lambda)^T \otimes I_m)C$ for some matrix $C \in \mathrm{M}_{\mu m \times \nu n}(\mathbb{R})$. Hence, regarding (3.4) once more, we may thus characterize the blocks $\mathcal{L}_{12}(\lambda)$ and $\mathcal{L}_{21}(\lambda)$ explicitly as

$$\mathcal{L}_{21}(\lambda) = C_1\left(L_\epsilon(\lambda) \otimes I_n\right) \quad \text{and} \quad \mathcal{L}_{12}(\lambda) = \left(L_\eta(\lambda)^T \otimes I_m\right)C_2 \quad (3.6)$$

for matrices $C_1 \in \mathrm{M}_{\epsilon n \times \epsilon n}(\mathbb{R})$ and $C_2 \in \mathrm{M}_{\eta m \times \eta m}(\mathbb{R})$. Notice that the expressions of $\mathcal{L}_{12}(\lambda)$ and $\mathcal{L}_{21}(\lambda)$ in (3.6) depend on the choice of ϵ and η but are totally independent of the matrix polynomial $A(\lambda)$.

Now let $A(\lambda) = \sum_{i=0}^{k} A_i \lambda^i \in \mathrm{M}_{m \times n}(\mathbb{R}[\lambda])$ be fixed and of degree $k = \epsilon + \eta + 1$. Considering again the upper-left block $\mathcal{L}_{11}(\lambda)$ in (3.4) we have $\Theta(\mathcal{L}_{11}(\lambda)) = \alpha A(\lambda)$ with $\alpha \in \mathbb{R}$. To determine all the possible solutions for $\mathcal{L}_{11}(\lambda)$ of this equation, first assume $\alpha \equiv 1$ and observe that the matrix pencil $\Sigma_{\eta,A}(\lambda)$ given by

$$\Sigma_{\eta,A}(\lambda) := \begin{bmatrix} \lambda A_k + A_{k-1} & A_{k-2} & \cdots & A_\eta \\ & & & A_{\eta-1} \\ & 0_{\eta m \times \epsilon n} & & \vdots \\ & & & A_0 \end{bmatrix} \in \mathrm{M}_{(\eta+1)m \times (\epsilon+1)n}(\mathbb{R}_1[\lambda])$$

satisfies

$$\Theta(\Sigma_{\eta,A}(\lambda)) = A(\lambda). \tag{3.7}$$

This can easily be checked by a direct calculation, see also [34, Thm. 5.4]. Moreover, for any other $(\eta + 1)m \times (\epsilon + 1)n$ matrix pencil $Q(\lambda)$ satisfying $\Theta(Q(\lambda)) = \alpha A(\lambda)$ for any $\alpha \in \mathbb{R}$ we obtain

$$\Theta\big(\alpha \Sigma_{\eta,A}(\lambda) - Q(\lambda)\big) = \alpha \Theta\big(\Sigma_{\eta,A}(\lambda)\big) - \Theta\big(Q(\lambda)\big)$$
$$= \alpha A(\lambda) - \alpha A(\lambda) = 0_{m \times n}.$$

Thus, interpreting Θ as a function mapping $(\eta + 1)m \times (\epsilon + 1)n$ matrix pencils $\mathcal{L}_{11}(\lambda)$ to $m \times n$ matrix polynomials $\Theta(\mathcal{L}_{11}(\lambda))$, Θ is linear (that is, a vector space homomorphism). The next important observation is that $\mathrm{range}(\Theta) = \mathrm{M}_{m \times n}(\mathbb{R}_k[\lambda])$ holds where here and in the following $\mathrm{range}(\Theta)$ denotes the image/range of Θ.

To accept this, first assume that $A(\lambda) = \sum_{i=0}^{k} A_i \lambda^i \in \mathrm{M}_{m \times n}(\mathbb{R}[\lambda])$ has exactly degree $k = \epsilon + \eta + 1$. Then we have $\Theta(\Sigma_{\eta,A}(\lambda)) = A(\lambda)$, i.e. in particular $A(\lambda) \in \mathrm{range}(\Theta)$ holds. Now suppose $A(\lambda) = \sum_{i=0}^{d} A_i \lambda^i$ has degree $d < k = \epsilon + \eta + 1$. Then, setting

$$A_{d+1} = \cdots = A_k = 0_{m \times n} \quad \text{and} \quad A_\star(\lambda) := \sum_{i=0}^{k} A_i \lambda^i$$

we obtain $\Theta(\Sigma_{\eta,A_\star}(\lambda)) = A(\lambda)$. Thus $A(\lambda) \in \mathrm{range}(\Theta)$ once more holds. Since $\deg(\Theta(\mathcal{L}(\lambda))) \leq \epsilon + \eta + 1$ always holds for any matrix pencil $\mathcal{L}(\lambda)$, we additionally have $A(\lambda) \notin \mathrm{range}(\Theta)$ whenever $\deg(A(\lambda)) > \epsilon + \eta + 1$. In particular, we conclude that $\mathrm{range}(\Theta) = \mathrm{M}_{m \times n}(\mathbb{R}_k[\lambda])$, i.e. Θ is surjective. Notice that this result holds independently of the particular choice for ϵ and η.

Now the homomorphism theorem may be applied to the linear mapping Θ and yields that the $\mathbb{R}$-vector spaces $\mathrm{range}(\Theta) = \mathrm{M}_{m \times n}(\mathbb{R}_k[\lambda])$ and $\mathrm{M}_{(\eta+1)m \times (\epsilon+1)n}(\mathbb{R}_1[\lambda])$ modulo $\mathrm{null}(\Theta)$ are isomorphic. Therefore, we may determine the dimension of $\mathrm{null}(\Theta)$ and obtain $\dim(\mathrm{null}(\Theta)) = (\eta(\epsilon + 1) + (\eta + 1)\epsilon)mn$.

In our next step, we will characterize $\mathrm{null}(\Theta)$. To this end, note that the set $\mathcal{N}_{\epsilon,\eta}$ of all $(\eta + 1)m \times (\epsilon + 1)n$ matrix pencils $\mathcal{M}(\lambda)$ of the form

$$\mathcal{M}(\lambda) = B_1\big(L_\epsilon(\lambda) \otimes I_n\big) + \big(L_\eta(\lambda)^T \otimes I_m\big)B_2 \tag{3.8}$$

with arbitrary matrices $B_1 \in \mathrm{M}_{(\eta+1)m \times \epsilon n}(\mathbb{R})$ and $B_2 \in \mathrm{M}_{\eta m \times (\epsilon+1)n}(\mathbb{R})$ form a real vector space completely contained in $\mathrm{null}(\Theta)$ (as $\Theta(\mathcal{M}(\lambda)) = 0$ holds independently of the particular choice of B_1 and B_2). Thus, $\mathcal{N}_{\epsilon,\eta} \subseteq \mathrm{null}(\Theta)$. Moreover, we claim that the mapping $(B_1, B_2) \mapsto \mathcal{M}(\lambda)$ as in (3.8) is injective. The verification of this claim will give us $\mathrm{null}(\Theta) = \mathcal{N}_{\epsilon,\eta}$.

To see this, assume that $\mathcal{M}(\lambda) = 0$ holds in (3.8) for some B_1, B_2. Now, multiplying (3.8) with $\Lambda_\epsilon(\lambda)^T \otimes I_n$ yields $0 = (L_\eta(\lambda)^T \otimes I_m)B_2(\Lambda_\epsilon(\lambda)^T \otimes I_n)$ since $L_\epsilon(\lambda) \otimes I_n$ and $\Lambda_\epsilon(\lambda) \otimes I_n$ are dual minimal bases, i.e. $(L_\epsilon(\lambda) \otimes I_n)(\Lambda_\epsilon(\lambda)^T \otimes I_n) = 0$. Since $\mathcal{N}_r(L_\eta(\lambda)^T \otimes I_m) = \{0\}$ as $L_\eta(\lambda)^T \otimes I_m$ has full normal rank, it immediately follows that $B_2(\Lambda_\epsilon(\lambda)^T \otimes I_n) = 0$ has to hold. In turn this implies $B_2 = 0$ due to the basis property of the monomials. Now (3.8) becomes $0 = B_1(L_\epsilon(\lambda) \otimes I_n)$ which implies (since $\mathcal{N}_r(L_\epsilon(\lambda)^T \otimes I_n) = \{0\}$) $B_1 = 0$. In conclusion, $\mathcal{M}(\lambda) = 0$ in (3.8) implies $B_1 = 0$ and $B_2 = 0$ and our claim is proved.

Therefore, we conclude that $\dim(\mathcal{N}_{\epsilon,\eta}) = (\eta(\epsilon + 1) + (\eta + 1)\epsilon)mn$ which in turn implies $\mathcal{N}_{\epsilon,\eta} = \mathrm{null}(\Theta)$. Since Θ is a homomorphism, we may express any solution $Q(\lambda)$ of $\Theta(Q(\lambda)) = \alpha A(\lambda)$ as $Q(\lambda) = \alpha \Sigma_{\eta,A}(\lambda) + N(\lambda)$ with $N(\lambda) \in \mathrm{null}(\Theta) = \mathcal{N}_{\epsilon,\eta}$. Together with Proposition 3.2 and (3.6) we obtain the following characterization of the block Kronecker ansatz space $\mathbb{G}_{\eta+1}(A)$.

Theorem 3.1 (Characterization of $\mathbb{G}_{\eta+1}(A)$, [42, Thm. 1]). *Let* $A(\lambda) \in \mathrm{M}_{m \times n}(\mathbb{R}[\lambda])$ *be of degree* $k = \eta + \epsilon + 1$. *Then* $\mathbb{G}_{\eta+1}(A)$ *is a vector space over* $\mathbb{R}$ *having dimension*

$$\dim\big(\mathbb{G}_{\eta+1}(A)\big) = (\epsilon n + \eta m)^2 + (\epsilon + \eta)mn + 1. \tag{3.9}$$

Any matrix pencil $\mathcal{L}(\lambda) \in \mathbb{G}_{\eta+1}(A)$ *may be characterized as*

$$\mathcal{L}(\lambda) = \left[\begin{array}{c|c} \alpha \Sigma_{\eta,A}(\lambda) + B_1(L_\epsilon(\lambda) \otimes I_n) + (L_\eta(\lambda)^T \otimes I_m)B_2 & (L_\eta(\lambda)^T \otimes I_m)C_2 \\ \hline C_1(L_\epsilon(\lambda) \otimes I_n) & 0 \end{array} \right] \tag{3.10}$$

for some scalar $\alpha \in \mathbb{R}$ *and some matrices* $B_1 \subset \mathrm{M}_{(\eta+1)m \times \epsilon n}(\mathbb{R})$, $B_2 \in \mathrm{M}_{\eta m \times (\epsilon+1)n}(\mathbb{R})$, $C_1 \in \mathrm{M}_{\epsilon n \times \epsilon n}(\mathbb{R})$ *and* $C_2 \in \mathrm{M}_{\eta m \times \eta m}(\mathbb{R})$.

The dimension of $\mathbb{G}_{\eta+1}(A)$ is just the sum of the dimensions of the constant matrices B_1, B_2, C_1, C_2 in expression (3.10) plus one for the scalar α. Moreover, notice that (3.9) is consistent with the choice $\epsilon = \eta = 0$ for $A(\lambda)$ of degree $k = 1$ since, in this case, $\mathbb{G}_1(A)$ simply consists of all multiples $\alpha A(\lambda), \alpha \in \mathbb{R}$, of $A(\lambda)$.

It is important to note that, for any $A(\lambda) \in \mathrm{M}_{m \times n}(\mathbb{R}[\lambda])$ of degree $\epsilon + \eta + 1$, any $\mathcal{L}(\lambda) \in \mathbb{G}_{\eta+1}(A)$ given in the form (3.10) can be factorized uniquely as

$$\mathcal{L}(\lambda) = \left[\begin{array}{c|c} I_{(\eta+1)m} & B_1 \\ \hline 0 & C_1 \end{array} \right] \left[\begin{array}{c|c} \alpha \Sigma_{\eta,A}(\lambda) & L_\eta(\lambda)^T \otimes I_m \\ \hline L_\epsilon(\lambda) \otimes I_n & 0 \end{array} \right] \left[\begin{array}{c|c} I_{(\epsilon+1)n} & 0 \\ \hline B_2 & C_2 \end{array} \right]. \tag{3.11}$$

We will call the factorization (3.11) of any $\mathcal{L}(\lambda) \in \mathbb{G}_{\eta+1}(A)$ its *natural factorization corresponding to* $\Sigma_{\eta,A}(\lambda)$. The linearity of Θ and the discussion preceding Theorem 3.1 show that $\Sigma_{\eta,A}(\lambda)$ in (3.10) or (3.11) can be replaced by any other $(\eta+1)m \times (\epsilon+1)n$

pencil $M(\lambda)$ such that $\Theta(M(\lambda)) = A(\lambda)$ holds. If any such $M(\lambda)$ different than $\Sigma_{\eta,A}(\lambda)$ is chosen, Theorem 3.1 (in particular the uniqueness result) is still valid. However, the matrices B_1 and B_2 in the characterization (3.10) will change whenever some different $M(\lambda)$ is used instead of $\Sigma_{\eta,A}(\lambda)$. The matrices C_1 and C_2 are not affected. Making use of this fact we obtain the following proposition. Notice that it is only valid in case $A(\lambda)$ is square.

Proposition 3.3. *Let* $A(\lambda) \in \mathrm{M}_{n \times n}(\mathbb{R}[\lambda])$ *be of degree* $k = \epsilon + \eta + 1$*. Then* $\mathcal{L}(\lambda) \in \mathbb{G}_{\eta+1}(A)$ *holds iff* $\mathcal{L}(\lambda)^{\mathcal{B}} \in \mathbb{G}_{k-\eta}(A)$*.*

Proof. Let $\mathcal{L}(\lambda) \in \mathbb{G}_{\eta+1}(A)$ be given as in (3.10). Then

$$\mathcal{L}(\lambda)^{\mathcal{B}} = \left[\begin{array}{c|c} \alpha\Sigma_{\eta,A}(\lambda)^{\mathcal{B}} + (L_\epsilon(\lambda)^T \otimes I_n)B_1^{\mathcal{B}} + B_2^{\mathcal{B}}(L_\eta(\lambda) \otimes I_n) & (L_\epsilon(\lambda)^T \otimes I_n)C_1^{\mathcal{B}} \\ \hline C_2^{\mathcal{B}}(L_\eta(\lambda) \otimes I_n) & 0 \end{array} \right].$$

Here, for instance, $((L_\eta(\lambda)^T \otimes I_n)C_2)^{\mathcal{B}} = C_2^{\mathcal{B}}(L_\eta(\lambda) \otimes I_n)$ holds since any nonzero $n \times n$ block of $L_\eta(\lambda)^T \otimes I_n$ either has the form $-I_n$ or λI_n and thus commutes with any $n \times n$ block in $C_2{}^2$, see also [80, Lem. 3.1.4]. Now $\mathcal{L}(\lambda)^{\mathcal{B}}$ can be factorized similarly to (3.11) as

$$\mathcal{L}(\lambda)^{\mathcal{B}} = \left[\begin{array}{c|c} I_{(\epsilon+1)n} & B_2^{\mathcal{B}} \\ \hline 0 & C_2^{\mathcal{B}} \end{array} \right] \left[\begin{array}{c|c} \alpha\Sigma_{\eta,A}(\lambda)^{\mathcal{B}} & L_\epsilon(\lambda)^T \otimes I_n \\ \hline L_\eta(\lambda) \otimes I_n & 0 \end{array} \right] \left[\begin{array}{c|c} I_{(\eta+1)n} & 0 \\ \hline B_1^{\mathcal{B}} & C_1^{\mathcal{B}} \end{array} \right]$$

which is the natural factorization of $\mathcal{L}(\lambda)^{\mathcal{B}}$ as an element of $\mathbb{G}_{\epsilon+1}(A) = \mathbb{G}_{k-\eta}(A)$ corresponding to $\Sigma_{\eta,A}(\lambda)^{\mathcal{B}}$. For $\mathbb{G}_{\epsilon+1}(A)$ is is straight forward to check that $\Theta(\Sigma_{\eta,A}(\lambda)^{\mathcal{B}}) = A(\lambda)$ holds. $\square$

Therefore, for any $\mathcal{L}(\lambda) \in \mathbb{G}_{\eta+1}(A)$ as in (3.11) and $A(\lambda) \in \mathrm{M}_{n \times n}(\mathbb{R}[\lambda])$ we have

$$\mathcal{L}(\lambda)^{\mathcal{B}} = \left[\begin{array}{c|c} I_{(\epsilon+1)n} & 0 \\ \hline B_2 & C_2 \end{array} \right]^{\mathcal{B}} \left[\begin{array}{c|c} \alpha\Sigma_{\eta,A}(\lambda) & L_\eta(\lambda)^T \otimes I_n \\ \hline L_\epsilon(\lambda) \otimes I_n & 0 \end{array} \right]^{\mathcal{B}} \left[\begin{array}{c|c} I_{(\eta+1)n} & B_1 \\ \hline 0 & C_1 \end{array} \right]^{\mathcal{B}} \qquad (3.12)$$

which is a pencil from $\mathbb{G}_{k-\eta}(A)$. In particular we have that $\mathbb{G}_{\eta+1}(A) \cong \mathbb{G}_{k-\eta}(A)$ are isomorphic if $A(\lambda)$ is square and block-transposition induces an isomorphism between $\mathbb{G}_{\eta+1}(A)$ and $\mathbb{G}_{k-\eta}(A)$. The mapping $\mathcal{L}(\lambda) \mapsto \mathcal{L}(\lambda)^{\mathcal{B}}$ from $\mathbb{G}_{\eta+1}(A)$ to $\mathbb{G}_{k-\eta}(A)$ (and vice versa) will be called *fundamental isomorphism* - according to [80, Thm. 3.1.6] - where the same name was chosen for the block-transposition map to establish the isomorphism $\mathbb{L}_1(A) \cong \mathbb{L}_2(A)$. The relation between $\mathbb{G}_{\eta+1}(A)$ and $\mathbb{G}_{k-\eta}(A)$ will be investigated deeply in Chapter 4.

Regarding Theorem 3.1 and the factorization in (3.11), the family $(\mathcal{F}_{\alpha,\eta,A}(\lambda))_{\alpha \in \mathbb{R}}$ of matrix pencils

$$\mathcal{F}_{\alpha,\eta,A}(\lambda) := \left[\begin{array}{c|c} \alpha\Sigma_{\eta,A}(\lambda) & L_\eta(\lambda)^T \otimes I_m \\ \hline L_\epsilon(\lambda) \otimes I_n & 0 \end{array} \right] \qquad (3.13)$$

---

[2]For arbitrary matrix polynomials $A(\lambda), B(\lambda)$ certainly $(A(\lambda)B(\lambda))^{\mathcal{B}} = B(\lambda)^{\mathcal{B}}A(\lambda)^{\mathcal{B}}$ does *not* hold.

plays a key role for the characterization of $\mathbb{G}_{\eta+1}(A)$. As in [42], these elements are called *anchor pencils* for $\mathbb{G}_{\eta+1}(A)$ since any $\mathcal{L}(\lambda) \in \mathbb{G}_{\eta+1}(A)$ can be obtained from $\mathcal{F}_{\alpha,\eta,A}(\lambda)$ (for some $\alpha \in \mathbb{R}$) by pre- and postmultiplication of $\mathcal{F}_{\alpha,\eta,A}(\lambda)$ by certain constant matrices. Notice that $(\mathcal{F}_{\alpha,\eta,A}(\lambda))_{\alpha \in \mathbb{R}}$ belong to the family of block Kronecker pencils, cf. Definition 2.16 or [34], and that any $\mathcal{F}_{\alpha,\eta,A}(\lambda)$ is always a strong linearization of $\alpha A(\lambda)$ if $\alpha \neq 0$, cf. (2.18). This fact will be useful several times in the sequel.

If $\mathcal{K}(\lambda) \in \mathbb{G}_{\eta+1}(A)$ is another pencil beside $\mathcal{L}(\lambda) \in \mathbb{G}_{\eta+1}(A)$ from (3.11), e.g.

$$\mathcal{K}(\lambda) = \left[\begin{array}{c|c} I_{(\eta+1)m} & \tilde{B}_1 \\ \hline 0 & \tilde{C}_1 \end{array}\right] \left[\begin{array}{c|c} \beta\Sigma_{\eta,A}(\lambda) & L_\eta(\lambda)^T \otimes I_m \\ \hline L_\epsilon(\lambda) \otimes I_n & 0 \end{array}\right] \left[\begin{array}{c|c} I_{(\epsilon+1)n} & 0 \\ \hline \tilde{B}_2 & \tilde{C}_2 \end{array}\right], \qquad (3.14)$$

with $\beta \in \mathbb{R}$ and some matrices $\tilde{B}_1, \tilde{B}_2, \tilde{C}_1, \tilde{C}_2$, then the sum $\mathcal{L}(\lambda) + \mathcal{K}(\lambda)$ is easily determined as

$$\left[\begin{array}{c|c} I_{(\eta+1)m} & B_1 + \tilde{B}_1 \\ \hline 0 & C_1 + \tilde{C}_1 \end{array}\right] \left[\begin{array}{c|c} (\alpha+\beta)\Sigma_{\eta,A}(\lambda) & L_\eta(\lambda)^T \otimes I_m \\ \hline L_\epsilon(\lambda) \otimes I_n & 0 \end{array}\right] \left[\begin{array}{c|c} I_{(\epsilon+1)n} & 0 \\ \hline B_2 + \tilde{B}_2 & C_2 + \tilde{C}_2 \end{array}\right].$$

Thus adding two matrix pencils $\mathcal{L}(\lambda)$ and $\mathcal{K}(\lambda)$ from $\mathbb{G}_{\eta+1}(A)$ is simply realized by adding the constant matrices (and the scalars) in their corresponding positions. The following example shows a typical matrix pencil from a block Kronecker ansatz space in the form (3.10) and (3.11).

Example 3.1 ([42, Ex. 1]). *Let* $A(\lambda) = \sum_{k=0}^{6} A_k \lambda^k \in M_{m \times n}(\mathbb{R})[\lambda]$ *be of degree* $k = 6$ *and consider the case* $\eta = 3, \epsilon = 2$. *According to (3.10) we may construct the following matrix pencil*

$$\mathcal{L}(\lambda) = \left[\begin{array}{cccc|ccc} \lambda A_6 + A_5 & A_4 & A_3 & 0 & -F & H \\ J & -(B + \lambda J) & A_2 & 0 & E + \lambda F & -\lambda H \\ -A_3 & \lambda B & A_1 & D & -\lambda E & 0 \\ \lambda A_3 & 0 & A_0 & -\lambda D & 0 & 0 \\ \hline C & -(G + \lambda C) & \lambda G & 0 & 0 & 0 \\ 0 & C & -\lambda C & 0 & 0 & 0 \end{array}\right]$$

with some matrices $B, J \in M_{m \times n}(\mathbb{R}), C, G \in M_{n \times n}(\mathbb{R})$ *and* $D, E, F, H \in M_{m \times m}(\mathbb{R})$. *It is not hard to see that* $\mathcal{L}(\lambda) \in \mathbb{G}_{\eta+1}(A) = \mathbb{G}_4(A)$ *since* $\mathcal{L}(\lambda)$ *may be expressed in the form (3.11) with* $\alpha = 1$ *and*

$$\begin{bmatrix} B_1 \\ C_1 \end{bmatrix} = \left[\begin{array}{cc} 0 & 0 \\ -J & 0 \\ 0 & 0 \\ 0 & 0 \\ \hline -C & G \\ 0 & -C \end{array}\right] \quad \text{and} \quad \begin{bmatrix} B_2 & C_2 \end{bmatrix} = \left[\begin{array}{ccc|ccc} 0 & 0 & 0 & 0 & F & -H \\ 0 & B & 0 & 0 & -E & 0 \\ A_3 & 0 & 0 & -D & 0 & 0 \end{array}\right].$$

As for the classical ansatz spaces $\mathbb{L}_1(A), \mathbb{L}_2(A)$ and $\mathbb{DL}(A)$, not every $\mathcal{L}(\lambda) \in$ $\mathbb{G}_{\eta+1}(A)$ is a strong linearization for $A(\lambda) \in M_{m\times n}(\mathbb{R}[\lambda])$ (regarding $\mathbb{L}_1(A)$ consult, e.g., [89, Sec. 4.1]). As the next theorem will reveal, $\mathcal{L}(\lambda)$ in Example 3.1 is guaranteed to be a strong linearization for $A(\lambda)$ if C, D, E and H are all nonsingular matrices. In case $m = n$ and $A(\lambda) \in M_{n\times n}(\mathbb{R}[\lambda])$ is regular, these three conditions turn out to be sufficient and necessary for $\mathcal{L}(\lambda)$ being a strong linearization for $A(\lambda)$. Surprisingly, the choice of J and B does, in either case, not have any effect in that regard. The following Theorem 3.2 is central for determining (strong) linearizations in $\mathbb{G}_{\eta+1}(A)$.

Theorem 3.2 (Linearization Condition, [42, Thm. 2]). *Let $A(\lambda) \in M_{m\times n}(\mathbb{R}[\lambda])$ be of degree $k = \eta + \epsilon + 1$ and let $\mathcal{L}(\lambda) \in \mathbb{G}_{\eta+1}(A)$ be as in* (3.11). *Then $\mathcal{L}(\lambda)$ is a strong linearization for $A(\lambda)$ if $\alpha \neq 0$ and*

$$\left[\begin{array}{c|c} I_{(\eta+1)m} & B_1 \\ \hline 0 & C_1 \end{array}\right] \in GL_{(\eta+1)m+\epsilon n}(\mathbb{R}) \ \ and \ \ \left[\begin{array}{c|c} I_{(\epsilon+1)n} & 0 \\ \hline B_2 & C_2 \end{array}\right] \in GL_{(\epsilon+1)n+\eta m}(\mathbb{R}). \quad (3.15)$$

Certainly (3.15) *is equivalent to* $\det(C_1), \det(C_2) \neq 0$, *i.e.* $C_1 \in M_{\epsilon n \times \epsilon n}(\mathbb{R})$ *and* C_2 $\in M_{\eta m \times \eta m}(\mathbb{R})$ *being nonsingular.*

Proof. Assuming the matrices

$$U := \left[\begin{array}{c|c} I_{(\eta+1)m} & B_1 \\ \hline 0 & C_1 \end{array}\right] \ \ and \ \ V := \left[\begin{array}{c|c} I_{(\epsilon+1)n} & 0 \\ \hline B_2 & C_2 \end{array}\right] \quad (3.16)$$

are nonsingular, $\mathcal{L}(\lambda)$ in (3.11) is strict equivalent to $\mathcal{F}_{\alpha,\eta,A}(\lambda)$ in (3.13), cf. Definition 2.12 (ii). According to (2.17) and (2.18), $\alpha \neq 0$ implies $\mathcal{F}_{\alpha,\eta,A}(\lambda)$ to be a strong linearization for $\alpha A(\lambda)$ (see also [34, Thm. 5.2]). Since $\alpha A(\lambda)$ and $A(\lambda)$ are strict equivalent in case $\alpha \neq 0$, $\mathcal{L}(\lambda)$ is a strong linearization for $A(\lambda)$ as well. $\qquad\square$

Remark 3.1 ([42, Rem. 2]). *Given the case of $A(\lambda)$ being regular, Theorem 3.2 becomes an equivalence. In fact, if $\mathcal{L}(\lambda) \in \mathbb{G}_{\eta+1}(A)$ as in* (3.11) *is a strong linearization for some regular $A(\lambda) \in M_{n\times n}(\mathbb{R}[\lambda])$ of degree $k = \epsilon + \eta + 1$, $\mathcal{L}(\lambda)$ is necessarily regular. This implies U and V in* (3.16) *to be nonsingular. Moreover, $\alpha \neq 0$ has to hold since*

$$\det(\mathcal{F}_{0,\eta,A}(\lambda)) = \det\left(\left[\begin{array}{c|c} 0 & L_\eta(\lambda)^T \otimes I_m \\ \hline L_\epsilon(\lambda) \otimes I_n & 0 \end{array}\right]\right) = 0$$

according to Definition 2.1 regardless of the choice of η and ϵ.

Note that any pencil $\mathcal{L}(\lambda) \in \mathbb{G}_{\eta+1}(A)$ that violates any of the conditions in Theorem 3.2 is automatically singular. For singular $A(\lambda) \in M_{m\times n}(\mathbb{R}[\lambda])$, (3.15) is in fact *not* necessary for $\mathcal{L}(\lambda) \in \mathbb{G}_{\eta+1}(A)$ to be a (strong) linearization for $A(\lambda)$; consult [31, Ex. 2] for a linearization $\mathcal{L}(\lambda) \in \mathbb{G}_1(A)$ that violates (3.15). The situation for singular matrix polynomials is analyzed more deeply in Section 7.3 (see also [45, Sec. 5]).

3.2 Advanced structural analysis of $\mathbb{G}_{\eta+1}(A)$

Let $A(\lambda) \in M_{m \times n}(\mathbb{R}[\lambda])$ be arbitrary. We now consider the question of how rich $\mathbb{G}_{\eta+1}(A)$ is assembled with (strong) linearizations. In case $m = n$, almost every pencil from $\mathbb{L}_1(A)$ and $\mathbb{L}_2(A)$ is a strong linearization for $A(\lambda)$ regardless whether $A(\lambda)$ is regular or singular, cf. [89, Thm. 4.7] and [31, Thm. 4.4]. Here, a similar statement holds for $\mathbb{G}_{\eta+1}(A)$ where $A(\lambda)$ can be square or rectangular. Thus, Theorem 3.3 below can be interpreted as an analogue to Theorem 4.7 from [89] for block Kronecker ansatz spaces.

Theorem 3.3 (Genericity of Linearizations, [42, Thm. 3]). *Let $A(\lambda) \in M_{m \times n}(\mathbb{R}[\lambda])$ be of degree $k = \epsilon + \eta + 1$. Then, for any $\eta, 0 \le \eta \le k - 1$, almost every matrix pencil in $\mathbb{G}_{\eta+1}(A)$ is a strong linearization for $A(\lambda)$.*

Proof. Theorem 3.3 follows immediately from Theorem 3.2 since $GL_{\epsilon n}(\mathbb{R}), GL_{\eta m}(\mathbb{R})$ and $\mathbb{R} \setminus \{0\}$ are dense subsets of $M_{\epsilon n \times \epsilon n}(\mathbb{R}), M_{\eta m \times \eta m}(\mathbb{R})$ and $\mathbb{R}$ respectively. $\square$

The following Theorem 3.4 was first stated and proven for the ansatz space $\mathbb{L}_1(A)$ in [89, Thm. 4.3] and reveals the central relationship between linearizations, strong linearizations and regular matrix pencils in $\mathbb{L}_1(A)$ in case $A(\lambda) \in M_{n \times n}(\mathbb{R}[\lambda])$ is regular. We now prove that this theorem substantially extends to the family of block Kronecker ansatz spaces $\mathbb{G}_{\eta+1}(A), 0 \le \eta \le k - 1$ (which shows the structural similarities between $\mathbb{L}_1(A)$ and the block Kronecker ansatz spaces).

Theorem 3.4 (Strong Linearization Theorem, [42, Thm. 4]). *Let $A(\lambda) \in M_{n \times n}(\mathbb{R}[\lambda])$ be regular and of degree $k = \epsilon + \eta + 1$ and let $\mathcal{L}(\lambda) \in \mathbb{G}_{\eta+1}(A)$. Then the following statements are equivalent:*

(i) $\mathcal{L}(\lambda)$ is a linearization for $A(\lambda)$.

(ii) $\mathcal{L}(\lambda)$ is a regular matrix pencil.

(iii) $\mathcal{L}(\lambda)$ is a strong linearization for $A(\lambda)$.

Proof. Since $(iii) \Rightarrow (i) \Rightarrow (ii)$ is obvious, we only need to show $(ii) \Rightarrow (iii)$. To this end, assume $\mathcal{L}(\lambda)$ in (3.11) to be regular. Then, since

$$0 \neq \det(\mathcal{L}(\lambda)) = \det\left(\left[\begin{array}{c|c} I_{(\eta+1)n} & B_1 \\ \hline 0 & C_1 \end{array}\right]\right) \det(\mathcal{F}_{\alpha,\eta,A}(\lambda)) \det\left(\left[\begin{array}{c|c} I_{(\epsilon+1)n} & 0 \\ \hline B_2 & C_2 \end{array}\right]\right),$$

this certainly requires the nonsingularity of U and V as in (3.16) and the regularity of $\mathcal{F}_{\alpha,\eta,A}(\lambda)$ as in (3.13), that is $\det(\mathcal{F}_{\alpha,\eta,A}(\lambda)) \neq 0$. However, assuming $\alpha = 0$, $\det(\mathcal{F}_{0,\eta,A}(\lambda)) = 0$ follows, cf. Remark 3.1. Therefore, the assumption of $\mathcal{L}(\lambda) \in \mathbb{G}_{\eta+1}(A)$ being regular also implies $\alpha \neq 0$ and thus the validity of all three conditions in Theorem 3.2. Thus, if $\mathcal{L}(\lambda)$ is regular, it is a strong linearization for $A(\lambda)$. $\square$

Certainly all conditions stated in Theorem 3.4 are, for regular $A(\lambda) \in M_{n \times n}(\mathbb{R}[\lambda])$, equivalent to the linearization condition from Theorem 3.2.

3.3   The recovery of eigenvectors

Beside the computation of eigenvalues of some $A(\lambda) \in M_{m \times n}(\mathbb{R}[\lambda])$ it is occasionally important to have the corresponding eigenspaces available, too. Thus, if the eigenproblem is solved for $A(\lambda)$ via any linearization $\mathcal{L}(\lambda) \in \mathbb{G}_{\eta+1}(A)$ it is desirable to be able to recover the eigenspaces $\mathrm{Eig}_r(A, \lambda_0)$ and $\mathrm{Eig}_\ell(A, \lambda_0)$ for any $\lambda_0 \in \sigma(A)$ from $\mathrm{Eig}_r(\mathcal{L}, \lambda_0)$ and $\mathrm{Eig}_\ell(\mathcal{L}, \lambda_0)$ at low costs. The following theorems show how this can be achieved.

We begin with the recovering of right and left nullspaces in case $A(\lambda)$ is singular.

Theorem 3.5. *Let* $A(\lambda) \in M_{m \times n}(\mathbb{R}[\lambda])$ *be singular and of degree* $k = \epsilon + \eta + 1$. *Moreover, let* $\mathcal{L}(\lambda) \in \mathbb{G}_{\eta+1}(A)$ *be of the form* (3.11) *so that* U *and* V *as in* (3.16) *are nonsingular. Then the following statements hold*

(i) *Let* $q(\lambda) \in \mathcal{N}_r(\mathcal{L}(\lambda))$. *Then* $q_\star(\lambda) := [q_{\epsilon n+1}(\lambda) \; \cdots \; q_{(\epsilon+1)n}]^T \in \mathcal{N}_r(A(\lambda))$.

(ii) *Let* $q(\lambda) \in \mathcal{N}_\ell(\mathcal{L}(\lambda))$. *Then* $q_\star(\lambda) := [q_{\eta m+1}(\lambda) \; \cdots \; q_{(\eta+1)m}]^T \in \mathcal{N}_\ell(A(\lambda))$.

Proof. We start by considering the general situation for $\mathcal{F}_{\alpha,\eta,A}(\lambda)$ from which the situation for $\mathcal{L}(\lambda) = U\mathcal{F}_{\alpha,\eta,A}(\lambda)V \in \mathbb{G}_{\eta+1}(A)$ can be deduced. This particular result can also be found as a consequence of [34, Thm. 7.6]. Here we give a direct and different proof. Moreover, we restrict our attention to vector polynomials.
(i) Let $\tilde{q}(\lambda) = [p_1(\lambda)^T \; p_2(\lambda)^T]^T \in \mathbb{R}[\lambda]^{(\epsilon+1)n+\eta m}$ with $p_1(\lambda) \in \mathbb{R}[\lambda]^{(\epsilon+1)n}$ be a vector polynomial from $\mathcal{N}_r(\mathcal{F}_{\alpha,\eta,A}(\lambda))$. Then, recalling the form of $\mathcal{F}_{\alpha,\eta,A}(\lambda)$ from (3.13), it holds that

$$\alpha\Sigma_{\eta,A}(\lambda)p_1(\lambda) + (L_\eta(\lambda)^T \otimes I_m)p_2(\lambda) = 0 \qquad \text{and} \qquad (3.17)$$
$$(L_\epsilon(\lambda) \otimes I_n)p_1(\lambda) = 0.$$

Since $\Lambda_\epsilon(\lambda) \otimes I_n$ is a dual minimal basis for $L_\epsilon(\lambda) \otimes I_n$, i.e. the columns of $\Lambda_\epsilon(\lambda)^T \otimes I_n$ are a basis of $\mathcal{N}_r(L_\epsilon(\lambda) \otimes I_n)$, any solution $p(\lambda)$ satisfying $(L_\epsilon(\lambda) \otimes I_n)p(\lambda) = 0$ can be expressed as $p(\lambda) = (\Lambda_\epsilon(\lambda)^T \otimes I_n)r(\lambda) = \Lambda_\epsilon(\lambda)^T \otimes r(\lambda)$ for some $r(\lambda) \in \mathbb{R}[\lambda]^n$. Backward substitution of the ansatz $p_1(\lambda) = \Lambda_\epsilon(\lambda)^T \otimes r(\lambda)$ in (3.17) (with yet unknown $r(\lambda)$) yields $\alpha\Sigma_{\eta,A}(\lambda)\big(\Lambda_\epsilon(\lambda)^T \otimes r(\lambda)\big) + \big(L_\eta(\lambda)^T \otimes I_m\big)p_2(\lambda) = 0$. Multiplying this expression from the left with $\Lambda_\eta(\lambda) \otimes I_m$ eliminates the second summand since $(\Lambda_\eta(\lambda) \otimes I_m)(L_\eta(\lambda)^T \otimes I_m) = 0$ and gives

$$\alpha\big(\Lambda_\eta(\lambda) \otimes I_m\big)\Sigma_{\eta,A}(\lambda)\big(\Lambda_\epsilon(\lambda)^T \otimes I_n\big)r(\lambda) = 0,$$

that is $\alpha\Theta(\Sigma_{\eta,A}(\lambda))r(\lambda) = \alpha A(\lambda)r(\lambda) = 0$. Therefore, we conclude that, whenever $\tilde{q}(\lambda) = [p_1(\lambda)^T \; p_2(\lambda)^T]^T \in \mathcal{N}_r(\mathcal{F}_{\alpha,\eta,A}(\lambda))$, then $p_1(\lambda) = \Lambda_\epsilon(\lambda)^T \otimes r(\lambda)$ holds for some $r(\lambda) \in \mathcal{N}_r(A(\lambda))$. In particular, since $\Lambda_\epsilon(\lambda) = [\, \lambda^\epsilon \; \cdots \; \lambda \; 1\,]$, the last n entries of $p_1(\lambda)$ coincide with $r(\lambda)$. With this observation at hand we now consider the general case $\mathcal{L}(\lambda) \in \mathbb{G}_{\eta+1}(A)$.

Assume $q(\lambda) \in \mathcal{N}_r(\mathcal{L}(\lambda))$ for some $\mathcal{L}(\lambda) = U\mathcal{F}_{\alpha,\eta,A}(\lambda)V \in \mathbb{G}_{\eta+1}(A)$ with U and V as in (3.16) being nonsingular and $\alpha \neq 0$. Then $s(\lambda) := V^{-1}q(\lambda) \in \mathcal{N}_r(\mathcal{F}_{\alpha,\eta,A}(\lambda))$

must hold due to the nonsingularity of U. Therefore, according to the previous discussion,

$$s_\star := [\, s_{\epsilon n+1}(\lambda) \;\cdots\; s_{(\epsilon+1)n}(\lambda) \,]^T \in \mathbb{R}[\lambda]^n$$

is an element from $\mathcal{N}_r(A(\lambda))$. However, due to the form of V we have

$$V^{-1} = \left[\begin{array}{c|c} I_{(\epsilon+1)n} & 0 \\ \hline \overline{B}_2 & \overline{C}_2 \end{array}\right]$$

for some matrices $\overline{B}_2 \in \mathrm{M}_{\eta m \times (\epsilon+1)n}(\mathbb{R})$ and $\overline{C}_2 \in \mathrm{M}_{\eta m \times \eta m}(\mathbb{R})$. Therefore, $s_\star(\lambda) = [\, q_{\epsilon n+1}(\lambda) \;\cdots\; q_{(\epsilon+1)n}(\lambda) \,]^T$. In addition, notice that all $(\epsilon+1)n$ first entries in $s(\lambda)$ and $q(\lambda)$ are the same due to the upper-left identity block in V^{-1}. This proves the first statement.

(*ii*) The result follows analogously to (*i*) considering $\mathcal{N}_\ell(\mathcal{L}(\lambda))$ instead of $\mathcal{N}_r(\mathcal{L}(\lambda))$ and the particular form of U and U^{-1}. $\square$

Since minimal bases for $A(\lambda)$ and $\mathcal{L}(\lambda) \in \mathbb{G}_{\eta+1}(A)$ do not play a central role in the sequel, a detailed study is omitted. A comprehensive treatment on their relationships can be found in [34]. For instance, the following Proposition 3.4 is a direct consequence of [34, Lem. 3.6, Lem. 7.1] and stated here for completeness. The proof is omitted.

Proposition 3.4. *Let $A(\lambda) \in \mathrm{M}_{m \times n}(\mathbb{R}[\lambda])$ be singular and of degree $k = \eta + \epsilon + 1$. In addition, let $\mathcal{L}(\lambda) \in \mathbb{G}_{\eta+1}(A)$ be a strong linearization for $A(\lambda)$ as in (3.11) so that U and V as in (3.16) are nonsingular. Then the following statements hold:*

(*i*) *If $0 \leq \epsilon_1 \leq \epsilon_2 \leq \cdots \leq \epsilon_t$ are the right minimal indices of $A(\lambda)$, then the right minimal indices of $\mathcal{L}(\lambda)$ are given as $\epsilon + \epsilon_1 \leq \epsilon + \epsilon_2 \leq \cdots \leq \epsilon + \epsilon_t$.*

(*ii*) *If $0 \leq \eta_1 \leq \eta_2 \leq \cdots \leq \eta_r$ are the left minimal indices of $A(\lambda)$, then the left minimal indices of $\mathcal{L}(\lambda)$ are given as $\eta + \eta_1 \leq \eta + \eta_2 \leq \cdots \leq \eta + \eta_r$.*

For regular matrix polynomials, the proof of Theorem 3.5 can be adapted considering $\mathcal{L}(\alpha)$ for some $\alpha \in \sigma_f(A)$ or $\mathrm{rev}\mathcal{L}(0)$ for $\infty \in \sigma(A)$ instead of working with the indeterminate λ. We obtain the following result:

Theorem 3.6 (Eigenvector Recovery, [42, Thm. 5]). *Let $A(\lambda) \in \mathrm{M}_{n \times n}(\mathbb{R}[\lambda])$ be regular and of degree $k = \epsilon + \eta + 1$. Moreover, let $\mathcal{L}(\lambda) \in \mathbb{G}_{\eta+1}(A)$ be a strong linearization for $A(\lambda)$. Then the following statements hold*

(*i*) *If $u \in \mathrm{Eig}_r(\mathcal{L}, \beta)$ ($u \in \mathbb{C}^{(\epsilon+1)n+\eta m}$) for some finite eigenvalue $\beta \in \sigma_f(A)$, then $u_\star = [\, u_{\epsilon n+1} \;\cdots\; u_{(\epsilon+1)n} \,]^T \in \mathrm{Eig}_r(A, \beta)$.*

(*ii*) *If $u \in \mathrm{Eig}_r(\mathcal{L}, \infty)$, then $u_\star := [\, u_1 \;\cdots\; u_n \,]^T \in \mathrm{Eig}_r(A, \infty)$.*

(*iii*) *If $y \in \mathrm{Eig}_\ell(\mathcal{L}, \beta)$ ($y \in \mathbb{C}^{(\eta+1)m+\epsilon n}$) for some finite eigenvalue $\beta \in \sigma_f(A)$, then $y_\star = [\, y_{\eta m+1} \;\cdots\; y_{(\eta+1)m} \,]^T \in \mathrm{Eig}_\ell(A, \beta)$.*

(iv) If $y \in \mathrm{Eig}_\ell(\mathcal{L}, \infty)$, then $y_\star = [\, y_1 \; \cdots \; y_m \,]^T \in \mathrm{Eig}_\ell(A, \infty)$.

Proof. The statements of *(i)* and *(iii)* follow by exactly the same reasoning as in the proof of Theorem 3.5 considering $\mathcal{L}(\beta)$ instead of $\mathcal{L}(\lambda)$ for $\beta \in \sigma_f(A)$.

For the proof of *(ii)* we first consider $\mathrm{rev}\mathcal{F}_{\alpha,\eta,A}(\lambda)$ which is, according to Theorem 3.2, a strong linearization for $\mathrm{rev}\alpha A(\lambda)$. Assuming that $\infty \in \sigma(A) = \sigma(\mathcal{L})$ we have $\mathrm{rev}\mathcal{F}_{\alpha,\eta,A}(0)\tilde{q} = 0$ for some $\tilde{q} = [p_1^T \; p_2^T]^T \in \mathbb{R}^{(\epsilon+1)n+\eta m}$. This gives the two equations

$$\mathrm{rev}\Sigma_{\eta,A}(0)p_1 + \mathrm{rev}(L_\eta(0)^T \otimes I_m)p_2 = 0 \quad \text{and}$$
$$\mathrm{rev}(L_\epsilon(0) \otimes I_n)p_1 = 0.$$

We see analogously as in the proof of Theorem 3.5 that $p_1 = \mathrm{rev}\Lambda_\epsilon(0)^T \otimes r$ has to hold for some $r \in \mathbb{R}^n$. In addition, the same argumentation as before shows that $r \in \mathrm{Eig}_r(A, \infty)$. Since $\mathrm{rev}\Lambda_\epsilon(0) = [1 \; 0 \; \cdots \; 0]$, r coincides with the first n entries of p_1. From here on, the rest of the proof proceeds in a similar manner as that of Theorem 3.5 using the form of V. The proof of *(iv)* works completely analogously. $\qquad\square$

Note that eigenvectors of $A(\lambda) \in \mathrm{M}_{m \times n}(\mathbb{R}[\lambda])$ and $\mathcal{L}(\lambda) \in \mathbb{G}_{\eta+1}(A)$ for $\lambda_0 = \infty$ are always real as $\mathrm{rev}\mathcal{F}_{\alpha,\eta,A}(0)$ and $\mathrm{rev}A(0)$ are real. This, of course, will in general not hold for eigenvectors for $\lambda_0 \in \sigma(A), \lambda_0 \in \mathbb{C}$.

3.4   Block Kronecker pencils and Fiedler-like linearizations

In this section we examine the relation between block Kronecker pencils, Fiedler-like linearization families and the block Kronecker ansatz spaces $\mathbb{G}_{\eta+1}(A)$. Block Kronecker pencils have been introduced in [34, Sec. 5] as a new class of linearizations, cf. Definition 2.16. In our context, the whole family of matrix pencils $(\mathcal{F}_{\alpha,\eta,A}(\lambda))_{\alpha \in \mathbb{R}}$ for $A(\lambda) \in \mathrm{M}_{m \times n}(\mathbb{R}[\lambda])$ of degree $k \geq 1$ and any $\eta, 0 \leq \eta \leq k - 1$, belongs to the class of block Kronecker pencils. On the other hand, every real (ϵ, n, η, m)-block Kronecker pencil $\mathcal{L}(\lambda)$ as in (2.17) belongs to the block Kronecker ansatz space $\mathbb{G}_{\eta+1}(A)$ for $A(\lambda) \in \mathrm{M}_{m \times n}(\mathbb{R}[\lambda])$ given by (2.18). In summary we have the following result characterizing the subspace of $\mathbb{G}_{\eta+1}(A)$ containing entirely block Kronecker pencils:

Proposition 3.5. *Let* $A(\lambda) \in \mathrm{M}_{m \times n}(\mathbb{R}[\lambda])$ *be of degree* $k = \eta + \epsilon + 1$. *Then the set of all* $((\eta+1)m + \epsilon n) \times ((\epsilon+1)n + \eta m)$ *matrix pencils* $\mathcal{L}(\lambda)$ *of the form*

$$\mathcal{L}(\lambda) = \left[\begin{array}{c|c} \Sigma_{\eta,A}(\lambda) + B_1(L_\epsilon(\lambda) \otimes I_n) + (L_\eta(\lambda)^T \otimes I_m)B_2 & L_\eta(\lambda)^T \otimes I_m \\ \hline L_\epsilon(\lambda) \otimes I_n & 0 \end{array} \right]. \quad (3.18)$$

coincides with the family of real (ϵ, n, η, m)-*block Kronecker pencils that are strong linearizations for* $A(\lambda)$.

The pencils as in (3.18) form an affine subspace of $\mathbb{G}_{\eta+1}(A)$ obtained from (3.11) with $\alpha = 1$, $C_1 = I_{\epsilon n}$, and $C_2 = I_{\eta m}$. The subfamily $(\mathcal{F}_{\alpha,\eta,A}(\lambda))_{\alpha \in \mathbb{R}}$ is obtained by replacing $\Sigma_{\eta,A}(\lambda)$ by $\alpha \Sigma_{\eta,A}(\lambda)$ and setting $B_1 = 0, B_2 = 0$. Furthermore, (ϵ, n, η, m)-block Kronecker pencils actually belong to the broader class of strong block minimal bases pencils which was introduced in [34, Sec. 3].

Definition 3.2 (Strong Block Minimal Bases Pencil, [34, Def. 3.1]). *A matrix pencil*

$$\mathcal{L}(\lambda) = \left[\begin{array}{c|c} M(\lambda) & K_2(\lambda)^T \\ \hline K_1(\lambda) & 0 \end{array} \right]$$

is called a block minimal bases pencil if $K_1(\lambda)$ and $K_2(\lambda)$ are both minimal bases. If, in addition, the row degrees of $K_1(\lambda)$ and $K_2(\lambda)$ are all equal to one, the row degrees of some minimal basis dual to $K_1(\lambda)$ are all equal and the row degrees of some minimal basis dual to $K_2(\lambda)$ are all equal, $\mathcal{L}(\lambda)$ is called a strong block minimal bases (SBMB) pencil.

According to Definition 2.10 it is easily checked that any strong linearization $\mathcal{L}(\lambda) \in \mathbb{G}_{\eta+1}(A)$ for some regular $A(\lambda) \in M_{m \times n}(\mathbb{R}[\lambda])$ is a SBMB-pencil. In fact $K_1(\lambda) = L_\epsilon(\lambda) \otimes I_n$ and $K_2(\lambda) = L_\eta(\lambda) \otimes I_m$ satisfy all necessary requirements stated in Definiton 3.2 (see also [34, Sec. 3]). Moreover, $C_1(L_\epsilon(\lambda) \otimes I_n)$ and $C_2^T(L_\eta(\lambda) \otimes I_m)$ in (3.10) satisfy these requirements as well and both are minimal bases dual to $\Lambda_\epsilon(\lambda) \otimes I_n$ and $\Lambda_\eta(\lambda) \otimes I_m$ iff C_1 and C_2 are nonsingular. In fact, the reverse statement is true as well. That is, any SBMB-pencil $\mathcal{L}(\lambda) \in \mathbb{G}_{\eta+1}(A)$ is a strong linearization for (any regular or singular) $A(\lambda)$. This follows from [34, Thm. 3.3] where the strong linearization property of SBMB-pencils was proven. For block Kronecker pencils this complies with Theorem 3.2, since, in this case, $\alpha \neq 0$ and $C_1 \in M_{\epsilon n \times \epsilon n}(\mathbb{R}[\lambda])$ and $C_2 \in M_{\eta m \times \eta m}(\mathbb{R}[\lambda])$ are both nonsingular. These results have been essentially used for the proof of Theorem 3.2. In conclusion, Theorem 3.1 provides an effective way for the explicit construction of strong block minimal bases pencils.

In [34, Thm. 4.5] it was shown that any (real) $kn \times kn$ Fiedler pencil $\mathcal{F}(\lambda)$ (recall Definition 2.15) corresponding to some $A(\lambda) \in M_{n \times n}(\mathbb{R}[\lambda])$ with degree k admits a permutation $\mathcal{L}(\lambda) := \Pi_\ell \mathcal{F}(\lambda) \Pi_r$ with two real $kn \times kn$ permutation matrices Π_ℓ, Π_r so that $\mathcal{L}(\lambda)$ has the form in (3.18) for certain values of ϵ, η and certain matrices B_1, B_2 (certainly, $\epsilon + \eta + 1 = k$ holds). Hereby, the values of ϵ and η can be determined by the bijection σ that defined the Fiedler pencil. In other words, any Fiedler pencil $\mathcal{F}(\lambda)$ for $A(\lambda)$ is - modulo permutation - an element of $\mathbb{G}_{\eta+1}(A)$ for some $\eta, 0 \leq \eta \leq k - 1$. Moreover, according to [34, Thm. 4.5], the permuted pencil $\mathcal{L}(\lambda) = \Pi_\ell \mathcal{F}(\lambda) \Pi_r$ of the form (3.18) always has a special structure in its upper-left $(\eta+1)n \times (\epsilon+1)n$ block[3]. Thus it is immediate that not every pencil $\mathcal{L}(\lambda)$ of the form (3.18) (with arbitrary matrices B_1, B_2) can be obtained via permutation of some Fiedler pencil $\mathcal{F}(\lambda)$.

A deeper investigation of all generalized versions of Fiedler pencils shall be omitted here since a detailed analysis can be found in, e.g., [17–21, 34, 120, 121]. However,

---

[3]This structure is called *staircase pattern*, see [34, Def. 4.3]. It is not further discussed here.

we would like to point out that the result about the connection between pencils from block Kronecker ansatz spaces and Fiedler pencils stated above can be extended to generalized types of Fiedler pencils. In particular, in [21, Thm. 7.1] it was shown that the correspondence stated above holds for proper generalized Fiedler pencils, too. So any proper generalized Fiedler pencil $\mathcal{F}(\lambda)$ can be transformed to the form (3.18) via $\Pi_\ell \mathcal{F}(\lambda) \Pi_r$ for two permutation matrices Π_ℓ, Π_r. For generalized Fiedler pencils with repetition, [21, Thm. 8.1] makes a slightly different statement. In particular, every generalized Fiedler pencil with repetition can be permuted into a special block form [21]. These permuted pencils have been called *extended block Kronecker pencils* in [21, Sec. 3.3] and are not necessarily of the form (3.18). However, in our context, the characterization of extended block Kronecker pencils is quite simple. Proposition 3.6 builds a bridge between [21] and our work.

Proposition 3.6. *Let $A(\lambda) \in M_{m \times n}(\mathbb{R}[\lambda])$ be of degree $k = \eta + \epsilon + 1$. The whole family of matrix pencils $\mathcal{L}(\lambda) \in \mathbb{G}_{\eta+1}(A)$ coincides with the set of real extended (ϵ, n, η, m)-block Kronecker pencils*

$$\mathcal{L}(\lambda) = \left[\begin{array}{c|c} M(\lambda) & (L_\eta(\lambda)^T \otimes I_m)C_2 \\ \hline C_1(L_\epsilon(\lambda) \otimes I_n) & 0 \end{array} \right]$$

where $M(\lambda) := \alpha \Sigma_{\eta,A}(\lambda) + B_1(L_\epsilon(\lambda) \otimes I_n) + (L_\eta(\lambda)^T \otimes I_m)B_2$ for some matrices B_1, B_2 and $\alpha \in \mathbb{R}$.

In particular, extended (ϵ, n, η, m)-block Kronecker pencils include the whole family of (ϵ, n, η, m)-block Kronecker pencils. Moreover, according to Theorem 3.1, the family of (real) extended (ϵ, n, η, m)-block Kronecker pencils corresponding to $A(\lambda) \in M_{m \times n}(\mathbb{R}[\lambda])$ can be identified with $\mathbb{G}_{\eta+1}(A)$. Although both structures are (in the real case) identical, they have been introduced in [21] and [42] independently and with different aims. While our aim is to analyze and characterize the vector spaces $\mathbb{G}_{\eta+1}(A)$ similarly to [80, 89] or [85], the authors of [21] are aiming to unify all Fiedler-like families of linearizations in one framework. Their findings on generalized Fiedler pencils with repetition can also be stated from the perspective of block Kronecker ansatz spaces as follows.

According to [21, Thm. 8.1] there exist two $kn \times kn$ permutation matrices Π_ℓ, Π_r for any generalized Fiedler pencil with repetition $\mathcal{F}(\lambda)$ so that $\Pi_\ell \mathcal{F}(\lambda)\Pi_r \in \mathbb{G}_{\eta+1}(A)$ for some $\eta \in \{0, 1, \ldots, k-1\}$. Another type of Fiedler-like linearizations are nonproper generalized Fiedler pencils [21, Def. 4.18]. For any nonproper generalized Fiedler pencil $\mathcal{F}(\lambda)$ corresponding to $A(\lambda)$ it was shown in [21, Thm. 7.3] that there always exist two $kn \times kn$ matrices L, R so that $L\mathcal{F}(\lambda)R \in \mathbb{G}_{\eta+1}(A)$ for some $\eta \in \{0, 1, \ldots, k-1\}$ although L and R are no permutation matrices in general. Therefore, we conclude this section with the observation that for any matrix polynomial $A(\lambda) \in M_{n \times n}(\mathbb{R}[\lambda])$ of degree $k \geq 2$ almost every type of Fiedler-like linearization $\mathcal{F}(\lambda)$ admits a permutation $\mathcal{L}(\lambda) = \Pi_\ell \mathcal{F}(\lambda)\Pi_r$ with two $kn \times kn$ permutation matrices or a transformation $\mathcal{L}(\lambda) = L\mathcal{F}(\lambda)R$ with $L, R \in M_{kn \times kn}(\mathbb{R})$ so that $\mathcal{L}(\lambda) \in \mathbb{G}_{\eta+1}(A)$ holds for some $\eta \in \{0, 1, \ldots, k-1\}$.

Chapter 4

Double block Kronecker ansatz spaces

" There is nothing so practical as a good theory. "

Kurt Lewin, [78]

This chapter is dedicated to the characterization of matrix pencils that belong to two or more block Kronecker ansatz spaces $\mathbb{G}_{j+1}(A)$, $j \in I \subseteq \{0, \dots, k-1\}$, simultaneously. In particular we analyze and characterize the double ansatz spaces $\mathbb{DG}_{\eta+1}(A) = \mathbb{G}_{\eta+1}(A) \cap \mathbb{G}_{k-\eta}(A)$ for any choice of η. Clearly, this study is strongly motivated by that of the double ansatz space $\mathbb{DL}(A)$, cf. [65, 89].

The spaces $\mathbb{DG}_{\eta+1}(A)$ and its elements are characterized in Section 4.1. A deeper analysis is then presented in Section 4.2 where the fundamental results from Chapter 3 and their implications are reformulated in this new context. In Section 4.3 the superpartition property of these intersection spaces is proven. In particular, it is shown that the spaces $\mathbb{DG}_{\eta+1}(A)$ form a nested subspace sequence for increasing values of η. This is a property being indigenous for $\mathbb{DG}_{\eta+1}(A), \eta > 0$, and has no equivalent in the classical setting on $\mathbb{L}_1(A), \mathbb{L}_2(A)$ and $\mathbb{DL}(A)$. The characterization and construction of block-symmetric and symmetric matrix pencils is considered in Section 4.5 and Section 4.6. This chapter concludes with a short summary of the results from this and the preceding chapter in Section 4.7.

4.1 Definition of $\mathbb{DG}_{\eta+1}(A)$ and basic properties

From here on we always assume $m = n$, i.e. we confine ourselves to square matrix polynomials $A(\lambda) \in M_{n \times n}(\mathbb{R}[\lambda])$. We begin with a formal definition of the *double block Kronecker ansatz spaces*.

Definition 4.1 (Double Ansatz Space, [42, Def. 2]). *Let $A(\lambda) \in M_{n \times n}(\mathbb{R}[\lambda])$ be of degree $k = \epsilon + \eta + 1$. Then we define*

$$\mathbb{DG}_{\eta+1}(A) := \mathbb{G}_{\eta+1}(A) \cap \mathbb{G}_{k-\eta}(A).$$

Given any $A(\lambda) \in M_{n \times n}(\mathbb{R}[\lambda])$ of degree $k = \epsilon + \eta + 1$, w.l.o.g. we will always assume $\eta \leq \epsilon = k - \eta - 1$ from now. This is reasonable since

$$\mathbb{DG}_{\eta+1}(A) = \mathbb{G}_{\eta+1}(A) \cap \mathbb{G}_{k-\eta}(A) = \mathbb{G}_{k-\epsilon}(A) \cap \mathbb{G}_{\epsilon+1}(A) = \mathbb{DG}_{\epsilon+1}(A). \quad (4.1)$$

Notice further that $\eta + 1 = k - \eta$ implies $k = 2\eta + 1$. Therefore, the special case $\mathbb{DG}_{\eta+1}(A) = \mathbb{G}_{\eta+1}(A) \cap \mathbb{G}_{\eta+1}(A)$ can only occur for $A(\lambda)$ having odd degree and $\eta = \epsilon = (k-1)/2$. Our first goal in this section will be to comprehensively characterize $\mathbb{DG}_{\eta+1}(A)$ for any $A(\lambda) \in M_{n \times n}(\mathbb{R}[\lambda])$ as this was done for $\mathbb{DL}(A)$ in [80, Sec. 3, 4].

Recalling Proposition 3.3 it is immediate that $\mathbb{DG}_{\eta+1}(A)$ is closed under block-transposition, that is $\mathcal{L}(\lambda) \in \mathbb{DG}_{\eta+1}(A)$ iff $\mathcal{L}(\lambda)^{\mathcal{B}} \in \mathbb{DG}_{\eta+1}(A)$. In particular, since $(\mathcal{L}(\lambda)^{\mathcal{B}})^{\mathcal{B}} = \mathcal{L}(\lambda)$ the mapping $\mathcal{L}(\lambda) \mapsto \mathcal{L}(\lambda)^{\mathcal{B}}$ is involutive on any $\mathbb{DG}_{\eta+1}(A)$. Do not overlook that this mapping on $\mathbb{DL}(A)$ is nothing but the identity, cf. [80, Thm. 3.4.2], since any pencil $\mathcal{L}(\lambda) \in \mathbb{DL}(A)$ is block-symmetric [65, Sec. 3.2]. It is a main difference to $\mathbb{DL}(A)$ that this property does not apply to $\mathbb{DG}_{\eta+1}(A)$ in general. This is illustrated in the following example.

Example 4.1 ([42, Ex. 3]). *Let $A(\lambda) = \sum_{i=0}^{6} A_i \lambda^i \in M_{n \times n}(\mathbb{R})[\lambda]$ have degree $k = 6$ and consider the case $\eta = 0, \epsilon = 5$. Then*

$$\mathcal{L}(\lambda) = \left[\begin{array}{c|ccccc} \lambda A_6 + A_5 & A_4 & A_3 & A_2 & A_1 & A_0 \\ \hline A_4 & A_3 - \lambda A_4 & A_2 - \lambda A_3 & A_1 - \lambda A_2 & A_0 - \lambda A_1 & -\lambda A_0 \\ A_3 & A_2 - \lambda A_3 & A_1 - \lambda A_2 & A_0 - \lambda A_1 & -\lambda A_0 & 0 \\ A_2 & A_1 - \lambda A_2 & A_0 - \lambda A_1 & -\lambda A_0 & 0 & 0 \\ A_1 & A_0 - \lambda A_1 & -\lambda A_0 & 0 & 0 & 0 \\ A_0 & -\lambda A_0 & 0 & 0 & 0 & 0 \end{array} \right] \quad (4.2)$$

is an element of $\mathbb{DG}_1(A) = \mathbb{G}_1(A) \cap \mathbb{G}_6(A)$. Moreover, $\mathcal{L}(\lambda)$ is block-symmetric. We will show in Theorem 6.2 (ii) that $\mathbb{DG}_1(A) = \mathbb{G}_1(A) \cap \mathbb{G}_k(A) \subseteq \mathbb{DL}(A)$ holds, so that $\mathbb{DG}_1(A)$ always consists entirely of block-symmetric pencils. However, consider now the case $\eta = 1, \epsilon = 4$ and the matrix pencil

$$\mathcal{K}(\lambda) = \left[\begin{array}{cc|cccc} \lambda A_6 + A_5 & A_4 & D & 0 & -B & -I_n \\ 0 & A_3 & A_2 - \lambda D & A_1 & A_0 + \lambda B & \lambda I_n \\ \hline 0 & A_2 & A_1 - \lambda A_2 & A_0 - \lambda A_1 & -\lambda A_0 & 0 \\ C & A_1 - \lambda C & A_0 - \lambda A_1 & -\lambda A_0 & 0 & 0 \\ 0 & A_0 & -\lambda A_0 & 0 & 0 & 0 \\ -I_n & \lambda I_n & 0 & 0 & 0 & 0 \end{array} \right] \quad (4.3)$$

with arbitrary matrices $B, C, D \in M_{n \times n}(\mathbb{R})$. It is readily checked that $\mathcal{K}(\lambda) \in \mathbb{DG}_2(A)$, i.e. $\mathcal{K}(\lambda)$ is an element of $\mathbb{G}_2(A)$ and $\mathbb{G}_5(A)$ simultaneously. Anyhow, it is obvious that $\mathcal{K}(\lambda)$ will (for any nontrivial choice of B, C and D) never be block-symmetric.

Example 4.1 shows that, unlike $\mathbb{DL}(A)$, double block Kronecker ansatz spaces $\mathbb{DG}_{\eta+1}(A)$ need not contain exclusively block-symmetric pencils. Theorem 4.1 below gives a comprehensive characterization of these spaces. For the purpose of a simpler characterization of $\mathbb{DG}_{\eta+1}(A)$, we replace $\mathcal{F}_{\alpha,\eta,A}(\lambda)$, used as the anchor pencil for the characterization of $\mathcal{L}(\lambda) \in \mathbb{G}_{\eta+1}(A)$ in Theorem 3.1 and (3.11), by $\mathcal{F}_{\alpha,\eta,A}^{\mathrm{DG}}(\lambda)$ introduced in Theorem 4.1. In particular, $\Sigma_{\eta,A}(\lambda)$ in $\mathcal{F}_{\alpha,\eta,A}(\lambda)$ is being replaced by $\Pi_{\eta,A}^{\mathrm{DG}}(\lambda)$ which is defined by (4.4) and (4.5) below. To this end recall that any $(\eta+1)n \times (\epsilon+1)n$ pencil $M(\lambda)$ with $\Theta(M(\lambda)) = A(\lambda)$ can be used in (3.11) instead of $\Sigma_{\eta,A}(\lambda)$ to characterize the elements from $\mathbb{G}_{\eta+1}(A)$. Keeping $\eta \leq \epsilon$ in mind we define

$$\Sigma_{\eta,A}^{\mathrm{DG}}(\lambda) = \begin{bmatrix} \lambda A_k + A_{k-1} & & 0 \\ & \ddots & \\ 0 & & \lambda A_{\tau+2} + A_{\tau+1} \end{bmatrix} \in \mathrm{M}_{(\eta+1)n \times (\eta+1)n}(\mathbb{R}_1[\lambda]) \qquad (4.4)$$

with $\tau := \epsilon - \eta - 1$ and set $\Pi_{\eta,A}^{\mathrm{DG}}(\lambda) := \left[\Sigma_{\eta,A}^{\mathrm{DG}}(\lambda) \; \mathcal{R}_{\eta,A} \right] \in \mathrm{M}_{(\eta+1)n \times (\epsilon+1)n}(\mathbb{R}[\lambda])$ with

$$\mathcal{R}_{\eta,A} = \left[\begin{array}{c} 0_{\eta n \times (\epsilon-\eta)n} \\ \hline A_\tau \; \cdots \; A_0 \end{array} \right] \in \mathrm{M}_{(\eta+1)n \times (\epsilon-\eta)n}. \qquad (4.5)$$

Notice that the definition of $\Sigma_{\eta,A}^{\mathrm{DG}}(\lambda)$ is different to the one given in [42, Sec. 4]. This change has been made in order to achieve a simpler setting for the discussion upcoming in Section 4.5 and does essentially not affect the following results.

Now, it is checked straight forward that $\Theta(\Pi_{\eta,A}^{\mathrm{DG}}(\lambda)) = A(\lambda)$ still holds. Moreover, for $\eta \leq \epsilon$ we define the block Hankel matrix

$$\mathcal{H}_{\epsilon-\eta}(A) = \begin{bmatrix} -A_\tau & \cdots & -A_1 & -A_0 \\ \vdots & \iddots & \iddots & \\ -A_1 & -A_0 & & \\ -A_0 & & & \end{bmatrix} \in \mathrm{M}_{(\epsilon-\eta)n \times (\epsilon-\eta)n}(\mathbb{R}). \qquad (4.6)$$

Notice that this block Hankel structure already showed up in the construction of block-symmetric linearizations in [65, 80]. We obtain the following theorem:

Theorem 4.1 (Characterization of $\mathbb{DG}_{\eta+1}(A)$, [42, Thm. 6]). *Let* $A(\lambda) \in \mathrm{M}_{n \times n}(\mathbb{R}[\lambda])$ *be of degree* $k = \eta + \epsilon + 1$ *and assume* $\eta \leq \epsilon$. *Then* $\mathbb{DG}_{\eta+1}(A)$ *is a vector space over* $\mathbb{R}$ *having dimension*

$$\dim\big(\mathbb{DG}_{\eta+1}(A)\big) = 2k\eta n^2 + 1. \qquad (4.7)$$

Any matrix pencil $\mathcal{L}(\lambda) \in \mathbb{DG}_{\eta+1}(A)$ *may be characterized as*

$$\mathcal{L}(\lambda) = \left[\begin{array}{c|c|c} I_{(\eta+1)n} & B_{11} & 0 \\ \hline 0 & C_{11} & \alpha\mathcal{H}_{\epsilon-\eta}(A) \\ \hline 0 & C_{21} & 0 \end{array} \right] \mathcal{F}_{\alpha,\eta,A}^{\mathrm{DG}}(\lambda) \left[\begin{array}{c|c} I_{(\epsilon+1)n} & 0 \\ \hline B_2 & C_2 \end{array} \right] \qquad (4.8)$$

with

$$\mathcal{F}_{\alpha,\eta,A}^{\mathrm{DG}}(\lambda) = \left[\begin{array}{c|c} \alpha\Pi_{\eta,A}^{\mathrm{DG}}(\lambda) & L_\eta^T(\lambda) \otimes I_n \\ \hline L_\epsilon(\lambda) \otimes I_n & 0 \end{array} \right] \qquad (4.9)$$

for some $\alpha \in \mathbb{R}$ *and some matrices* $B_{11} \in M_{(\eta+1)n \times \eta n}(\mathbb{R})$, $C_{11} \in M_{(\epsilon-\eta)n \times \eta n}(\mathbb{R})$, $C_{21} \in M_{\eta n \times \eta n}(\mathbb{R})$, $B_2 \in M_{\eta n \times (\epsilon+1)n}(\mathbb{R})$ *and* $C_2 \in M_{\eta n \times \eta n}(\mathbb{R})$. *Moreover, unless* $\eta = \epsilon$, $\mathbb{DG}_{\eta+1}(A)$ *is a proper subspace of both* $\mathbb{G}_{\eta+1}(A)$ *and* $\mathbb{G}_{k-\eta}(A)$.

The first proof of Theorem 4.1 appeared in [42, Sec. 4]. It was kept rather compact without any deeper explanation and formalism. Here we present a fully developed proof of Theorem 4.1 including all the details. Since the proof is rather lengthy and technical, the whole upcoming section is devoted to it.

4.1.1   Proof of Theorem 4.1

First notice that in case $\epsilon = \eta$ there is nothing to show since $\mathbb{DG}_{\eta+1}(A) = \mathbb{G}_{\eta+1}(A)$. Thus, the characterization from Theorem 3.1 applies. Indeed, as $\epsilon = \eta$ implies $k = 2\eta + 1$, the dimension given in (3.9) is the same as the one given in (4.7). Moreover, (4.8) coincides with (3.11) for $\mathcal{F}_{\alpha,\eta,A}(\lambda)$ being switched to $\mathcal{F}^{\mathrm{DG}}_{\alpha,\eta,A}(\lambda)$.

Now assume $A(\lambda) \in M_{n \times n}(\mathbb{R}[\lambda])$ to be of degree $k = \epsilon + \eta + 1$ with $\eta < \epsilon$ and let $\mathcal{L}^{\star}(\lambda) \in \mathbb{DG}_{\eta+1}(A)$. Certainly, since $\mathbb{DG}_{\eta+1}(A) = \mathbb{G}_{\eta+1}(A) \cap \mathbb{G}_{k-\eta}(A)$, $\mathcal{L}^{\star}(\lambda)$ can be characterized as an element of $\mathbb{G}_{\eta+1}(A)$ and of $\mathbb{G}_{k-\eta}(A)$. As an element of $\mathbb{G}_{\eta+1}(A)$, $\mathcal{L}'(\lambda) := \mathcal{L}^{\star}(\lambda)$ can be expressed as

$$\mathcal{L}'(\lambda) = \left[\begin{array}{c|c} I_{(\eta+1)n} & B_1 \\ \hline 0 & C_1 \end{array}\right] \left[\begin{array}{c|c} \alpha[\,\Sigma^{\mathrm{DG}}_{\eta,A}(\lambda)\ \mathcal{R}_{\eta,A}] & L_\eta(\lambda)^T \otimes I_n \\ \hline L_\epsilon(\lambda) \otimes I_n & 0 \end{array}\right] \left[\begin{array}{c|c} I_{(\epsilon+1)n} & 0 \\ \hline B_2 & C_2 \end{array}\right], \quad (4.10)$$

with $[\,\Sigma^{\mathrm{DG}}_{\eta,A}(\lambda)\ \mathcal{R}_{\eta,A}] = \Pi^{\mathrm{DG}}_{\eta,A}(\lambda)$ satisfying $\Theta(\Pi^{\mathrm{DG}}_{\eta,A}(\lambda)) = A(\lambda)$ and some matrices B_1, B_2, C_1, C_2 of appropriate sizes. Notice that for (4.10) we replaced $\Sigma_{\eta,A}(\lambda)$ by $[\,\Sigma^{\mathrm{DG}}_{\eta,A}(\lambda)\ \mathcal{R}_{\eta,A}]$. Moreover, $\mathcal{L}^{\star}(\lambda)$ can also be expressed as an element $\mathcal{L}''(\lambda) := \mathcal{L}^{\star}(\lambda)$ of $\mathbb{G}_{\epsilon+1}(A) = \mathbb{G}_{k-\eta}(A)$,

$$\mathcal{L}''(\lambda) = \left[\begin{array}{c|c} I_{(\epsilon+1)n} & D_1 \\ \hline 0 & E_1 \end{array}\right] \left[\begin{array}{c|c} \beta \left[\begin{array}{c} \Sigma^{\mathrm{DG}}_{\eta,A}(\lambda) \\ \mathcal{R}^{\mathcal{B}}_{\eta,A} \end{array}\right] & L_\epsilon(\lambda)^T \otimes I_n \\ \hline L_\eta(\lambda) \otimes I_n & 0 \end{array}\right] \left[\begin{array}{c|c} I_{(\eta+1)n} & 0 \\ \hline D_2 & E_2 \end{array}\right], \quad (4.11)$$

with

$$\Theta\left(\left[\begin{array}{c} \Sigma^{\mathrm{DG}}_{\eta,A}(\lambda) \\ \mathcal{R}^{\mathcal{B}}_{\eta,A} \end{array}\right]\right) = \Theta\left(\left(\Pi^{\mathrm{DG}}_{\eta,A}(\lambda)\right)^{\mathcal{B}}\right) = A(\lambda), \quad \text{and matrices } D_1, D_2, E_1, E_2.$$

Our goal is to derive necessary and sufficient conditions and relations on the unknowns $\alpha, \beta,\ B_1, C_1, B_2, C_2, D_1, E_1, D_2$ and E_2 so that (4.10) and (4.11) are simultaneously satisfied. To this end, recall that we assume that $\mathcal{L}'(\lambda) = \mathcal{L}''(\lambda) = \mathcal{L}^{\star}(\lambda)$ holds.

First of all, via block-transposition of the ansatz equation (3.1) and Proposition 3.3 it is immediate, that $\mathcal{L}^{\star}(\lambda) \in \mathbb{DG}_{\eta+1}(A)$ can only hold if $\alpha = \beta$, so this will be assumed from now on. As it will turn out to be convenient, we will use the notation $\mathcal{L}'_{ij}(\lambda)$ and $\mathcal{L}''_{ij}(\lambda), 1 \leq i, j \leq 2$, for the blocks in the natural partition of $\mathcal{L}^{\star}(\lambda)$ with respect to $\mathbb{G}_{\eta+1}(A)$ and $\mathbb{G}_{k-\eta}(A)$, respectively. Now consider Figure 4.1.

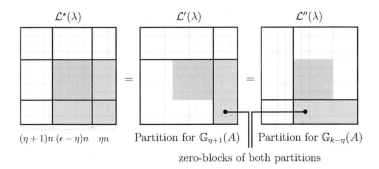

$$\mathcal{L}^\star(\lambda) \qquad\qquad \mathcal{L}'(\lambda) \qquad\qquad \mathcal{L}''(\lambda)$$

$(\eta+1)n \ (\epsilon-\eta)n \ \ \eta n$ Partition for $\mathbb{G}_{\eta+1}(A)$ ‖ Partition for $\mathbb{G}_{k-\eta}(A)$

zero-blocks of both partitions

FIGURE 4.1: $\mathcal{L}^\star(\lambda)$ in its natural 3×3 partition. This partition
may be interpreted as the overlay of the natural partitions of
elements in $\mathbb{DG}_{\eta+1}(A)$ and $\mathbb{DG}_{k-\eta}(A)$.

Here $\mathcal{L}^\star(\lambda)$ is displayed with its natural partitions corresponding to $\mathbb{G}_{\eta+1}(A)$
(middle sketch) and $\mathbb{G}_{k-\eta}(A)$ (right sketch). The left sketch in Figure 4.1 shows
both partitions simultaneously, that is, the natural 3×3 partition of a pencil from
$\mathbb{DG}_{\eta+1}(A)$. We subsequently denote the (i,j)-block, $1 \leq i,j \leq 3$, of this 3×3
partition by $\mathcal{L}^\star(i,j)$. Certainly, $\mathcal{L}''_{22}(\lambda) = 0$ has to hold for pencils in $\mathbb{G}_{k-\eta}(A)$ which
a priori implies for $\mathcal{L}'(\lambda)$ that

$$\mathcal{L}'(\lambda) = \left[\begin{array}{c|c|c} I_{(\eta+1)n} & B_{11} & B_{12} \\ \hline 0 & C_{11} & C_{12} \\ \hline 0 & C_{21} & 0 \end{array}\right]\left[\begin{array}{c|c} \alpha[\,\Sigma^{\mathbb{DG}}_{\eta,A}(\lambda)\ \mathcal{R}_{\eta,A}] & L^T_\eta(\lambda) \otimes I_n \\ \hline L_\epsilon(\lambda) \otimes I_n & 0 \end{array}\right]\left[\begin{array}{c|c} I_{(\epsilon+1)n} & 0 \\ \hline B_{21}\,|\,B_{22} & C_2 \end{array}\right],$$

(4.12)

where the newly appearing zero-block in (4.12) (in comparison to (4.10)) has dimension
$\eta n \times (\epsilon-\eta)n$. Here, for later convenience, we have already partitioned B_1, C_1
and B_2 (as $B_2 = [\,B_{21}\ B_{22}\,]$ with $B_{21} \in \mathrm{M}_{\eta n \times (\eta+1)n}(\mathbb{R})$). We now determine the
explicit form of $\mathcal{L}'(\lambda)$ in (4.12) with respect to the partition from Figure 4.1 (left
sketch). Thus, partitioning B_{12} and C_{12} as $B_{12} = [\,b\ \bar{B}_{12}\,], b \in \mathrm{M}_{(\eta+1)n \times n}(\mathbb{R})$, and
$C_{12} = [\,c\ \bar{C}_{12}\,], c \in \mathrm{M}_{(\epsilon-\eta)n \times n}(\mathbb{R})$, $\mathcal{L}'(\lambda)$ in (4.12) can be written explicitly as

$$\mathcal{L}'(\lambda) = \left[\begin{array}{c|c|c} Q_1(\lambda) - B & Q_2(\lambda) + \lambda\hat{B} & (L_\eta(\lambda)^T \otimes I_n)C_2 \\ \hline C_{11}(L_\eta(\lambda) \otimes I_n) - C & \bar{C}_{12}(L_{\epsilon-\eta-1}(\lambda) \otimes I_n) + \lambda\hat{C} & 0 \\ \hline C_{21}(L_\eta(\lambda) \otimes I_n) & 0 & 0 \end{array}\right]$$

(4.13)

with $C := [\,0\ \cdots\ 0\ c\,], B := [\,0\ \cdots\ 0\ b\,]$ and $\hat{C} := [\,c\ 0\ \cdots\ 0\,]$, $\hat{B} := [\,b\ 0\ \cdots\ 0\,]$ and

$$Q_1(\lambda) = \alpha\Sigma^{\mathbb{DG}}_{\eta,A}(\lambda) + (L_\eta(\lambda)^T \otimes I_n)B_{21} + B_{11}(L_\eta(\lambda) \otimes I_n)$$
$$Q_2(\lambda) = \alpha\mathcal{R}_{\eta,A} + (L_\eta(\lambda)^T \otimes I_n)B_{22} + \bar{B}_{12}(L_{\epsilon-\eta-1}(\lambda) \otimes I_n).$$

Notice that, in case $\epsilon - \eta = 1$, the term $\bar{B}_{12}(L_{\epsilon-\eta-1}(\lambda) \otimes I_n)$ in the expression of $Q_2(\lambda)$ is not existent (as $B_{12} = b$ holds and there is no matrix $\bar{B}_{12}$). Moreover, following the same strategy with $\mathcal{L}''(\lambda)$ instead of $\mathcal{L}'(\lambda)$, $\mathcal{L}'_{22}(\lambda) = 0$ implies that $\mathcal{L}''(\lambda)$ from (4.11) actually takes the form

$$\mathcal{L}''(\lambda) = \left[\begin{array}{c|c} I_{(\epsilon+1)n} & \begin{array}{c} D_{11} \\ \hline D_{12} \\ \hline E_1 \end{array} \end{array}\right] \left[\begin{array}{c|c} \alpha\begin{array}{c} \Sigma_{\eta,A}^{\mathrm{DG}}(\lambda) \\ \mathcal{R}_{\eta,A}^{\mathcal{B}} \end{array} & L_\epsilon(\lambda)^T \otimes I_n \\ \hline L_\eta(\lambda) \otimes I_n & 0 \end{array}\right] \left[\begin{array}{c|c|c} I_{(\eta+1)n} & 0 & 0 \\ \hline D_{21} & \bar{E}_2 & E_{21} \\ \hline D_{22} & E_{22} & 0 \end{array}\right]$$

with $D_1 = \begin{bmatrix} D_{11} \\ D_{12} \end{bmatrix}$, $D_{11} \in \mathrm{M}_{(\eta+1)n \times \eta n}(\mathbb{R})$ (and D_2, E_2 partitioned for later use), and the newly appearing zero-block having dimension $(\epsilon - \eta)n \times \eta n$. Notice that this is an analogous result compared to the form derived for $\mathcal{L}'(\lambda)$ in (4.12). Furthermore, with

$$D_{22} = \left[\frac{d}{\bar{D}_{22}}\right], d \in \mathrm{M}_{n \times (\eta+1)n}(\mathbb{R}), \quad \text{and} \quad E_{22} = \left[\frac{e}{\bar{E}_{22}}\right], e \in \mathrm{M}_{n \times (\epsilon-\eta)n}(\mathbb{R}),$$

$\mathcal{L}''(\lambda)$ can - analogously to $\mathcal{L}'(\lambda)$ in (4.13) - explicitly be characterized as

$$\mathcal{L}''(\lambda) = \left[\begin{array}{c|c|c} \tilde{Q}_1(\lambda) - D & (L_\eta(\lambda)^T \otimes I_n)\tilde{E}_2 - E & (L_\eta(\lambda)^T \otimes I_n)E_{21} \\ \hline \tilde{Q}_2(\lambda) + \lambda\hat{D} & (L_{\epsilon-\eta-1}(\lambda)^T \otimes I_n)\bar{E}_{22} + \lambda\hat{E} & 0 \\ \hline E_1(L_\eta(\lambda) \otimes I_n) & 0 & 0 \end{array}\right]$$

(4.14)

with $D = \begin{bmatrix} 0\cdots 0 \\ d \end{bmatrix}$, $E = \begin{bmatrix} 0\cdots 0 \\ e \end{bmatrix}$, $\hat{D} = \begin{bmatrix} d \\ 0\cdots 0 \end{bmatrix}$ and $\hat{E} = \begin{bmatrix} e \\ 0\cdots 0 \end{bmatrix}$ and

$$\tilde{Q}_1(\lambda) = \alpha\Sigma_{\eta,A}^{\mathrm{DG}}(\lambda) + D_{11}(L_\eta(\lambda) \otimes I_n) + (L_\eta(\lambda)^T \otimes I_n)D_{21}$$
$$\tilde{Q}_2(\lambda) = \alpha\mathcal{R}_{\eta,A}^{\mathcal{B}} + D_{12}(L_\eta(\lambda) \otimes I_n) + (L_{\epsilon-\eta-1}(\lambda)^T \otimes I_n)\bar{D}_{22}.$$

Once again the term $(L_{\epsilon-\eta-1}(\lambda)^T \otimes I_n)\bar{D}_{22}$ in the expression of $\tilde{Q}_2(\lambda)$ dissapears for $\epsilon - \eta = 1$. Now recall that the partitions shown for $\mathcal{L}'(\lambda)$ in (4.13) and for $\mathcal{L}''(\lambda)$ in (4.14) have to coincide in case $\mathcal{L}^\star(\lambda) = \mathcal{L}'(\lambda) = \mathcal{L}''(\lambda)$. Thus, comparing the blocks in (4.13) and (4.14) will reveal necessary and sufficient conditions for $\mathcal{L}'(\lambda) = \mathcal{L}''(\lambda)$ to hold. First, consider $\mathcal{L}^\star(1,1)$ which gives the equation

$$\alpha\Sigma_{\eta,A}^{\mathrm{DG}}(\lambda) + (L_\eta(\lambda)^T \otimes I_n)B_{21} + B_{11}(L_\eta(\lambda) \otimes I_n)$$
$$= \alpha\Sigma_{\eta,A}^{\mathrm{DG}}(\lambda) + D_{11}(L_\eta(\lambda) \otimes I_n) + (L_\eta(\lambda)^T \otimes I_n)D_{21} = \mathcal{L}^\star(1,1).$$

(4.15)

Moreover, as we assumed $\mathcal{L}^\star(\lambda) \in \mathbb{DG}_{\eta+1}(A)$ to be given, $\mathcal{L}^\star(1,1)$ is known. Now, the scalar α and the matrices D_{11} and D_{21} are uniquely determined by $\mathcal{L}^\star(1,1)$. Equation (4.15) is thus always satisfied iff $B_{21} = D_{21}$ and $B_{11} = D_{11}$ and no further restrictions have to be imposed on B_{11} and B_{21} on the basis of (4.15). However, so far we still need to analyze the conditions on the remaining five nonzero blocks from (4.13) and (4.14) in order to ensure that these choices are in no conflict with other conditions. Next, consider the block $\mathcal{L}^{\star}(1,2)$ displayed in Figure 4.1 for which the following equation (derived from (4.13) and (4.14)) has to hold:

$$
\alpha\mathcal{R}_{\eta,A} + (L_\eta(\lambda)^T \otimes I_n)B_{22} + \bar{B}_{12}(L_{\epsilon-\eta-1}(\lambda) \otimes I_n) + \lambda\hat{B}
$$
$$
= (L_\eta(\lambda)^T \otimes I_n)\tilde{E}_2 - E = \mathcal{L}^{\star}(1,2). \tag{4.16}
$$

Since $\mathcal{L}^{\star}(\lambda)$ and thus $\mathcal{L}^{\star}(1,2) \in M_{(\eta+1)n\times(\epsilon-\eta)n}(\mathbb{R}[\lambda])$ was assumed to be given, it is easy to see that (4.16) can only hold for $B_{22} = \tilde{E}_2$, $-E = \alpha\mathcal{R}_{\eta,A}$ (implying $e = -[\,A_r \cdots A_0\,]$) and $\bar{B}_{12} = 0, \hat{B} = 0$. The latter two conditions hold iff $B_{12} = 0$, i.e. we must have $B_{12} = 0$. Notice that this is the only restriction we obtain from (4.16) and the form (4.13).

Therefore, so far, we have shown that $\mathcal{L}^{\star}(\lambda) = \mathcal{L}'(\lambda) \in \mathbb{D}\mathbb{G}_{\eta+1}(A)$ - considered as an element of $\mathbb{G}_{\eta+1}(A)$ - must be of the form

$$
\mathcal{L}'(\lambda) = \left[\begin{array}{c|c|c} I_{(\eta+1)n} & B_{11} & 0 \\ \hline 0 & C_{11} & C_{12} \\ \hline 0 & C_{21} & 0 \end{array}\right] \left[\begin{array}{c|c} \Pi^{\mathbb{D}\mathbb{G}}_{\eta,A}(\lambda) & L_\eta^T(\lambda) \otimes I_n \\ \hline L_\epsilon(\lambda) \otimes I_n & 0 \end{array}\right] \left[\begin{array}{c|c|c} I_{(\epsilon+1)n} & 0 \\ \hline B_{21} & B_{22} & C_2 \end{array}\right] \tag{4.17}
$$

where the newly appearing zero-block has dimension $(\eta+1)n \times (\epsilon-\eta)n$. Analogously, the equation derived from (4.13) and (4.14) for $\mathcal{L}^{\star}(2,1)$ is

$$
\alpha\mathcal{R}^{\mathcal{B}}_{\eta,A} + D_{12}(L_\eta(\lambda) \otimes I_n) + (L_{\epsilon-\eta-1}(\lambda)^T \otimes I_n)\bar{D}_{22} + \lambda\hat{D}
$$
$$
= C_{11}(L_\eta(\lambda) \otimes I_n) - C = \mathcal{L}^{\star}(2,1). \tag{4.18}
$$

This only holds iff $-C = \alpha\mathcal{R}^{\mathcal{B}}_{\eta,A}$, $C_{11} = D_{12}$ and both $\bar{D}_{22}$ and $\hat{D}$ are zero. The latter two conditions imply $D_{22} = 0$. Thus, seen as an element from $\mathbb{G}_{k-\eta}(A)$, $\mathcal{L}^{\star}(\lambda)$ must have an expression of the form

$$
\mathcal{L}''(\lambda) = \left[\begin{array}{c|c} I_{(\epsilon+1)n} & B_{11} \\ \hline & C_{11} \\ \hline & E_1 \end{array}\right] \left[\begin{array}{c|c} \alpha\left(\Pi^{\mathbb{D}\mathbb{G}}_{\eta,A}(\lambda)\right)^{\mathcal{B}} & L_\epsilon(\lambda)^T \otimes I_n \\ \hline L_\eta(\lambda) \otimes I_n & 0 \end{array}\right] \left[\begin{array}{c|c|c} I_{(\eta+1)n} & 0 & 0 \\ \hline B_{21} & B_{22} & E_{21} \\ \hline 0 & E_{22} & 0 \end{array}\right]. \tag{4.19}
$$

Notice that we still have not considered conditions on E_1, E_{21} and E_{22}. We proceed by considering the blocks $\mathcal{L}^{\star}(1,3)$ and $\mathcal{L}^{\star}(3,1)$. From (4.13) and (4.14) we obtain the equation

$$
\left(L_\eta(\lambda)^T \otimes I_n\right)C_2 = \left(L_\eta(\lambda)^T \otimes I_n\right)E_{21} \tag{4.20}
$$

for $\mathcal{L}^{\star}(1,3)$ giving the condition $C_2 = E_{21}$. Thus one matrix is fixed once the other has been chosen (while there is no restriction on the form of the chosen matrix itself). For $\mathcal{L}^{\star}(3,1)$ we get

$$
C_{21}\left(L_\eta(\lambda) \otimes I_n\right) = E_1\left(L_\eta(\lambda) \otimes I_n\right) \tag{4.21}
$$

yielding $C_{21} = E_1$. Thus, assuming that $\mathcal{L}^\star(\lambda) \in \mathbb{G}_{\eta+1}(A)$ is also an element from $\mathbb{G}_{k-\eta}(A)$, the expression of $\mathcal{L}^\star(\lambda) = \mathcal{L}''(\lambda)$ must be of the form

$$\mathcal{L}''(\lambda) = \left[\begin{array}{c|c} I_{(\epsilon+1)n} & \begin{array}{c} B_{11} \\ \hline C_{11} \\ \hline C_{21} \end{array} \end{array}\right] \left[\begin{array}{c|c} \alpha\left(\Pi_{\eta,A}^{\mathbb{DG}}(\lambda)\right)^{\mathcal{B}} & \begin{array}{c|c} L_\epsilon(\lambda)^T \otimes I_n \end{array} \\ \hline L_\eta(\lambda) \otimes I_n & 0 \end{array}\right] \left[\begin{array}{c|c|c} I_{(\eta+1)n} & 0 & 0 \\ \hline B_{21} & B_{22} & C_2 \\ \hline 0 & E_{22} & 0 \end{array}\right]$$
$$(4.22)$$

So far, the only nonzero block that has not been considered is $\mathcal{L}^\star(2,2)$. Therefore, the question remains whether some conditions have to be imposed on C_{12} and E_{22}.

The corresponding equation derived from (4.13) and (4.14) is

$$\bar{C}_{12}(L_{\epsilon-\eta-1}(\lambda) \otimes I_n) - \alpha\lambda \begin{bmatrix} A_\tau & 0 & \cdots & 0 \\ \vdots & \vdots & & \vdots \\ A_0 & 0 & \cdots & 0 \end{bmatrix}$$
$$= (L_{\epsilon-\eta-1}(\lambda)^T \otimes I_n)\bar{E}_{22} - \alpha\lambda \begin{bmatrix} A_\tau & \cdots & A_0 \\ 0 & \cdots & 0 \\ \vdots & & \vdots \\ 0 & \cdots & 0 \end{bmatrix}.$$
$$(4.23)$$

It is straight forward to check that (4.23) has a unique solution. In fact, (4.23) holds iff $\bar{C}_{12}$ and $\bar{E}_{22}$ are chosen as

$$\bar{C}_{12} = \left[\begin{array}{c|c|c} \alpha\mathcal{H}_\tau(A) \\ \hline 0_{n\times n} & \cdots & 0_{n\times n} \end{array}\right] \quad \text{and} \quad \bar{E}_{22} = \left[\begin{array}{c|c} \alpha\mathcal{H}_\tau(A) & \begin{array}{c} 0_{n\times n} \\ \vdots \\ 0_{n\times n} \end{array} \end{array}\right]$$

where $\tau = \epsilon-\eta-1$ as before. Notice that, incorporating the previous findings for c and e, this implies $C_{12} = [\,c\ \bar{C}_{12}\,] = \mathcal{H}_{\epsilon-\eta}(A) = \mathcal{H}_{\tau+1}(A) = \begin{bmatrix} e \\ \bar{E}_{22} \end{bmatrix} = E_{22}$. This completes the proof since we have shown that, whenever some pencil $\mathcal{L}^\star(\lambda) \in \mathbb{DG}_{\eta+1}(A)$, then it can be expressed as stated in (4.8).

We conclude this section by briefly summarizing our proof: We assumed that $\mathcal{L}^\star(\lambda)$ was given as an element of $\mathbb{DG}_{\eta+1}(A) = \mathbb{G}_{\eta+1}(A) \cap \mathbb{G}_{k-\eta}(A)$ and deduced that expressions of the form (4.10) and (4.11) must exist. The condition that both should simultaneously hold imposed several restrictions. First of all we showed that some blocks in (4.10) and (4.11) need to be zero. Otherwise, the property of $\mathcal{L}^\star(\lambda)$ being an element of both ansatz spaces would not be satisfied. This led us to the forms (4.17) and (4.19). We derived the relations that have to hold between the expressions (4.17) and (4.19) for all constant matrices showing up in (4.10) and (4.11). We found that in fact no restrictions on the form on the matrices (which are allowed to be nonzero) have to be imposed. Finally, we considered the middle block $\mathcal{L}^\star(2,2)$ which revealed the necessary block-Hankel structure introduced in (4.6). This completed the proof showing that any $\mathcal{L}^\star(\lambda) \in \mathbb{DG}_{\eta+1}(A)$ must have an expression of the form claimed in Theorem 4.1.

4.2　Advanced structural analysis of $\mathbb{DG}_{\eta+1}(A)$

In this section we derive some statements analogous to those of Chapter 3. In particular, Theorem 4.1 directly admits the following corollary:

Corollary 4.1 (Non-Emptiness of $\mathbb{DG}_{\eta+1}(A)$, [42, Cor. 1]). *Let* $A(\lambda) \in M_{n \times n}(\mathbb{R}[\lambda])$ *be of degree* $k = \eta \mid \epsilon + 1$ *and assume* $\eta \leq \epsilon$. *Then* $\mathbb{DG}_{\eta+1}(A) \neq \{0\}$.

Note that $\deg(A(\lambda)) = \epsilon + \eta + 1$ particularly implies $A(\lambda) \neq 0$. Therefore, according to Corollary 4.1, $\mathbb{DG}_{\eta+1}(A)$ is always nontrivial. The following Corollary 4.2 is an immediate consequence from the proof of Theorem 4.1 and states how any pencil $\mathcal{L}(\lambda) \in \mathbb{DG}_{\eta+1}(A)$, given in its natural factorization according to $\Pi_{\eta,A}^{\mathrm{DG}}(\lambda)$ (as an element from $\mathbb{G}_{\eta+1}(A)$), can easily be reformulated as a pencil from $\mathbb{G}_{k-\eta}(A)$ in its natural factorization with respect to $(\Pi_{\eta,A}^{\mathrm{DG}}(\lambda))^{\mathcal{B}}$.

Corollary 4.2. *Let* $A(\lambda) \in M_{n \times n}(\mathbb{R}[\lambda])$ *be of degree* $k = \eta + \epsilon + 1$ *with* $\eta \leq \epsilon$ *and assume* $\mathcal{L}(\lambda) \in \mathbb{DG}_{\eta+1}(A)$ *is given in the form* (4.8) *as an element from* $\mathbb{G}_{\eta+1}(A)$. *Then, setting* $B_2 = [\, B_{21} \; B_{22} \,]$ *appropriately partitioned, the pencil* $\mathcal{L}(\lambda) \in \mathbb{G}_{k-\eta}(A)$ *can be expressed as* $\mathcal{L}(\lambda) =$

$$
\left[\begin{array}{c|c} I_{(\epsilon+1)n} & B_{11} \\ & \overline{C_{11}} \\ \hline 0 & C_{21} \end{array}\right]
\left[\begin{array}{c|c} \alpha\left(\Pi_{\eta,A}^{\mathrm{DG}}(\lambda)\right)^{\mathcal{B}} & L_\epsilon(\lambda)^T \otimes I_n \\ \hline L_\eta(\lambda) \otimes I_n & 0 \end{array}\right]
\left[\begin{array}{c|c|c} I_{(\eta+1)n} & 0 & 0 \\ \hline B_{21} & B_{22} & C_2 \\ \hline 0 & \alpha H_{\epsilon-\eta}(A) & 0 \end{array}\right].
$$

Notice that the block-transposition of $\Pi_{\eta,A}^{\mathrm{DG}}(\lambda)$ in Corollary 4.2 can also be applied to $\mathcal{F}_{\alpha,\eta,A}^{\mathrm{DG}}(\lambda)$. Thus, the above expression of $\mathcal{L}(\lambda)$ is equivalent to

$$
\mathcal{L}(\lambda) = \left[\begin{array}{c|c} I_{(\epsilon+1)n} & B_{11} \\ & \overline{C_{11}} \\ \hline 0 & C_{21} \end{array}\right]
\left(\mathcal{F}_{\alpha,\eta,A}^{\mathrm{DG}}(\lambda)\right)^{\mathcal{B}}
\left[\begin{array}{c|c|c} I_{(\eta+1)n} & 0 & 0 \\ \hline B_{21} & B_{22} & C_2 \\ \hline 0 & \alpha H_{\epsilon-\eta}(A) & 0 \end{array}\right]. \qquad (4.24)
$$

From Theorem 4.1 and Corollary 4.2 it can be seen that, expressing some $\mathcal{L}(\lambda) \in \mathbb{DG}_{\eta+1}(A)$ as an element of $\mathbb{G}_{\eta+1}(A)$ and $\mathbb{G}_{k-\eta}(A)$, essentially only the block Hankel matrix $\mathcal{H}_{\epsilon-\eta}(A)$ changes its place and the anchor pencil $\mathcal{F}_{\alpha,\eta,A}^{\mathrm{DG}}(\lambda)$ is being block transposed. In particular, in case $\epsilon = \eta$ it holds that $(\mathcal{F}_{\alpha,\eta,A}^{\mathrm{DG}}(\lambda))^{\mathcal{B}} = \mathcal{F}_{\alpha,\eta,A}^{\mathrm{DG}}(\lambda)$ and the Hankel block $\mathcal{H}_{\epsilon-\eta}(A)$ does not exist. Furthermore, notice that expressing $\mathcal{L}(\lambda) \in \mathbb{G}_{\eta+1}(A)$ as an element from $\mathbb{G}_{k-\eta}(A)$ as in (4.24) is not equal to the computation of $\mathcal{L}(\lambda)^{\mathcal{B}}$ according to (3.12). This establishes once more that $\mathcal{L}(\lambda) \mapsto \mathcal{L}(\lambda)^{\mathcal{B}}$ is not the identity mapping on $\mathbb{DG}_{\eta+1}(A)$.

A pure block Kronecker pencil as introduced in [34] (see also Definition 2.16) can never be an element of some ansatz space $\mathbb{DG}_{\eta+1}(A)$ for any matrix polynomial $A(\lambda) \in M_{n \times n}(\mathbb{R}[\lambda])$ unless $\epsilon = \eta$, i.e. $\eta + 1 = k - \eta$ for $\deg(A(\lambda)) = k$. Figuratively speaking, we need some connection between $\mathbb{G}_{\eta+1}(A)$ and $\mathbb{G}_{k-\eta}(A)$ to make a pencil $\mathcal{L}(\lambda)$ an element of both spaces. The $(2,2)$-block of $\mathcal{L}(\lambda) \in \mathbb{DG}_{\eta+1}(A)$ from (4.8) considered in the double partition as displayed in Figure 4.1 is denoted $\mathcal{C}(\mathcal{L})$ subsequently and takes on this task. It has the Hankel-structure

$$
\mathcal{C}(\mathcal{L}) = \alpha \begin{bmatrix}
A_{\tau-1} - \lambda A_\tau & A_{\tau-2} - \lambda A_{\tau-1} & \cdots & A_0 - \lambda A_1 & -\lambda A_0 \\
A_{\tau-2} - \lambda A_{\tau-1} & A_{\tau-3} - \lambda A_{\tau-2} & \cdot^{\displaystyle\cdot} & \cdot^{\displaystyle\cdot} & \\
\vdots & & \cdot^{\displaystyle\cdot} & \cdot^{\displaystyle\cdot} & \\
A_0 - \lambda A_1 & -\lambda A_0 & & 0 & \\
-\lambda A_0 & & & &
\end{bmatrix}
\tag{4.25}
$$

with $\tau := \epsilon - \eta - 1$ and is called the *core part* of $\mathcal{L}(\lambda)$ from now on. Modulo scalar multiplication, $\mathcal{C}(\mathcal{L})$ is the same for each pencil $\mathcal{L}(\lambda) \in \mathbb{DG}_{\eta+1}(A)$, so it does specifically not depend on $\mathcal{L}(\lambda)$ but on ϵ, η and $A(\lambda)$. In case $\epsilon = \eta$, $\mathcal{C}(\mathcal{L})$ vanishes entirely and no further restrictions remain for $\mathbb{DG}_{\eta+1}(A)$, i.e. we have $\mathbb{DG}_{\eta+1}(A) = \mathbb{G}_{\eta+1}(A)$. It is only in this very situation where we obtain pure block Kronecker pencils. Finally, note that $\mathcal{C}(\mathcal{L}) = \mathcal{C}(\mathcal{L})^{\mathcal{B}}$ always holds. This block-symmetry turns out to be an important property of pencils in double block Kronecker ansatz spaces and is further studied in the next section.

Regarding linearizations, the following fact can immediately be derived from the characterization of $\mathbb{DG}_{\eta+1}(A)$ in Theorem 4.1 or Corollary 4.2 and Theorem 3.2.

Theorem 4.2 (Linearization Condition, [42, Thm. 7]). *Let $A(\lambda) \in M_{n\times n}(\mathbb{R}[\lambda])$ be regular and of degree $k = \eta + \epsilon + 1$. Moreover, let $\mathcal{L}(\lambda) \in \mathbb{DG}_{\eta+1}(A)$ be given in the form (4.8) or (4.24). Then the following statements are equivalent:*

(i) *$\mathcal{L}(\lambda)$ is a (strong) linearization for $A(\lambda)$.*

(ii) *$A_0 \in GL_n(\mathbb{R}), C_{21} \in GL_{\eta n}(\mathbb{R}), C_2 \in GL_{\eta n}(\mathbb{R})$ and $\alpha \in \mathbb{R} \setminus \{0\}$.*

Theorem 4.2 is actually just a correspondingly adjusted version of Theorem 3.2 taking the special structure of pencils in $\mathbb{DG}_{\eta+1}(A)$, cf. (4.8), into account. In particular, $A_0 \in GL_n(\mathbb{R})$ reflects the nonsingularity of $\mathcal{H}_{\epsilon-\eta}(A)$. Therefore, in case $\epsilon = \eta$, the equivalence in Theorem 4.2 holds without the requirement $A_0 \in GL_n(\mathbb{R})$ due to the disappearance of the $\mathcal{H}_{\epsilon-\eta}(A)$-block. In this situation, the implication $(ii) \Rightarrow (i)$ also holds for singular matrix polynomials $A(\lambda) \in M_{n\times n}(\mathbb{R}[\lambda])$. On the other hand, the equivalence of Theorem 4.2 is void for singular $A(\lambda)$ if $\epsilon \neq \eta$ as the second statement will never hold (the trailing coefficient A_0 is always singular). This does a priori not mean that there can not be any linearizations for $A(\lambda)$ in this case (see Remark 3.1 and the references therein).

Consider once again Theorem 4.2. The compliance of the irrevocable condition $A_0 \in GL_n(\mathbb{R})$ certainly depends exclusively on the matrix polynomial $A(\lambda)$. On the other hand, the conditions $C_{21}, C_2 \in GL_{\eta n}(\mathbb{R})$ are satisfied for almost every matrix in $M_{\eta n \times \eta n}(\mathbb{R})$. In accordance with Theorem 3.2, we obtain the following general density property.

Corollary 4.3 (Genericity of Linearizations, [42, Cor. 2]). *Let $A(\lambda) \in M_{n\times n}(\mathbb{R}[\lambda])$. Then almost every matrix pencil in $\mathbb{DG}_{\eta+1}(A)$ is a strong linearization for $A(\lambda)$ as long as the trailing coefficient A_0 of $A(\lambda)$ is nonsingular.*

4.3 The superpartition principle

Although double block Kronecker ansatz spaces $\mathbb{DG}_{\eta+1}(A), \eta \geq 0$, usually do not contain solely block-symmetric pencils $\mathcal{L}(\lambda) = \mathcal{L}(\lambda)^{\mathcal{B}}$ (reconsider Example 4.1), they possess a remarkable feature that we call *superpartition property*[1]. To its motivation, consider the following example.

Example 4.2 ([42, Ex. 5]). *Let $A(\lambda) = \sum_{i=0}^{6} A_i \lambda^i \in M_{n \times n}(\mathbb{R})[\lambda]$ be of degree $k = 6$. Consider as in Example 4.1 the case $\eta = 1, \epsilon = 4$ and the corresponding matrix pencil*

$$
\mathcal{K}(\lambda) = \left[
\begin{array}{cc|ccc|c}
\lambda A_6 + A_5 & A_4 & D & 0 & -B & -I_n \\
0 & A_3 & A_2 - \lambda D & A_1 & A_0 + \lambda B & \lambda I_n \\
0 & A_2 & A_1 - \lambda A_2 & A_0 - \lambda A_1 & -\lambda A_0 & 0 \\
C & A_1 - \lambda C & A_0 - \lambda A_1 & -\lambda A_0 & 0 & 0 \\
0 & A_0 & -\lambda A_0 & 0 & 0 & 0 \\
-I_n & \lambda I_n & 0 & 0 & 0 & 0
\end{array}
\right]
\tag{4.26}
$$

for arbitrary matrices $B, C, D \in M_{n \times n}(\mathbb{R})$. As already discussed, $\mathcal{K}(\lambda) \in \mathbb{DG}_2(A) = \mathbb{G}_2(A) \cap \mathbb{G}_5(A)$. The natural partitions of $\mathcal{K}(\lambda)$ corresponding to $\mathbb{G}_2(A)$ and $\mathbb{G}_5(A)$ are indicated in (4.26). Now consider $\mathcal{K}(\lambda)$ in the slightly modified partitioned form

$$
\mathcal{K}(\lambda) = \left[
\begin{array}{ccc|c|cc}
\lambda A_6 + A_5 & A_4 & D & 0 & -B & -I_n \\
0 & A_3 & A_2 - \lambda D & A_1 & A_0 + \lambda B & \lambda I_n \\
0 & A_2 & A_1 - \lambda A_2 & A_0 - \lambda A_1 & -\lambda A_0 & 0 \\
\hline
C & A_1 - \lambda C & A_0 - \lambda A_1 & -\lambda A_0 & 0 & 0 \\
\hline
0 & A_0 & -\lambda A_0 & 0 & 0 & 0 \\
-I_n & \lambda I_n & 0 & 0 & 0 & 0
\end{array}
\right].
\tag{4.27}
$$

It is readily checked that $\mathcal{K}(\lambda)$, partitioned as in (4.27), may alternatively be taken as an element of $\mathbb{G}_3(A)$ or $\mathbb{G}_4(A)$ (i.e. $\eta = 2$, $\epsilon = 3$). In other words, $\mathcal{K}(\lambda) \in \mathbb{DG}_3(A)$.

Theorem 4.3 below states that the phenomenon highlighted in Example 4.2 always holds. The main reason behind this fact turns out to be the block-symmetric core part $\mathcal{C}(\mathcal{L})$, cf. (4.25), of pencils $\mathcal{L}(\lambda) \in \mathbb{DG}_{\eta+1}(A)$.

Theorem 4.3 (Superpartition Property, [42, Thm. 8]). *Let $A(\lambda) \in M_{n \times n}(\mathbb{R}[\lambda])$ be of degree $k = \epsilon + \eta + 1$ and assume $\eta \leq \epsilon$. Then $\mathcal{L}(\lambda) \in \mathbb{DG}_{\eta+1}(A)$ implies that*

$$
\mathcal{L}(\lambda) \in \mathbb{G}_{\eta+i}(A) \qquad \text{for all } i = 1, 2, \ldots, k - 2\eta.
$$

4.3.1 Proof of Theorem 4.3

Before proving Theorem 4.3 we make a statement on the zero-structure of pencils $\mathcal{L}(\lambda)$ in $\mathbb{DG}_{\eta+1}(A) = \mathbb{G}_{\eta+1}(A) \cap \mathbb{G}_{k-\eta}(A)$. It follows directly from the characterization in Theorem 4.1 (see also [42, Rem. 6]).

---

[1]This property was also recognized by the authors of [21], cf. Remark 3.3. and Theorem 3.10. therein.

To show that $\mathcal{L}(\lambda) \in \mathbb{DG}_{\eta+1}(A)$ implies $\mathcal{L}(\lambda) \in \mathbb{DG}_{\eta+i}(A)$ for all $i = 1, \ldots, k - 2\eta$ we a priori have to guarantee that the $(2,2)$-block $\mathcal{L}_{22}(\lambda)$ of $\mathcal{L}(\lambda)$ - naturally partitioned according to some space $\mathbb{DG}_{\eta+i}(A)$ - is always entirely zero. To see that this holds, consider exemplarily some $A(\lambda)$ of degree $k = 7$ with $\eta = 1$. According to Theorem 4.1 any pencil $\mathcal{L}(\lambda) \in \mathbb{DG}_2(A)$ schematically has the form

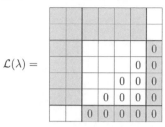

$$\mathcal{L}(\lambda) =$$

with the indicated unalterable zero-structure and the 3×3 partition as in Figure 4.1 (all empty $n \times n$ squares may be nonzero). For any $\mathcal{K}(\lambda) \in \mathbb{DG}_\ell(A)$, $\ell = 3, 4, 5, 6$, the following sketches indicate the 2×2 natural partitions of $\mathcal{K}(\lambda)$ for the spaces $\mathbb{DG}_\ell(A)$ displayed with the zero-structure of $\mathcal{L}(\lambda)$ as above:

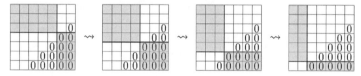

This example shows, that the zero-structure of any $\mathcal{L}(\lambda) \in \mathbb{DG}_2(A)$ is exactly of the form that it covers all the $(2,2)$-zero blocks of pencils in $\mathbb{DG}_\ell(A)$ with $2 \leq \ell \leq 6$. The situation is exactly the same for all other choices of η and all other degrees of $A(\lambda)$. We now turn out attention to the proof of Theorem 4.3.

Let $\mathcal{L}(\lambda) \in \mathbb{DG}_{\eta+1}(A)$ and consider again the form of $\mathcal{L}(\lambda)$ given in (4.8), i.e.

$$\mathcal{L}(\lambda) = \left[\begin{array}{c|c|c} I_{(\eta+1)n} & B_{11} & 0 \\ \hline 0 & C_{11} & \alpha\mathcal{H}_{\epsilon-\eta}(A) \\ \hline 0 & C_{21} & 0 \end{array} \right] \left[\begin{array}{c|c} \alpha\Pi_{\eta,A}^{\mathbb{DG}}(\lambda) & L_\eta(\lambda)^T \otimes I_n \\ \hline L_\epsilon(\lambda) \otimes I_n & 0 \end{array} \right] \left[\begin{array}{c|c} I_{(\epsilon+1)n} & 0 \\ \hline B_2 & C_2 \end{array} \right].$$

It is convenient to express $\mathcal{L}(\lambda)$ as $\mathcal{L}(\lambda) = \alpha\mathcal{K}_1(\lambda) + \mathcal{K}_2(\lambda)$ in a way that $\mathcal{K}_1(\lambda)$ is responsible for the recovering of $A(\lambda)$ in (3.1) whereas $\mathcal{K}_2(\lambda)$ satisfies (3.1) with $\alpha = 0$. In particular, we may set $\mathcal{K}_1(\lambda)$ as

$$\mathcal{K}_1(\lambda) = \left[\begin{array}{c|c|c} \Sigma_{\eta,A}^{\mathbb{DG}}(\lambda) & \mathcal{R}_{\eta,A} & 0 \\ \hline \mathcal{R}_{\eta,A}^{\mathcal{B}} & \mathcal{H}_\tau(A)(L_{\epsilon-\eta-1}(\lambda) \otimes I_n) - \lambda \left[\begin{array}{c|c} A_\tau & \\ \vdots & 0 \\ A_0 & \end{array} \right] & 0 \\ \hline 0 & 0 & 0 \end{array} \right] \qquad (4.28)$$

with $\tau = \epsilon - \eta - 1$. Notice that the $(2,2)$-block in $\mathcal{K}_1(\lambda)$ equals $\mathcal{C}(\mathcal{L})$, cf. (4.25). With $Q(\lambda) := (L_\eta(\lambda)^T \otimes I_n)B_{21} + B_{11}(L_\eta(\lambda) \otimes I_n)$ we obtain for $\mathcal{K}_2(\lambda)$:

$$\mathcal{K}_2(\lambda) = \left[\begin{array}{c|c|c} Q(\lambda) & (L_\eta(\lambda)^T \otimes I_n)B_{22} & (L_\eta(\lambda)^T \otimes I_n)C_2 \\ \hline C_{11}(L_\eta(\lambda) \otimes I_n) & 0 & 0 \\ \hline C_{21}(L_\eta(\lambda) \otimes I_n) & 0 & 0 \end{array}\right], \qquad (4.29)$$

so that $\mathcal{L}(\lambda) = \alpha\mathcal{K}_1(\lambda) + \mathcal{K}_2(\lambda)$. Obviously, $\mathcal{K}_1(\lambda), \mathcal{K}_2(\lambda) \in \mathbb{DG}_{\eta+1}(A)$ and $\mathcal{K}_1(\lambda)$ is block-symmetric (however $\mathcal{K}_2(\lambda) \in \mathbb{DG}_{\eta+1}(A)$ is not). Explicitly, $\mathcal{K}_1(\lambda)$ may be expressed as

$$\mathcal{K}_1(\lambda) = \left[\begin{array}{c|c|c} \Sigma^{\mathbb{DG}}_{\eta,A}(\lambda) & \mathcal{R}_{\eta,A} & 0 \\ \hline \mathcal{R}^{\mathcal{B}}_{\eta,A} & \begin{array}{cccc} A_{\tau-1} - \lambda A_\tau & A_{\tau-2} - \lambda A_{\tau-1} & \cdots & A_0 - \lambda A_1 \quad -\lambda A_0 \\ A_{\tau-2} - \lambda A_{\tau-1} & A_{\tau-3} - \lambda A_{\tau-2} & \iddots & \iddots \\ \vdots & \iddots & \iddots & \\ A_0 - \lambda A_1 & -\lambda A_0 & & \\ -\lambda A_0 & & & \end{array} & 0 \\ \hline 0 & 0 & 0 \end{array}\right].$$

Now, due to the form of $\mathcal{K}_1(\lambda)$ it is straight forward to check, that $\alpha\mathcal{K}_1(\lambda)$ can be (naturally) partitioned according to any $\mathbb{G}_{\eta+1}(A), i = 1, \ldots, k - 2\eta$, and still satisfies the corresponding ansatz equation (3.1) for α. This property certainly only holds because of the particular form of $\mathcal{C}(\mathcal{L})$ for any pencil $\mathcal{L}(\lambda) \in \mathbb{G}_{\eta+1}(A)$.

Now consider the pencil $\mathcal{K}_2(\lambda)$. Once more it can be checked straight forward that, due to the form of $\mathcal{K}_2(\lambda)$, the ansatz equation (3.1) for $\mathbb{G}_{\eta+i}(A), i = 1, \ldots, k - 2\eta$ applied to $\mathcal{K}_2(\lambda)$ always yields the zero $kn \times kn$ matrix on its right-hand-side. That is, the scalar α in (3.1) in always zero. So, in other words,

$$\left[\begin{array}{c|c} \Lambda_{\eta+i}(\lambda) \otimes I_n & 0 \\ \hline 0 & I_{(\epsilon-i)n} \end{array}\right] \mathcal{K}_2(\lambda) \left[\begin{array}{c|c} \Lambda_{\epsilon-i}(\lambda)^T \otimes I_n & 0 \\ \hline 0 & I_{(\eta+i)n} \end{array}\right] = \left[\begin{array}{c|c} 0 & 0 \\ \hline 0 & 0_{(\epsilon-i)n \times (\eta+i)n} \end{array}\right]$$

holds for all $i = 1, \ldots, k - 2\eta$. In conclusion, $\mathcal{L}(\lambda) = \alpha\mathcal{K}_1(\lambda) + \mathcal{K}_2(\lambda)$ satisfies the ansatz equation (3.1) for $\mathbb{G}_{\eta+i}(A), i = 1, \ldots, k - 2\eta$ giving each time $\alpha A(\lambda)$ in the upper-left block on its right-hand-side. This completes the proof.

4.4 The inclusion relation

According to Example 4.2 it is not surprising that Theorem 4.3 holds. The property of a matrix pencil $\mathcal{L}(\lambda) \in \mathbb{G}_{\eta+1}(A)$ being additionally an element of $\mathbb{G}_{k-\eta}(A)$ imposes several restrictions on the form of $\mathcal{L}(\lambda)$. In particular, whereas some bordering blocks in the 3×3 partition as in Figure 4.1 underly the condition of having no contribution in one space and being completely reproducible in the other (see Theorem 4.1), the core part $\mathcal{C}(\mathcal{L})$ of the pencil has to be adequate for both spaces, $\mathbb{G}_{\eta+1}(A)$ and $\mathbb{G}_{k-\eta}(A)$.

This lucky circumstance determines the (block-symmetric) form of $C(\mathcal{L})$ as in (4.25) completely, no matter how η and ϵ are chosen. In turn this implies that

$$\Theta(\mathcal{L}_{11}^{(i)}(\lambda)) = \alpha A(\lambda)$$

always holds, where $\mathcal{L}_{11}^{(i)}(\lambda)$ denotes the upper-left block in the natural partition of $\mathcal{L}(\lambda)$ according to any $\mathbb{G}_{\eta+i}(A), i = 1, \ldots, k - 2\eta$. An algorithm to reformulate a pencil $\mathcal{L}(\lambda) \in \mathbb{DG}_{\eta+1}(A)$ as an element of $\mathbb{DG}_{\eta+i}(A)$ for all $i = 1, \ldots, k - 2\eta$ was presented in [42]. The following important observation is now immediate.

Corollary 4.4 (Inclusion Property, [42, Cor. 3]). *Let $A(\lambda) \in M_{n\times n}(\mathbb{R}[\lambda])$ be of degree $k = \epsilon + \eta + 1$. Then we have*

$$\mathbb{DG}_1(A) \subsetneqq \mathbb{DG}_2(A) \subsetneqq \cdots \subsetneqq \mathbb{DG}_{\lceil \frac{k}{2} \rceil}(A). \tag{4.30}$$

For $\mathbb{DG}_\gamma(A)$ for $\gamma = \lceil k/2 \rceil, \ldots, k - 1$ nothing new is obtained according to (4.1) as $\mathbb{DG}_{\eta+1}(A) = \mathbb{DG}_{\epsilon+1}(A)$.

4.5　The block-symmetric ansatz spaces $\mathbb{BG}_{\eta+1}(A)$

This section is dedicated to the basic study of block-symmetric pencils in double ansatz spaces $\mathbb{DG}_{\eta+1}(A)$. Vector spaces of block-symmetric linearizations have also been investigated in [22, 65, 80]. For motivation, consider once more the matrix pencil $\mathcal{K}(\lambda)$ from (4.3) in the following example.

Example 4.3 ([42, Rem. 8]). *In contrast to our experience with $\mathbb{DL}(A)$, in general not all pencils $\mathcal{L}(\lambda) \in \mathbb{DG}_{\eta+1}(A)$ satisfy $\mathcal{L}(\lambda) = \mathcal{L}(\lambda)^{\mathcal{B}}$, cf. Example 4.1. Nevertheless, considering $\mathcal{K}(\lambda)$ from Example 4.1 it is not hard to see how a block-symmetric pencil $\widetilde{\mathcal{K}}(\lambda) \in \mathbb{DG}_2(A)$ can be obtained. For $\widetilde{\mathcal{K}}(\lambda)$ we modify the $(1,1)$-block according to its natural partition to be block-symmetric and adjust the bordering blocks:*

$$\widetilde{\mathcal{K}}(\lambda) = \left[\begin{array}{cc|ccc|c} \lambda A_6 + A_5 & \frac{1}{2}A_4 & D & C & -B & -I_n \\ \frac{1}{2}A_4 & A_3 & A_2 - \lambda D & A_1 - \lambda C & A_0 + \lambda B & \lambda I_n \\ \hline D & A_2 - \lambda D & A_1 - \lambda A_2 & A_0 - \lambda A_1 & -\lambda A_0 & 0 \\ C & A_1 - \lambda C & A_0 - \lambda A_1 & -\lambda A_0 & 0 & 0 \\ -B & A_0 + \lambda B & -\lambda A_0 & 0 & 0 & 0 \\ \hline -I_n & \lambda I_n & 0 & 0 & 0 & 0 \end{array} \right]. \tag{4.31}$$

In fact, we still have $\widetilde{\mathcal{K}}(\lambda) \in \mathbb{DG}_2(A)$ and $\widetilde{\mathcal{K}}(\lambda)$ now is block-symmetric.

The following definition now seems natural:

Definition 4.2 (Block-symmetric Ansatz Space, [42, Def. 3]). *Let $A(\lambda) \in M_{n\times n}(\mathbb{R}[\lambda])$ be of degree $k = \epsilon + \eta + 1$ and assume $\eta \le \epsilon$. Then we define*

$$\mathbb{BG}_{\eta+1}(A) = \left\{ \mathcal{L}(\lambda) \in \mathbb{DG}_{\eta+1}(A) \;\middle|\; \mathcal{L}(\lambda) = \mathcal{L}(\lambda)^{\mathcal{B}} \right\}.$$

As Example 4.2 immediately reveals, in general $\mathbb{BG}_{\eta+1}(A) \subsetneq \mathbb{DG}_{\eta+1}(A)$ holds. Recall that for $\mathcal{L}(\lambda) \in \mathbb{L}_1(A)$ it holds that $\mathcal{L}(\lambda)$ is block-symmetric iff $\mathcal{L}(\lambda) \in \mathbb{DL}(A)$ [80, Thm. 3.4.2]. However, in our setting, block-symmetric pencils $\mathcal{L}(\lambda) \in \mathbb{DG}_{\eta+1}(A)$ form a proper subspace of $\mathbb{DG}_{\eta+1}(A)$ for $0 < \eta \le \lceil \frac{k}{2} \rceil$ (which will be proven in Theorem 4.4 below) and therefore a nowhere dense subset in $\mathbb{DG}_{\eta+1}(A)$. Since any $\mathcal{L}(\lambda) \in \mathbb{DG}_1(A) \subset \mathbb{DL}(A)$ is block-symmetric, cf. Example 4.1 and Theorem 4.1, $\mathbb{DG}_{\eta+1}(A) = \mathbb{BG}_{\eta+1}(A)$ holds indeed in case $\eta = 0$. With the characterization of $\mathbb{DG}_{\eta+1}(A)$ from Theorem 4.1 at hand, the ansatz spaces $\mathbb{BG}_{\eta+1}(A), \eta \ge 0$, can easily be characterized as follows:

Theorem 4.4 (Characterization of $\mathbb{BG}_{\eta+1}(A)$, [42, Thm. 9]). *Let* $A(\lambda) \in \mathrm{M}_{n \times n}(\mathbb{R}[\lambda])$ *be of degree* $k = \epsilon + \eta + 1$ *and assume* $\eta \le \epsilon$. *Then* $\mathbb{BG}_{\eta+1}(A)$ *is a vector space over* $\mathbb{R}$ *having dimension*

$$\dim\big(\mathbb{BG}_{\eta+1}(A)\big) = k\eta n^2 + 1.$$

Any matrix pencil $\mathcal{L}(\lambda) \in \mathbb{BG}_{\eta+1}(A)$ *may be characterized as*

$$\mathcal{L}(\lambda) = \left[\begin{array}{c|c|c} I_{(\eta+1)n} & B_{11} & 0 \\ \hline 0 & C_{11} & \alpha\mathcal{H}_{\epsilon-\eta}(A) \\ \hline 0 & C_{21} & 0 \end{array}\right] \mathcal{F}^{\mathrm{DG}}_{\alpha,\eta,A}(\lambda) \left[\begin{array}{c|c|c} I_{(\epsilon+1)n} & & 0 \\ \hline B_{11}^{\mathcal{B}} & C_{11}^{\mathcal{B}} & C_{21}^{\mathcal{B}} \end{array}\right] \quad (4.32)$$

with arbitrary matrices $B_{11} \in \mathrm{M}_{(\eta+1)n \times \eta n}(\mathbb{R})$, $C_{11} \in \mathrm{M}_{(\epsilon-\eta)n \times \eta n}(\mathbb{R})$, $C_{21} \in \mathrm{M}_{\eta n \times \eta n}(\mathbb{R})$ *and* $\alpha \in \mathbb{R}$. *Moreover, unless* $\eta = 0$, $\mathbb{BG}_{\eta+1}(A)$ *is a proper subspace of both* $\mathbb{DG}_{\eta+1}(A)$ *and* $\mathbb{DG}_{k-\eta}(A)$.

Proof. Theorem 4.4 can be deduced directly from $\mathcal{L}(\lambda) = \mathcal{L}(\lambda)^{\mathcal{B}}$ and the previous results. As in (3.12) (see also Proposition 3.3) we obtain from $\mathcal{L}(\lambda)$ in (4.8)

$$\mathcal{L}(\lambda)^{\mathcal{B}} = \left[\begin{array}{c|c} I_{(\epsilon+1)n} & 0 \\ \hline B_{21} & B_{22} & C_2 \end{array}\right]^{\mathcal{B}} \mathcal{F}^{\mathrm{DG}}_{\alpha,\eta,A}(\lambda)^{\mathcal{B}} \left[\begin{array}{c|c|c} I_{(\eta+1)n} & B_{11} & 0 \\ \hline 0 & C_{11} & \alpha\mathcal{H}_{\epsilon-\eta}(A) \\ \hline 0 & C_{21} & 0 \end{array}\right]^{\mathcal{B}}$$

which more explicitly can be written as

$$\mathcal{L}(\lambda)^{\mathcal{B}} = \left[\begin{array}{c|c} I_{(\epsilon+1)n} & B_{21}^{\mathcal{B}} \\ \hline & B_{22}^{\mathcal{B}} \\ \hline 0 & C_2^{\mathcal{B}} \end{array}\right] \mathcal{F}^{\mathrm{DG}}_{\alpha,\eta,A}(\lambda)^{\mathcal{B}} \left[\begin{array}{c|c|c} I_{(\eta+1)n} & 0 & 0 \\ \hline B_{11}^{\mathcal{B}} & C_{11}^{\mathcal{B}} & C_{21}^{\mathcal{B}} \\ \hline 0 & \alpha H_{\epsilon-\eta}(A) & 0 \end{array}\right]. \quad (4.33)$$

We now take Corollary 4.2 into account (in particular (4.24)) which tells us how $\mathcal{L}(\lambda) \in \mathbb{G}_{\eta+1}(A)$ can be expressed as an element of $\mathbb{G}_{k-\eta}(A)$ and obtain

$$\mathcal{L}(\lambda) = \left[\begin{array}{c|c} I_{(\epsilon+1)n} & B_{11} \\ \hline & C_{11} \\ \hline 0 & C_{21} \end{array}\right] \mathcal{F}^{\mathrm{DG}}_{\alpha,\eta,A}(\lambda)^{\mathcal{B}} \left[\begin{array}{c|c|c} I_{(\eta+1)n} & 0 & 0 \\ \hline B_{21} & B_{22} & C_2 \\ \hline 0 & \alpha H_{\epsilon-\eta}(A) & 0 \end{array}\right]. \quad (4.34)$$

Now recall that $\mathcal{L}(\lambda)$ was assumed to be block-symmetric, i.e. it holds that $\mathcal{L}(\lambda) = \mathcal{L}(\lambda)^{\mathcal{B}}$. According to Theorem 3.1 we conclude from the expressions (4.33) and (4.34) that $B_{21} = B_{11}^{\mathcal{B}}, B_{22} = \mathcal{C}_{11}^{\mathcal{B}}$ and $C_2 = C_{21}^{\mathcal{B}}$ hold. The statement on the dimension of $\mathbb{BG}_{\eta+1}(A)$ follows straight forward from (4.7) in Theorem 4.1 since the matrices B_{21}, B_{22} and C_2 are fixed once B_{11}, C_{11} and C_{21} have been determined. □

To find or construct block-symmetric pencils $\mathcal{L}(\lambda) \in \mathbb{DG}_{\eta+1}(A)$ several aspects have to be considered. These are summarized as follows: As in the previous discussion, let $\mathcal{L}(\lambda)$ be partitioned as a 3×3 block-matrix as in Figure 4.1. First and foremost (4.31) reveals, that we have to take care of the bordering blocks in Figure 4.1 in order to enforce $\mathcal{L}(\lambda) \in \mathbb{DG}_{\eta+1}(A)$ on being block-symmetric. This is controlled by choosing $B_{21} = B_{11}^{\mathcal{B}}, B_{22} = \mathcal{C}_{11}^{\mathcal{B}}$ and $C_2 = C_{21}^{\mathcal{B}}$. Secondly, the upper left square diagonal block certainly has to be block-symmetric as well. This is guaranteed by choosing $\Sigma_{\eta,A}^{\mathrm{DG}}(\lambda)$ in the upper-left block of the anchor pencil $\mathcal{F}_{\alpha,\eta,A}^{\mathrm{DG}}(\lambda)$ which is in fact block-symmetric. Of course, adding the expression $B_{11}(L_\eta(\lambda) \otimes I_n) + (L_\eta(\lambda)^T \otimes I_n)B_{11}^{\mathcal{B}}$ to $\alpha\Sigma_{\eta,A}^{\mathrm{DG}}(\lambda)$ keeps on the block-symmetry. Finally, we do not have to take care of the core part $\mathcal{C}(\mathcal{L})$ of the pencil $\mathcal{L}(\lambda)$ which is, for pencils in $\mathbb{DG}_{\eta+1}(A)$, block-symmetric anyway.

Since block-symmetry is a property that has been intensively studied, we present an algorithm that computes block-symmetric pencils from $\mathbb{BG}_{\eta+1}(A)$. The conditions mentioned above were thereby taken into account.

Construction Procedure for block-symmetric Pencils
Let $A(\lambda) = \sum_{i=0}^{k} A_i \lambda^i \in \mathrm{M}_{n \times n}(\mathbb{R})[\lambda]$ be of degree $k = \epsilon + \eta + 1$.
  1. Compute the matrix $\Pi_{\eta,A}^{\mathrm{DG}}(\lambda)$ as in Section 4.1.
  2. Compute the matrix

$$C_1 = \left[\begin{array}{c|c} C_{11} & \alpha\mathcal{H}_{\epsilon-\eta}(A) \\ \hline C_{21} & 0_{\eta n \times (\epsilon-\eta)n} \end{array} \right] \in \mathrm{M}_{\epsilon n \times \epsilon n}(\mathbb{R})$$

with arbitrary matrices $C_{11} \in \mathrm{M}_{(\epsilon-\eta)n \times \eta n}(\mathbb{R})$ and $C_{21} \in \mathrm{M}_{\eta n \times \eta n}(\mathbb{R})$.
  3. Choose an arbitrary matrix $B_{11} \in \mathrm{M}_{(\eta+1)n \times \eta n}(\mathbb{R})$ and set

$$B_1 = \begin{bmatrix} B_{11} & 0_{(\eta+1)n \times (\epsilon-\eta)n} \end{bmatrix} \qquad C_2 = C_{21}^{\mathcal{B}} \qquad B_2 = \begin{bmatrix} B_{11}^{\mathcal{B}} & C_{11}^{\mathcal{B}} \end{bmatrix}. \qquad (4.35)$$

  4. Construct the $kn \times kn$ matrix pencil $\mathcal{L}(\lambda) \in \mathbb{DG}_{\eta+1}(A)$:

$$\mathcal{L}(\lambda) = \left[\begin{array}{c|c} I_{(\eta+1)n} & B_1 \\ \hline 0 & C_1 \end{array} \right] \left[\begin{array}{c|c} \alpha\Pi_{\eta,A}^{\mathrm{DG}}(\lambda) & L_\eta^T(\lambda) \otimes I_n \\ \hline L_\epsilon(\lambda) \otimes I_n & 0 \end{array} \right] \left[\begin{array}{c|c} I_{(\epsilon+1)n} & 0 \\ \hline B_2 & C_2 \end{array} \right]. \qquad (4.36)$$

Since $\Sigma_{\eta,A}^{\mathbb{BG}}(\lambda)$ is block-symmetric by construction, (4.35) ensures the block-symmetry of $\mathcal{L}(\lambda)$ in total according to Theorem 4.4. The next results about $\mathbb{BG}_{\eta+1}(A)$ are immediate consequences of Theorem 4.4 and Corollary 4.4.

Corollary 4.5 (Linearization Condition, [42, Cor. 4]). *Let $A(\lambda) \in \mathrm{M}_{n \times n}(\mathbb{R}[\lambda])$ be regular and of degree $k = \eta + \epsilon + 1$ and assume $\eta \leq \epsilon$. Moreover, let $\mathcal{L}(\lambda) \in \mathbb{BG}_{\eta+1}(A)$ be given in the form (4.32). Then the following statements are equivalent:*

  (i) $\mathcal{L}(\lambda)$ is a (strong) linearization for $A(\lambda)$.

(ii) $A_0 \in \mathrm{GL}_n(\mathbb{R}), C_{21} \in \mathrm{GL}_{\eta n}(\mathbb{R})$ and $\alpha \in \mathbb{R} \setminus \{0\}$.

For $\epsilon = \eta$ the equivalence in Corollary 4.5 holds again without the condition $A_0 \in \mathrm{GL}_n(\mathbb{R})$ in the second statement (due to the disappearance of the $\mathcal{H}$-block). In this case, $(ii) \Rightarrow (i)$ in Corollary 4.5 is also valid for singular matrix polynomials (according to Theorem 3.2). Moreover, certainly Corollary 4.3 applies. That is, whenever $A_0 \in \mathrm{GL}_n(\mathbb{R})$ or $\epsilon = \eta$, almost every pencil $\mathcal{L}(\lambda) \in \mathbb{B}\mathbb{G}_{\eta+1}(A)$ is a block-symmetric strong linearization for $A(\lambda) \in \mathrm{M}_{n \times n}(\mathbb{R}[\lambda])$ (regardless of $A(\lambda)$ being regular or singular). Moreover, the inclusion property from the previous section still holds:

Proposition 4.1 (Inclusion Property, [42, Lem. 3]). *Let* $A(\lambda) \in \mathrm{M}_{n \times n}(\mathbb{R}[\lambda])$ *be of degree* $k = \epsilon + \eta + 1$. *Then we have*

$$\mathbb{B}\mathbb{G}_1(A) \subsetneqq \mathbb{B}\mathbb{G}_2(A) \subsetneqq \cdots \subsetneqq \mathbb{B}\mathbb{G}_{\lceil \frac{k}{2} \rceil}(A). \tag{4.37}$$

Notice once more that $\mathbb{B}\mathbb{G}_{\eta+1}(A) = \mathbb{B}\mathbb{G}_{\epsilon+1}(A)$ holds for any $\eta, 0 \leq \eta \leq k - 1$, in accordance with (4.1). To illustrate the construction procedure for block-symmetric pencils consider the following simple example.

Example 4.4 ([42, Ex. 7]). *Let* $A(\lambda) = \sum_{i=0}^{7} A_i \lambda^i \in \mathrm{M}_{n \times n}(\mathbb{R})[\lambda]$ *be of degree* $k = 7$. *First consider the case* $\eta = 1$ *and* $\epsilon = k - \eta - 1 = 5$. *The construction procedure easily gives*

$$\mathcal{H}_4 = \begin{bmatrix} -A_3 & -A_2 & -A_1 & -A_0 \\ -A_2 & -A_1 & -A_0 & \\ -A_1 & -A_0 & & \\ -A_0 & & & \end{bmatrix} \in \mathrm{M}_{4n \times 4n}(\mathbb{R})$$

and $\Pi_{1,A}^{\mathbb{B}\mathbb{G}}(\lambda) = \begin{bmatrix} \lambda A_7 + A_6 & 0_n & 0_n & 0_n & 0_n & 0_n \\ 0_n & \lambda A_5 + A_4 & A_3 & A_2 & A_1 & A_0 \end{bmatrix}$. *Choose* $B_{11} = 0, C_{11} = 0$ *and* $C_{21} = I_n$. *Then, computing* $\mathcal{L}(\lambda)$ *from (4.36) with* $\alpha = 1$ *yields*

$$\mathcal{L}(\lambda) = \left[\begin{array}{cc|cccc|c} \lambda A_7 + A_6 & 0 & 0 & 0 & 0 & 0 & -I_n \\ 0 & \lambda A_5 + A_4 & A_3 & A_2 & A_1 & A_0 & \lambda I_n \\ \hline 0 & A_3 & A_2 - \lambda A_3 & A_1 - \lambda A_2 & A_0 - \lambda A_1 & -\lambda A_0 & 0 \\ 0 & A_2 & A_1 - \lambda A_2 & A_0 - \lambda A_1 & -\lambda A_0 & 0 & 0 \\ 0 & A_1 & A_0 - \lambda A_1 & -\lambda A_0 & 0 & 0 & 0 \\ 0 & A_0 & -\lambda A_0 & 0 & 0 & 0 & 0 \\ \hline -I_n & \lambda I_n & 0 & 0 & 0 & 0 & 0 \end{array} \right]$$

which is indeed a block-symmetric $7n \times 7n$ *matrix pencil. Thus* $\mathcal{L}(\lambda) \in \mathbb{B}\mathbb{G}_2(A)$. *Note that the choice of* B_{11} *and* C_{11} *has no influence on* $\mathcal{L}(\lambda)$ *for being a linearization. In fact, the nonsingularity of* A_0 *and* C_{21} *is the decisive factor, while choosing* B_{11} *and* C_{11} *to be singular matrices does not affect the linearization property of* $\mathcal{L}(\lambda)$ *at* all. So $\mathcal{L}(\lambda)$ is a strong linearization for any $A(\lambda)$ with $A_0 \in \mathrm{GL}_n(\mathbb{R})$. Now consider $\eta = 2$ and $\epsilon = k - \eta - 1 = 4$. Then

$$\mathcal{H}_2 = \begin{bmatrix} -A_1 & -A_0 \\ -A_0 & 0 \end{bmatrix} \in \mathrm{M}_{2n \times 2n}(\mathbb{R}).$$

Now choose $C_{11} = \begin{bmatrix} -A_7 & -A_6 \\ -A_5 & -A_4 \end{bmatrix}$ and $C_{21} = \begin{bmatrix} -A_3 & -A_2 \\ -A_1 & -A_0 \end{bmatrix}$. The computation in (4.36) gives $\mathcal{L}(\lambda) =$

$$\left[\begin{array}{ccc|cc|cc} \lambda A_7 + A_6 & 0 & 0 & A_7 & A_5 & A_3 & A_1 \\ 0 & \lambda A_5 + A_4 & 0 & A_6 - \lambda A_7 & A_4 - \lambda A_5 & A_2 - \lambda A_3 & A_0 - \lambda A_1 \\ 0 & 0 & \lambda A_3 + A_2 & A_1 - \lambda A_6 & A_0 - \lambda A_4 & -\lambda A_2 & -\lambda A_0 \\ \hline A_7 & A_6 - \lambda A_7 & A_1 - \lambda A_6 & A_0 - \lambda A_1 & -\lambda A_0 & 0 & 0 \\ A_5 & A_4 - \lambda A_5 & A_0 - \lambda A_4 & -\lambda A_0 & 0 & 0 & 0 \\ \hline A_3 & A_2 - \lambda A_3 & -\lambda A_2 & 0 & 0 & 0 & 0 \\ A_1 & A_0 - \lambda A_1 & -\lambda A_0 & 0 & 0 & 0 & 0 \end{array} \right]$$

which is a block-symmetric matrix pencil from $\mathbb{BG}_3(A)$.

4.6   A note on symmetric linearizations for symmetric matrix polynomials

If $A(\lambda) \in \mathrm{M}_{n \times n}(\mathbb{R})[\lambda]$ is regular and symmetric, the pencils from $\mathbb{DL}(A)$ provide a convenient way to linearize $A(\lambda)$, cf. [65]. In fact, $A(\lambda)$ is symmetric iff any pencil $\mathcal{L}(\lambda) \in \mathbb{DL}(A)$ is block-symmetric and symmetric [80, Thm. 5.1.2].

Now reconsider the pencil $\widetilde{\mathcal{K}}(\lambda)$ from (4.31) which was an element of $\mathbb{BG}_2(A)$ and $\mathbb{BG}_3(A)$, i.e. $\widetilde{\mathcal{K}}(\lambda)$ is block-symmetric. However, assuming that the matrix coefficients $A_6, \ldots, A_0$ are all symmetric (i.e. $A(\lambda) = A(\lambda)^T$ holds) does not enforce $\widetilde{\mathcal{K}}(\lambda)$ to be symmetric. To achieve this, the (arbitrarily chosen) matrices B, C and D have to be symmetric, too. In conclusion, for $A(\lambda) = A(\lambda)^T$ the ansatz space $\mathbb{BG}_{\eta+1}(A)$ does in general not consist entirely of symmetric pencils contrary to $\mathbb{DL}(A)$. This section summarizes how the results from Section 4.3 and Section 4.5 apply to symmetric matrix polynomials. However, we will only briefly deal with this subject since the situation is very similar to the discussion for $\mathbb{DG}_{\eta+1}(A)$ and $\mathbb{BG}_{\eta+1}(A)$.

To construct symmetric linearizations, we consider $\mathbb{DG}_{\eta+1}(A)$ and make the following definition similar to that of $\mathbb{BG}_{\eta+1}(A)$:

Definition 4.3 (Symmetric Ansatz Space). *Let $A(\lambda) \in \mathrm{M}_{n \times n}(\mathbb{R}[\lambda])$ be of degree $k = \epsilon + \eta + 1$ and symmetric, i.e. $A(\lambda) = A(\lambda)^T$. Moreover, assume $\eta \leq \epsilon$. Then we define*

$$\mathbb{SG}_{\eta+1}(A) = \left\{ \mathcal{L}(\lambda) \in \mathbb{DG}_{\eta+1}(A) \;\middle|\; \mathcal{L}(\lambda) = \mathcal{L}(\lambda)^T \right\}.$$

It is easily seen that for $\tilde{\mathcal{K}}(\lambda)$ in (4.31) to satisfy $\tilde{\mathcal{K}}(\lambda)^T = \tilde{\mathcal{K}}(\lambda)$ it is not only necessary but also sufficient to choose B, C and D as symmetric matrices. The generalization of this fact is therefore easily proved according to Theorem 4.4 (using transposition instead of block-transposition). The characterization of $\mathbb{SG}_{\eta+1}(A)$ is thus stated below without proof.

Theorem 4.5 (Characterization of $\mathbb{SG}_{\eta+1}(A)$). *Let $A(\lambda) \in M_{n \times n}(\mathbb{R}[\lambda])$ be of degree $k = \epsilon + \eta + 1$ and assume $A(\lambda) = A(\lambda)^T$. Moreover, let $\eta \leq \epsilon$. Then $\mathbb{SG}_{\eta+1}(A)$ is a vector space over $\mathbb{R}$ having dimension*

$$\dim\big(\mathbb{SG}_{\eta+1}(A)\big) = k\eta n^2 + 1.$$

Any matrix pencil $\mathcal{L}(\lambda) \in \mathbb{SG}_{\eta+1}(A)$ may be characterized as

$$\mathcal{L}(\lambda) = \left[\begin{array}{c|c|c} I_{(\eta+1)n} & B_{11} & 0 \\ \hline 0 & C_{11} & \alpha\mathcal{H}_{\epsilon-\eta}(A) \\ \hline 0 & C_{21} & 0 \end{array}\right] \mathcal{F}^{\mathrm{DG}}_{\alpha,\eta,A}(\lambda) \left[\begin{array}{c|c|c} I_{(\epsilon+1)n} & & 0 \\ \hline B_{11}^T & C_{11}^T & C_{21}^T \end{array}\right] \tag{4.38}$$

with arbitrary matrices $B_{11} \in M_{(\eta+1)n \times \eta n}(\mathbb{R})$, $C_{11} \in M_{(\epsilon-\eta)n \times \eta n}(\mathbb{R})$, $C_{21} \in M_{\eta n \times \eta n}(\mathbb{R})$ and $\alpha \in \mathbb{R}$.

The superpartition and inclusion properties according to Theorem 4.3 and Corollary 4.4 follow immediately for $\mathbb{SG}_{\eta+1}(A)$ as before (the proofs are analogous and thus omitted). Moreover, the linearization condition for $\mathbb{SG}_{\eta+1}(A)$ is exactly the same as stated in Corollary 4.5 for $\mathbb{BG}_{\eta+1}(A)$. Notice further that $\mathcal{H}_{\epsilon-\eta}(A) = \mathcal{H}_{\epsilon-\eta}(A)^T$ holds automatically if $A(\lambda)$ is symmetric and that $\mathcal{F}^{\mathrm{DG}}_{\alpha,\eta,A}(\lambda)^{\mathcal{B}} = \mathcal{F}^{\mathrm{DG}}_{\alpha,\eta,A}(\lambda)^T$. For $A(\lambda) = A(\lambda)^T$, all pencils $\mathcal{L}(\lambda) \in \mathbb{DG}_{\eta+1}(A)$ as in (4.38) form a subspace of $\mathbb{DG}_{\eta+1}(A)$ which is for $n > 1$ in general not equal to $\mathbb{BG}_{\eta+1}(A)$, although is has the same dimension. However, it is easily seen that there is one exception: the case $\eta = 0$ where $\mathbb{SG}_1(A)$ is again a subset of $\mathbb{DL}(A)$ (see also Lemma 6.1).

As already mentioned, all results from Section 4.5 apply in the same or a similar fashion to $\mathbb{SG}_{\eta+1}(A)$. Therefore, we opt out of discussing these results for $\mathbb{SG}_{\eta+1}(A)$ in detail and refer the reader to Section 4.5. For each single result, the proof can always be adapted in a straight forward manner to give the desired statement on $\mathbb{SG}_{\eta+1}(A)$.

4.7   Conclusions

In this and the preceding section we introduced the concept of block Kronecker ansatz spaces. Beside the classical ansatz spaces $\mathbb{L}_1(A), \mathbb{L}_2(A)$ and $\mathbb{DL}(A)$ introduced in [89] this is a new ansatz space framework that is suitable for the construction of strong linearizations $\mathcal{L}(\lambda) \in \mathbb{G}_{\eta+1}(A)$ for square or rectangular matrix polynomials $A(\lambda) \in M_{m \times n}(\mathbb{R}[\lambda])$. We analyzed these large vector spaces in the style of [89] showing that, in general, almost every matrix pencil $\mathcal{L}(\lambda) \in \mathbb{G}_{\eta+1}(A)$ is a strong linearization for $A(\lambda)$ regardless whether $A(\lambda)$ is regular or singular. Moreover, we showed that the strong linearization theorem [89, Thm. 4.3] extends from $\mathbb{L}_1(A)$ and $\mathbb{L}_2(A)$ to $\mathbb{G}_{\eta+1}(A)$.

As for $\mathbb{DL}(A)$ we draw attention to the intersection spaces $\mathbb{DG}_{\eta+1}(A) = \mathbb{G}_{\eta+1}(A) \cap \mathbb{G}_{k-\eta}(A)$. The fact that $\mathbb{DG}_{\eta+1}(A)$ does not contain solely block-symmetric matrix pencils was shown as one main difference to $\mathbb{DL}(A)$. In addition we characterized the double ansatz spaces $\mathbb{DG}_{\eta+1}(A)$ comprehensively and showed that, under the condition $\epsilon = \eta$, $\mathbb{DG}_{\eta+1}(A)$ also contains strong linearizations for singular matrix polynomials. The latter is in fact a property that is not shared by $\mathbb{DL}(A)$. Furthermore, we presented an algorithm for the construction of block-symmetric matrix pencils in $\mathbb{DG}_{\eta+1}(A)$ and introduced the subspaces $\mathbb{BG}_{\eta+1}(A), \mathbb{SG}_{\eta+1}(A) \subseteq \mathbb{DG}_{\eta+1}(A)$. In particular, we characterized all block-symmetric and symmetric matrix pencils in $\mathbb{BG}_{\eta+1}(A)$ and $\mathbb{SG}_{\eta+1}(A)$, respectively. For all types of ansatz spaces we gave conditions to identify strong linearizations in these spaces and showed that, in case $A(\lambda)$ has a nonsingular trailing coefficient A_0, almost every pencil in $\mathbb{DG}_{\eta+1}(A), \mathbb{BG}_{\eta+1}(A)$ and $\mathbb{SG}_{\eta+1}(A)$ is a strong linearization for $A(\lambda)$. Moreover, we have proved the superpatition property of the spaces $\mathbb{DG}_{\eta+1}(A), \mathbb{BG}_{\eta+1}(A)$ and $\mathbb{SG}_{\eta+1}(A)$ which implied that these spaces form nested subspace sequences for increasing values of η - a property that has no analogue for $\mathbb{L}_1(A), \mathbb{L}_2(A)$ or $\mathbb{DL}(A)$. We pointed out that the eigenvector recovery formulas for pencils in $\mathbb{G}_{\eta+1}(A)$ are similar as for pure block Kronecker pencils, which form an affine subspace of $\mathbb{G}_{\eta+1}(A)$ for each choice of η. In particular, using the results from [34] and [21] we mentioned that most of the matrix pencils in Fielder-like linearization families may be transformed (by a permutation) so that they become elements of some $\mathbb{G}_{\eta+1}(A)$. Therefore, block Kronecker ansatz spaces provide - modulo permutation - an ansatz space framework for most Fielder-like families of linearizations.

Finally we would like to mention that the concepts presented in this and the preceding chapter admit a direct extension to matrix polynomials over the field of complex numbers. However, we decided to confine ourselves to matrix polynomials over $\mathbb{R}$. Moreover, according to the results from [34], the minimal bases $L_k(\lambda)$ and $\Lambda_k(\lambda)$ used for the construction and characterization of the block Kronecker ansatz spaces can certainly be replaced by other dual minimal bases. As long as the fundamental results from [34, Sec. 3] hold for any another particular choice of dual minimal basis, there is no restriction to define and study block Kronecker ansatz spaces in this new setting.

Chapter 5

Solving generalized eigenproblems

> *Es ist nicht genug zu wissen, man muss auch anwenden; es ist nicht genug zu wollen, man muss auch tun.*

Johann Wolfgang von Goethe (1749 – 1832), [54, Kap. 71]

In this section we discuss a few ways how a polynomial eigenproblem corresponding to $A(\lambda) \in M_{n \times n}(\mathbb{R}[\lambda])$ can be solved via a linearization $\mathcal{L}(\lambda) \in \mathbb{G}_{\eta+1}(A)$. Thereby we focus on methods that enables us to compute a few (finite) eigenvalues of $\mathcal{L}(\lambda)$ near a desired target $\tau \in \mathbb{C}$.

In Section 5.1 we show how a generalized eigenvalue problem corresponding to a matrix pencil $\mathcal{L}(\lambda) = \lambda A + B$ can be transformed into an ordinary eigenproblem using a standard "shift-and-invert" strategy [6, Sec. 3.3]. Eigenvalue problems arising in practice are usually of large size and there is often only an interest in a few eigenvalues from a specific region in the complex plane. Therefore, Krylov subspace methods are in general a suitable tool to solve these problems. A brief discussion on Krylov subspace methods is presented in Section 5.1.1. The shift-and-invert reformulation of a GEP in cooperation with a Krylov subspace method in general requires the solution of linear systems $\mathcal{L}(\tau)x = z$ for some $\tau \in \mathbb{C}$. We show in Section 5.2 how these systems can be solved very efficiently if $\mathcal{L}(\lambda)$ belongs to a block Kronecker ansatz space $\mathbb{G}_{\eta+1}(A)$. These results are applied in Section 5.3 where we consider structured generalized eigenvalue problems. In particular, we discuss the EVEN-IRA algorithm from [94] for the solution of T-even generalized eigenproblems in Section 5.3.1. In particular, we show how a structure-preserving linearizations for T-even matrix polynomials can be obtained from the block Kronecker ansatz spaces so that the results from Section 5.2 can be used beneficially for their solution with EVEN-IRA. This is illustrated by some numerical experiments in Section 5.3.2. Finally, we briefly describe in Section 5.4 how the results from Section 5.2 can be incorporated in eigenvalues solvers for generalized symmetric eigenvalue problems.

5.1   Shift-and-invert reformulation

Let $A(\lambda) \in M_{n \times n}(\mathbb{R}[\lambda])$ be a regular matrix polynomial of degree $k \geq 2$ and assume that $\mathcal{L}(\lambda) = \lambda A + B \in M_{kn \times kn}(\mathbb{R})[\lambda]$ is some linearization for $A(\lambda)$ (not necessarily from a block Kronecker ansatz space $\mathbb{G}_{\eta+1}(A)$).

The most widely used technique to solve the generalized eigenproblem corresponding to $\mathcal{L}(\lambda)$ is the QZ algorithm dating back to 1973 [99]. Based on the generalized Schur decomposition this algorithm is appropriate for the computation of the entire spectrum of the GEP $\lambda Aw = -Bw$ and requires about $46(kn)^3$ floating point operations [57]. A nice survey on the QZ algorithm and related techniques can be found in [71]. Beside the QZ algorithm there are other techniques that can be applied if only a small part of the spectrum shall be found. For instance, rational Krylov subspace methods are suitable for the approximation of eigenvalues close to appropriately chosen shift parameters [111]. The divide-and-conquer approaches, e.g. [5], can also be applied to generalized eigenproblems although they are often vulnerable to bad convergence behavior or stability issues, cf. [100]. Finally, a modification of the FEAST algorithm appropriate for non-Hermitian generalized eigenproblems was given in [69]. More information on generalized (unstructured) eigenvalue problems and solution techniques can be found in [6, 57, 71] and the references therein. Our approach in this section is the reformulation of the GEP corresponding to $\mathcal{L}(\lambda)$ into an ordinary eigenvalue problem. To this end, let some parameter $\tau \in \mathbb{C}$ with $\tau \notin \sigma_f(\mathcal{L}) = \sigma_f(A)$ be given.

Recall that the finite eigenvalues of $\mathcal{L}(\lambda) \in M_{kn \times kn}(\mathbb{R}[\lambda])$ are those scalars $\lambda_0 \in \mathbb{C}$ with $\mathrm{rank}(\mathcal{L}(\lambda_0)) < kn$. If the generalized eigenproblem $\mathcal{L}(\lambda)w = 0$ is rewritten as $\lambda Aw = -Bw$, the subtraction of τAw on both sides gives $(\lambda - \tau)Aw = -(\tau A + B)w = -\mathcal{L}(\tau)w$. Moreover, since τ is no eigenvalue of $A(\lambda)$, the (complex) matrix $\mathcal{L}(\tau)$ is nonsingular. This yields the standard eigenvalue problem

$$\mathcal{L}(\tau)^{-1}Aw = -\frac{1}{\lambda - \tau}w =: \mu w. \tag{5.1}$$

Notice that this "shift-and-invert" strategy for (generalized) eigenvalue problems is common and can be found in many textbooks, e.g. [6, Sec. 3.3]. Some further remarks are in order:

(i) According to (5.1) any eigenvalue $\mu_0 \neq 0$ of $\mathcal{L}(\tau)^{-1}A$ determines an eigenvalue $\lambda_0 = \tau - 1/\mu_0$ of $\mathcal{L}(\lambda)$. On the other hand, any finite eigenvalue λ_0 of $\mathcal{L}(\lambda)$ gives rise to a eigenvalue $\mu_0 = -1/(\lambda_0 - \tau)$ of $\mathcal{L}(\tau)^{-1}A$.

(ii) The matrix $\mathcal{L}(\tau)^{-1}A$ is singular iff $A(\lambda)$ has infinite eigenvalues (which is the case iff A is singular). The correspondence in (5.1) extends to this situation with the convention that $-1/(\infty - \tau) = 0$, i.e. an infinite eigenvalue of $\mathcal{L}(\lambda)$ gives a zero eigenvalue of $\mathcal{L}(\tau)^{-1}A$.

(iii) Approximations μ_0 to eigenvalues of $\mathcal{L}(\tau)^{-1}A$ with large absolute value give small terms for $1/\mu_0$. Thus the eigenvalue approximation $\lambda_0 = \tau - 1/\mu_0$ yields eigenvalues of $\mathcal{L}(\lambda)$ near τ if μ_0 has large magnitude.

In applications, the situation for $A(\lambda)$ most often occurring is when $\deg(A(\lambda)) = k \ll n$. Nevertheless, in general $n \gg 1$, so the size $kn \times kn$ of the matrix $\mathcal{L}(\tau)^{-1}A$ (i.e. of the linearization $\mathcal{L}(\lambda)$ for $A(\lambda)$) in (5.1) is usually significantly large. Furthermore, in most practical considerations there is in fact not an interest in computing the entire spectrum of $A(\lambda)$ but in finding approximations to only a few eigenvalues of $A(\lambda)$ in some specific region of the complex plane (i.e. the eigenvalues with smallest absolute value, the eigenvalue in the left half plane etc.). According to these two observations, Krylov subspace methods are in general a valid tool to solve problems of the form (5.1). We will briefly highlight a few properties of Krylov subspace methods in the next section (in particular, related to the problem (5.1)).

5.1.1 Aspects on Krylov subspace methods

Let $M \in M_{n \times n}(\mathbb{C})$ and $x \in \mathbb{C}^n$. The m-th Krylov subspace corresponding to M and x is usually denoted $\mathcal{K}_m(M, x)$ and is defined as the span of the vectors $x, Mx, M^2x, \ldots, M^{m-1}x$. In a great many of numerical algorithms for the solution of eigenvalue problems Krylov subspaces play a key role.

Methods based on Krylov subspaces generally make use of the fact that, if $v_1, \ldots, v_m \in \mathbb{C}^n$ is an orthonormal basis for $\mathcal{K}_m(M, x)$ and $V = [\, v_1 \cdots v_m \,] \in M_{n \times m}(\mathbb{C})$, the eigenvalues of the *orthogonal projection* $V^H M V \in M_{m \times m}(\mathbb{C})$ of M onto $\mathcal{K}_m(M, x)$ often provide good approximations to some eigenvalues of M [117, Thm. 4.3]. Furthermore, the approximation accuracy will usually increase as m increases. This approach is often called the Rayleigh-Ritz method and the eigenvalues and eigenvectors of $V^H M V$ are called Ritz values and Ritz vectors for M [57]. Certainly, if $\mathcal{K}_m(M, x) = \mathcal{K}_{m+1}(M, x)$ the Ritz values of $V^H M V$ are exact eigenvalues of M since $\mathcal{K}_m(M, x)$ is an invariant subspace for M. Numerical algorithms which exploit the fact that the eigenvalues of M are approximated by its Ritz values (in any direct or more sophisticated way) are usually subsumed under the term *Krylov subspace methods*. In the following we will sketch two aspects which are common to many Krylov subspace methods. More information about these algorithms can be found in most textbooks on the numerical solution of eigenvalue problems, e.g., in [6, 112, 117].

(i) First of all, Krylov subspace methods are particularly valuable for large dimensional (and sparse) problems. In particular, if n is large and $m \ll n$ is chosen to be (comparatively) small, the eigenvalue problem corresponding to $V^H M V$ is again of small size $m \times m$ and can be solved via any appropriate method within a complexity depending on m (rather than on n). In consequence, only m eigenvalue approximations can be found from the spectrum of $V^H M V$. In general, the approximated part of $\sigma(M)$ from $\sigma(V^H M V)$ are those eigenvalues of largest magnitude [112].

(ii) Another key feature of Krylov subspace methods is that $V^H A V$ can be found according to the Arnoldi method efficiently with the only knowledge required about M being the result of matrix-vector-computations Mx for arbitrary vectors $x \in \mathbb{C}^n$ [6, Sec. 7.5]. Therefore, as long as a routine for the computation of Mx is available, Krylov subspace methods usually do not require direct access to the matrix M. For large-dimensional problems this can be advantageous since, if a routine for the matrix-vector-products Mx exists, the storage of M is not necessary.

Commonly, the matrix-vector-computations Mx are among the most expensive steps in Krylov subspace algorithms. If the m eigenvalues of $V^H M V$ do not provide eigenvalue approximations for M which are accurate enough, instead of increasing m a suitable restart strategy is usually incorporated in those methods. Two of the main restart schemes of practical relevance are the implicitly restarted Arnoldi method (IRAM), cf. [114, 115], and the Krylov-Schur algorithm [116].

Now consider once more the eigenvalue problem $\mathcal{L}(\tau)^{-1} A w = \mu w$ from (5.1). As, in general, Krylov subspace methods are suitable for finding eigenvalues of large absolute value (cf. (i)), the *shift parameter* τ in (5.1) can be used to control which eigenvalues of $\mathcal{L}(\tau)^{-1} A$ will be of large magnitude. In particular, a shift $\tau \in \mathbb{C}$ will effect the eigenvalues of $\mathcal{L}(\lambda)$ which lie near τ to be "large" eigenvalues of $\mathcal{L}(\tau)^{-1} A$. In this way, Krylov subspace methods are able to find those eigenvalues lying around τ in the complex plane. Moreover, the profitable usage of Krylov subspace methods for the eigenvalue problem $\mathcal{L}(\tau)^{-1} A w = \mu w$ depends crucially on how efficient matrix-vector-products with $\mathcal{L}(\tau)^{-1} A$ can be computed (cf. (ii)). As the inverse $\mathcal{L}(\tau)^{-1}$ shall not be computed explicitly, the products $(\mathcal{L}(\tau)^{-1} A) w$ can be found by first computing $z = Aw$ and then solving $\mathcal{L}(\tau) x = z$.

In the next section we discuss how the solution to $\mathcal{L}(\tau) x = z$ can be obtained in a very efficient way whenever the linearization $\mathcal{L}(\lambda)$ is from some block Kronecker ansatz space $\mathbb{G}_{\eta+1}(A)$. These results can be used within almost any Krylov subspace method (and related methods, see Section 5.3.1) to perform the matrix-vector-products with $\mathcal{L}(\tau)^{-1} A$ in (5.1) and $\mathcal{L}(\lambda) \in \mathbb{G}_{\eta+1}(A)$. Results similar to those from Section 5.2 already appeared in [43, Sec. 2] in the context of the Even-Ira algorithm and T-even polynomial eigenproblems (this is addressed in detail in Section 5.3.1).

5.2   Solving linear systems $\mathcal{L}(\tau) x = z$

Let $A(\lambda) \in \mathrm{M}_{n \times n}(\mathbb{R}[\lambda])$ of degree $k = \epsilon + \eta + 1$ be regular and assume $\tau \in \mathbb{C}, \tau \notin \sigma_f(A)$. Before we consider the general situation of solving $\mathcal{L}(\tau) x = z$ for any linearization $\mathcal{L}(\lambda) \in \mathbb{G}_{\eta+1}(A)$, we analyze the simpler situation where $\mathcal{L}(\lambda) = \mathcal{F}_{\alpha,\eta,A}(\lambda)$, i.e.

$$\mathcal{F}_{\alpha,\eta,A}(\lambda) = \left[\begin{array}{c|c} \alpha \Sigma_{\eta,A}(\lambda) & L_\eta(\lambda)^T \otimes I_m \\ \hline L_\epsilon(\lambda) \otimes I_n & 0 \end{array} \right]$$

with $\alpha \neq 0$ (cf. (3.13)). Recall that $\mathcal{F}_{\alpha,\eta,A}(\lambda)$ is a linearization for $A(\lambda)$ and that $\mathcal{F}_{\alpha,\eta,A}(\tau)$ is nonsingular. Therefore, given any vector $v \in \mathbb{C}^{kn}$, the linear system $\mathcal{F}_{\alpha,\eta,A}(\tau) w = v$ has a unique solution $w \in \mathbb{C}^{kn}$. The question how this solution can be efficiently computed taking the structure of $\mathcal{F}_{\alpha,\eta,A}(\tau)$ into account is addressed next. To this end, consider the system $\mathcal{F}_{\alpha,\eta,A}(\tau)w = v$ in the form

$$\mathcal{F}_{\alpha,\eta,A}(\tau)w = \left[\begin{array}{c|c} \alpha\Sigma_{\eta,A}(\tau) & L_\eta(\tau)^T \otimes I_n \\ \hline L_\epsilon(\tau) \otimes I_n & 0 \end{array}\right] \begin{bmatrix} w_1 \\ w_2 \end{bmatrix} = \begin{bmatrix} v_1 \\ v_2 \end{bmatrix},$$

that is,

$$\alpha\Sigma_{\eta,A}(\tau)w_1 + (L_\eta(\tau)^T \otimes I_n)w_2 = v_1 \tag{5.2}$$

$$(L_\epsilon(\tau) \otimes I_n)w_1 = v_2, \tag{5.3}$$

where w and v are partitioned with $w_1 \in \mathbb{C}^{(\epsilon+1)n}$ and $v_1 \in \mathbb{C}^{(\eta+1)n}$. First consider the system (5.3). As this is an underdetermined system, it has many solutions in general. If $\hat{w}_1$ is one particular solution, all solutions w_1 to (5.3) can be obtained as $\hat{w}_1 + \tilde{w}_1$ for arbitrary $\tilde{w}_1 \in \text{null}(L_\epsilon(\tau) \otimes I_n)$. As for any $\tau \in \mathbb{C}$ the $\epsilon n \times (\epsilon+1)n$ matrix $L_\epsilon(\tau) \otimes I_n$ has rank ϵn, the nullspace of $L_\epsilon(\tau) \otimes I_n$ is always n-dimensional. Moreover, since $(L_\epsilon(\tau) \otimes I_n)(\Lambda_\epsilon(\tau)^T \otimes I_n) = 0$ holds, this nullspace is spanned by the n columns of $\Lambda_\epsilon(\tau)^T \otimes I_n$. Therefore, the solutions w_1 to (5.3) are all vectors of the form $w_1 = \hat{w}_1 + (\Lambda_\epsilon(\tau)^T \otimes I_n)r$ where $r \in \mathbb{C}^n$ is arbitrary.

Now consider (5.3) in the form

$$\begin{bmatrix} -I_n & \tau I_n & & & \\ & -I_n & \tau I_n & & \\ & & \ddots & \ddots & \\ & & & -I_n & \tau I_n \end{bmatrix} \begin{bmatrix} w_{1,1} \\ w_{1,2} \\ \vdots \\ w_{1,\epsilon+1} \end{bmatrix} = \begin{bmatrix} v_{2,1} \\ v_{2,2} \\ \vdots \\ v_{2,\epsilon} \end{bmatrix}, \quad \text{with } w_{1,j}, v_{2,j} \in \mathbb{C}^n. \tag{5.4}$$

Notice that $w_{1,1}, \ldots, w_{1,\epsilon}$ can be found recursively once $w_{1,\epsilon+1}$ is known. On the other hand, any choice of $w_{1,\epsilon+1}$ uniquely determines a solution for (5.4). In particular, with $w_{1,\epsilon+1} \equiv 0$ we have $w_{1,\epsilon} = -v_{2,\epsilon}$ and

$$w_{1,j} = \tau w_{1,j+1} - v_{2,j}, \qquad \text{for all } j = \epsilon - 1, \ldots, 1. \tag{5.5}$$

The unique solution to (5.4) with $w_{1,\epsilon+1} \equiv 0$ can thus be computed in $2(\epsilon-1)n$ floating point operations. In the following computations this solution is always denoted by $\hat{w}_1$.

As we know that any solution w_1 for (5.3) has the form $w_1 = \hat{w}_1 + (\Lambda_\epsilon(\tau)^T \otimes I_n)r$ for some $r \in \mathbb{C}^n$, we may use this characterization to replace w_1 in (5.2) by this expression and consider (5.2), (5.3) as a system in the unknowns $w_2 \in \mathbb{C}^{\eta m}$ and $r \in \mathbb{C}^n$. Now suppose that $w_1^\star \in \mathbb{C}^{(\epsilon+1)n}$ and $w_2^\star \in \mathbb{C}^{\eta m}$ are the unique solutions to (5.2), (5.3) and $w_1^\star = \hat{w}_1 + (\Lambda_\epsilon(\tau)^T \otimes I_n)r^\star$. In particular, (5.2) now reads

$$\alpha\Sigma_{\eta,A}(\tau)\big(\hat{w}_1 + (\Lambda_\epsilon(\tau)^T \otimes I_n)r^\star\big) + \big(L_\eta(\tau)^T \otimes I_n\big)w_2^\star = v_1. \tag{5.6}$$

Then, multiplying (5.6) with $\Lambda_\eta(\tau) \otimes I_n$ from the left eliminates the second term since $(\Lambda_\eta(\tau) \otimes I_n)(L_\eta(\tau)^T \otimes I_n) = 0$ and yields after some reordering

$$\big(\Lambda_\eta(\tau) \otimes I_n\big)\big(v_1 - \alpha\Sigma_{\eta,A}(\tau)\hat{w}_1\big) = \big(\Lambda_\eta(\tau) \otimes I_n\big)\alpha\Sigma_{\eta,A}(\tau)\big(\Lambda_\epsilon(\tau)^T \otimes I_n\big)r^\star$$
$$= \alpha A(\tau)r^\star. \tag{5.7}$$

To see that (5.7) holds, recall that $(\Lambda_\eta(\tau) \otimes I_n)\alpha\Sigma_{\eta,A}(\tau)(\Lambda_\epsilon(\tau)^T \otimes I_n) = \alpha A(\tau)$ according to (3.7). Therefore, whenever (5.6) holds for $w_2^\star$ and $r^\star$, then (5.7) holds for $r^\star$. On the other hand, notice that $r^\star \in \mathbb{C}^n$ certainly has to be the only solution to (5.7) since the matrix $\alpha A(\tau)$ is nonsingular (recall that we assumed $\tau \notin \sigma_f(A)$). As a consequence, the correct $r^\star$ in (5.6) can directly (and independently of $w_2^\star$) be found by solving the linear system

$$\alpha A(\tau)r = \left(\Lambda_\eta(\tau)^T \otimes I_n\right)\left(v_1 - \alpha\Sigma_{\eta,A}(\tau)\widehat{w}_1\right) \tag{5.8}$$

for r. The system (5.8) can be solved by any appropriate method. For instance, if (5.8) is solved via the LU decomposition of $A(\tau)$, this yields a computational complexity of about $\mathcal{O}(2n^3/3)$ if no additional structure in the matrix $A(\tau)$ can be exploited.

Having computed $r^\star$ from (5.8), $w_1^\star$ is given by $w_1^\star = \widehat{w}_1 + (\Lambda_\epsilon(\tau)^T \otimes I_n)r^\star$, so it remains to find $w_2^\star$. As (5.6) holds for $r^\star$ and $w_2^\star$ and since $r^\star$ has already been found, (5.6) becomes a system with one unknown, namely $w_2^\star$. In particular, $w_2^\star \in \mathbb{C}^{\eta m}$ is the unique vector that satisfies

$$\begin{bmatrix} -I_n & & & \\ \tau I_n & -I_n & & \\ & \tau I_n & \ddots & \\ & & \ddots & -I_n \\ & & & \tau I_n \end{bmatrix}\begin{bmatrix} w_{2,1}^\star \\ \vdots \\ w_{2,\eta}^\star \end{bmatrix} = v_1 - \alpha\Sigma_{\eta,A}(\tau)w_1^\star =: \begin{bmatrix} \widetilde{v}_{1,1} \\ \vdots \\ \widetilde{v}_{1,\eta+1} \end{bmatrix}. \tag{5.9}$$

The system (5.9) can be solved by forward substitution. In fact, we obtain $w_{2,1}^\star = -\widetilde{v}_{1,1}$ and

$$w_{2,j}^\star = \tau w_{2,j-1}^\star - \widetilde{v}_{1,j} \qquad \text{for } j = 2,\ldots,\eta \tag{5.10}$$

whereas the last condition $\tau w_{2,\eta}^\star = \widetilde{v}_{1,\eta+1}$ of the overdetermined system (5.9) will automatically hold and need not be considered (this is since we know that $w_2^\star$ satisfies (5.6) with the correct $r^\star$ that already has been found). Analogously to the recursion for $\widehat{w}_1$ in (5.5), the computation of $w_2^\star$ according to (5.10) requires $2(\eta-1)n$ flops.

Remark 5.1. *Notice that the solution of $\mathcal{F}_{\alpha,\eta,A}(\tau)w = v$ does not require the explicit construction of the matrix $\mathcal{F}_{\alpha,\eta,A}(\tau)$. In particular, an efficient implementation of this method can exploit the fact that in all intermediate steps according to (5.5) to (5.10) only matrix-vector-operations of size $n \times n$ have to be performed.*

Now assume that $\mathcal{L}(\lambda) \in \mathbb{G}_{\eta+1}(A)$ is some linearization for $A(\lambda)$. Then $\mathcal{L}(\lambda)$ can be expressed in the form

$$\mathcal{L}(\tau) = U\mathcal{F}_{\alpha,\eta,A}(\tau)V$$
$$= \left[\begin{array}{c|c} I_{(\eta+1)n} & B_1 \\ \hline 0 & C_1 \end{array}\right]\left[\begin{array}{c|c} \alpha\Sigma_{\eta,A}(\tau) & L_\eta(\tau)^T \otimes I_m \\ \hline L_\epsilon(\tau) \otimes I_n & 0 \end{array}\right]\left[\begin{array}{c|c} I_{(\epsilon+1)n} & 0 \\ \hline B_2 & C_2 \end{array}\right]$$

for certain matrices B_1, B_2, C_1, C_2 (with C_1 and C_2 being nonsingular) and some $\alpha \in \mathbb{R}, \alpha \neq 0$. If $\tau \in \mathbb{C}$ is no eigenvalue of $A(\lambda)$, then $\mathcal{L}(\tau)$ is nonsingular and, given any vector $z \in \mathbb{C}^{kn}$, the linear system $\mathcal{L}(\tau)x = z$ has a unique solution. This solution can be found by solving the three $kn \times kn$ systems $Uv = z$, $\mathcal{F}_{\alpha,\eta,A}(\tau)w = v$ and $Vx = w$ one after another. The solutions to $Uv = z$ and $Vx = w$ can be easily obtained as follows:

(i) For $Uv = z$ let v be partitioned as before and let z be given as $z = [\, z_1^T \; z_2^T \,]^T$ with $z_1 \in \mathbb{C}^{(\eta+1)n}$. Then the system $Uv = z$ consists of two equations, namely $I_{(\eta+1)n}v_1 + B_1v_2 = z_1$ and $C_1v_2 = z_2$. Thus, if $C_1v_2 = z_2$ is solved for v_2, then v_1 is given as $z_1 - B_1v_2$. The cost of computing v is essentially that of finding the solution v_2 to the $\epsilon n \times \epsilon n$ (complex) system $C_1v_2 = z_2$.

(ii) Now consider the system $Vx = w$ with w partitioned as before and $x = [\, x_1^T \; x_2^T \,]^T$ with $x_1 \in \mathbb{C}^{(\epsilon+1)n}$. Then we have the two equations $I_{(\epsilon+1)n}w_1 = x_1$ and $B_2w_1 + C_2w_2 = x_2$. As the first equation yields $w_1 = x_1$ we have to solve $C_2w_2 = x_2 - B_2x_1$ to obtain w_2. The computational costs for finding the solution to $Vx = w$ are dominated by the costs for solving the $\eta n \times \eta n$ (complex) system $C_2w_2 = x_2 - B_2x_1$.

If we assume that either $C_1 = I_{\epsilon n}$ and $C_2 = I_{\eta n}$ holds or that a routine for solving the systems in (i) and (ii) at low computational costs is available, the main costs for solving $\mathcal{L}(\tau)x = z$ rise from the solution of the system $\mathcal{F}_{\alpha,\eta,A}(\tau)w = v$. As the computational effort in (5.5) and (5.10) is significantly less than that of finding the solution in (5.8), we conclude that the solution to $\mathcal{L}(\tau)x = z$ can be found with overall complexity $\mathcal{O}(2n^3/3)$ whenever an LU decomposition of $A(\tau)$ is computed to solve the system in (5.8).

5.3 Structured polynomial eigenproblems

Matrix polynomials $A(\lambda)$ arising in applications are often structured. For instance, matrix polynomials $A(\lambda) = \sum_{j=0}^{k} A_j\lambda^j \in \mathrm{M}_{n \times n}(\mathbb{R})[\lambda]$ whose matrix coefficients $A_0, \ldots, A_k$ are alternately symmetric and skew-symmetric are called T-even or T-odd (see Table 2.1). Unfortunately, regardless of the choice of τ, the matrix $\mathcal{L}(\tau)^{-1}A$ in (5.1) will in general not inherit any related structure. For instance, even if $A(\lambda)$ and a linearization $\mathcal{L}(\lambda) = \lambda A + B$ for $A(\lambda)$ are both T-even, the standard eigenvalue problem corresponding to $\mathcal{L}(\tau)^{-1}A$ in (5.1) will not possess any significant structure. To solve the eigenvalue problem for a T-even linearization $\mathcal{L}(\lambda)$ for $A(\lambda)$ in a numerically reliable way, the Even-Ira algorithm was introduced in [94]. As a Krylov subspace method it is most appropriate for large and sparse problems and computes a part of the spectrum of $\mathcal{L}(\lambda)$ near a predefined target $\tau \in \mathbb{C}$ in the complex plane. This method generalized the SHIRA algorithm from [97] which was designed for Hamiltonian eigenproblems corresponding to $\mathcal{L}(\lambda) = A\lambda + B$ where

A and B are Hamiltonian and skew-Hamiltonian matrices[1], respectively. Related methods based on implicitly restarted Arnoldi methods can be found in, e.g., [68, 96, 128] (more references can be found in [94]). Whenever a generalized Hamiltonian eigenproblem becomes small and dense, several techniques for their solution have been developed in [9, 10, 73], mostly based on the computation of deflating subspaces. In particular, [73] presents an QR-like approach which is also suitable for generalized palindromic eigenproblems, i.e. $Ax = \lambda A^T x$. It is based on the palindromic Schur decomposition [73, Thm. 1].

The next section is based on the publication [43]. It shows how the results from Section 5.2 on the solution of linear systems $\mathcal{L}(\tau)x = z$ for linearizations $\mathcal{L}(\lambda) \in \mathbb{G}_{\eta+1}(A)$ and the EVEN-IRA algorithm from [94] can be used for the solution of T-even polynomial eigenproblems.

5.3.1 T-even eigenproblems and the Even-IRA algorithm

Let

$$A(\lambda) = \sum_{j=0}^{k} A_j \lambda^j = A_k \lambda^k + A_{k-1}\lambda^{k-1} + \cdots + A_1\lambda + A_0 \in \mathrm{M}_{n \times n}(\mathbb{R})[\lambda]$$

be a regular and T-even matrix polynomial of degree $k \geq 2$. That is, it holds that $A(\lambda)^T = A(-\lambda)$, which is equivalent to the fact that A_j is symmetric if j is even and A_j is skew-symmetric otherwise. Such matrix polynomials possess a spectrum that is symmetric with respect to the real and imaginary axis. Numerical methods respecting this spectral symmetry are particularly suitable for the eigenvalue computation of these problems. A linearization $\mathcal{L}(\lambda) = \lambda A + B \in \mathrm{M}_{kn \times kn}(\mathbb{R})[\lambda]$ for $A(\lambda)$ is called structure-preserving if $\mathcal{L}(\lambda)$ is T-even, too (i.e. $A^T = -A$, $B^T = B$). Before we consider the situation $\mathcal{L}(\lambda) \in \mathbb{G}_{\eta+1}(A)$ in detail, assume that $\mathcal{L}(\lambda)$ is an arbitrary T-even linearization for $A(\lambda)$.

The EVEN-IRA algorithm from [94] was introduced as a method for finding some eigenvalues of a T-even matrix pencil as $\mathcal{L}(\lambda)$. As pointed out earlier, the transformation of the generalized T-even eigenvalue problem $(\lambda A + B)x = 0$ corresponding to $\mathcal{L}(\lambda)$ into the form (5.1) will yield a standard eigenvalue problem $\mathcal{L}(\tau)^{-1}Ax = \mu x$ with no particular structure. The main idea behind the EVEN-IRA algorithm is a clever "double" shift-and-invert transformation similar to the one presented in (5.1). In fact, the authors from [94] consider the standard eigenvalue problem $Kx = \theta x$ for K given as

$$K := \mathcal{L}(\tau)^{-T}A\mathcal{L}(\tau)^{-1}A \in \mathrm{M}_{kn \times kn}(\mathbb{C}) \tag{5.11}$$

and show that the eigenvalues of $\mathcal{L}(\lambda)$ and K are related in a particular fashion. Indeed, a finite eigenvalue-eigenvector pair (λ_0, x_0) of $\mathcal{L}(\lambda)$ satisfying $\mathcal{L}(\lambda_0)x_0 = 0$ yields an eigenvalue-eigenvector pair (θ, x_0) of K with $\theta = 1/(\lambda_0^2 - \tau^2)$. Moreover, an infinite eigenvalue of $\mathcal{L}(\lambda)$ gives rise to a zero eigenvalue of K in a similar way

---

[1]More information on these type of matrices is provided in Chapters 8 to 10.

as discussed in Section 5.1. The main advantage in considering $Kx = \theta x$ instead of (5.1) is that any finite eigenvalue pair $\lambda_0, -\lambda_0 \in \sigma_f(A)$ is mapped to the same eigenvalue[2] $\theta = 1/(\lambda_0^2 - \tau^2)$ of K. Therefore, $\pm$ matching pairs of eigenvalues are preserved. Notice the following two important facts:

(i) Whenever some eigenvalue θ_0 of K has been found, it gives rise to a $\pm$ matching pair of two eigenvalues of $\mathcal{L}(\lambda)$, i.e.

$$\lambda_0 = \sqrt{(1/\theta_0) + \tau^2} \quad \text{and} \quad \widehat{\lambda}_0 = -\sqrt{(1/\theta_0) + \tau^2}.$$

(ii) The matrix $K \in \mathrm{M}_{kn \times kn}(\mathbb{C})$ will in general be complex but remains real if a purely real or purely complex shift is chosen. If $\tau = a + ib$ with $a \neq 0, b \neq 0$, is chosen, K can be slightly modified to stay within real arithmetic [94, Rem. 2.1]. Here we will assume that τ is either real or purely imaginary.

In [94] the authors suggest to apply the implicitly restarted Krylov-Schur method [116] to the matrix K to find some eigenvalues of $\mathcal{L}(\lambda)$. That is, if $v_1, \ldots, v_m$ is a orthonormal basis of $\mathcal{K}_m(K, x)$ for some $x \in \mathbb{R}^{kn}$ (computed by the Arnoldi method) and $V = [\, v_1 \; \cdots \; v_m \,] \in \mathrm{M}_{kn \times m}(\mathbb{R})$, some of the eigenvalues of K of largest magnitude are approximated by the eigenvalues of $V^H K V$. To improve the approximation accuracy, the process is repeated using the Krylov-Schur restart strategy. Despite all the details (which can be found in [94]) this approach is called the EVEN-IRA algorithm. Additional detailed information on this method and how eigenvectors may be captured can be found in [94, p. 4074ff]; a discussion on its implementation is presented in [94, Sec. 4].

For the efficient application of the EVEN-IRA algorithm for $\mathcal{L}(\lambda)$ it is important to efficiently compute the matrix-vector-products with K in (5.11) for arbitrary vectors $x \in \mathbb{R}^{kn}$. Furthermore, a matrix-vector-multiplication Kx requires the solution of two linear systems. As it was shown in Section 5.2 how linear systems $\mathcal{L}(\tau)x = z$ for linearizations from $\mathbb{G}_{\eta \restriction 1}(A)$ can be solved, our next goal is to construct a T-even linearization $\mathcal{L}(\lambda)$ for any T-even regular matrix polynomial $A(\lambda)$ so that the results from Section 5.2 can be applied. To this end, let $A(\lambda) = \sum_{j=0}^k A_j \lambda^j \in \mathrm{M}_{n \times n}(\mathbb{R})[\lambda]$ be a regular, T-even matrix polynomial of degree k and first assume the degree k of $A(\lambda)$ is odd, i.e. $k = 2d + 1, d \geq 1$. Now, for each $\ell \in \mathbb{N}$ let $\mathcal{I}_n^\ell$ denote the matrix

$$\mathcal{I}_n^\ell = \begin{bmatrix} (-1)^{\ell-1} I_n & & & \\ & (-1)^{\ell-2} I_n & & 0 \\ & & \ddots & \\ & & & (-1)^1 I_n \\ 0 & & & & (-1)^0 I_n \end{bmatrix} \in \mathrm{M}_{\ell n \times \ell n}(\mathbb{R}).$$

---

[2]Therefore, each eigenvalue of K necessarily has an even algebraic multiplicity [94, p. 4071].

With this definition at hand, we build the matrix pencil $\mathcal{L}_A(\lambda)$ defined below. Notice that it is a (strict equivalent) modification of $\mathcal{F}^{\mathrm{DG}}_{1,d,A}(\lambda) \in \mathbb{G}_{d+1}(A)$ from (4.9):

$$\mathcal{L}_A(\lambda) := \begin{bmatrix} \mathcal{I}_n^{d+1} & 0 \\ \hline 0 & I_{dn} \end{bmatrix} \begin{bmatrix} \Sigma^{\mathrm{DG}}_{d,A}(\lambda) & L_d^T(\lambda) \otimes I_n \\ \hline L_d(\lambda) \otimes I_n & 0 \end{bmatrix} \begin{bmatrix} I_{(d+1)n} & 0 \\ \hline 0 & -\mathcal{I}_n^d \end{bmatrix}$$

$$= \widehat{U} \mathcal{F}^{\mathrm{DG}}_{1,d,A}(\lambda) \widehat{V} = \lambda X + Y. \tag{5.12}$$

It is checked straight forward that $\mathcal{L}_A(\lambda)^T = \mathcal{L}_A(-\lambda)$ holds if $A(\lambda)^T = A(-\lambda)$, so $\mathcal{L}_A(\lambda)$ is T-even. Moreover, as $\mathcal{L}_A(\lambda)$ is strictly equivalent to $\mathcal{F}^{\mathrm{DG}}_{1,d,A}(\lambda)$, it is always a strong linearization for $A(\lambda)$. With $\mathcal{L}(\lambda) = \mathcal{L}_A(\lambda)$, the matrix K in (5.11) now takes the form $K = \mathcal{L}_A(\tau)^{-T} X \mathcal{L}_A(\tau)^{-1} X$ and the Krylov-Schur algorithm can be applied to K.

For the computation of matrix-vector-products

$$Kx = \left(\mathcal{L}_A(\tau)^{-T} X \mathcal{L}_A(\tau)^{-1} X \right) x \tag{5.13}$$

with $x \in \mathbb{R}^{kn}$ two linear systems involving $\mathcal{L}_A(\tau)$ and $\mathcal{L}_A(\tau)^T$ have to be solved. As the matrices $\widehat{U}$ and $\widehat{V}$ in (5.12) are easily seen to be involutions (i.e. $\widehat{U}^{-1} = \widehat{U}$ and $\widehat{V}^{-1} = \widehat{V}$ holds), $\mathcal{L}_A(\tau)x = z$ can be found by (i) computing $v = \widehat{U}z$, (ii) solving $\mathcal{F}^{\mathrm{DG}}_{1,d,A}(\tau)w = v$ and (iii) computing $\widehat{V}w = x$. The linear system $\mathcal{F}^{\mathrm{DG}}_{1,d,A}(\tau)w = v$ can be solved efficiently as described in Section 5.2 using an LU decomposition of $A(\tau)$. The second system $\mathcal{L}_A(\tau)^T x = (\widehat{V} \mathcal{F}^{\mathrm{DG}}_{1,d,A}(\tau)^T \widehat{U})x = z$ can be handled analogously noting that

$$\mathcal{F}^{\mathrm{DG}}_{1,d,A}(\tau)^T = \begin{bmatrix} \Sigma^{\mathrm{DG}}_{d,A}(-\tau) & L_d^T(\tau) \otimes I_n \\ \hline L_d(\tau) \otimes I_n & 0 \end{bmatrix}$$

since $A(\lambda)$ is T-even. Thus, the procedure described in Section 5.2 yields a linear system in (5.8) involving $A(-\tau)$ instead of $A(\tau)$ which has to be solved. However, as $A(-\tau) = A(\tau)^T$, if $A(\tau) = LU$ is an LU decomposition of $A(\tau)$, then $A(\tau)^T = U^T L^T$ is a decomposition of $A(\tau)^T$ from which a solution can be obtained as easily as before by forward and backward substitution.

In conclusion, whenever $A(\lambda) \in \mathrm{M}_{n \times n}(\mathbb{R})$ is some regular and T-even matrix polynomial of odd degree, the matrix pencil $\mathcal{L}_A(\lambda)$ from (5.12) is a structure-preserving (i.e. T-even) linearization for $A(\lambda)$ and the EVEN-IRA algorithm from [94, p. 4076ff] can be applied to $\mathcal{L}_A(\lambda)$. In particular, the matrix-vector-products with K as in (5.13) in the Krylov-Schur algorithm can be found as described above using the results on the solution of linear systems involving $\mathcal{L}_A(\tau)$ and $\mathcal{L}_A(\tau)^T$ from Section 5.2. This can be achieved in $\mathcal{O}(n^3)$ whenever an LU decomposition of $A(\tau)$ is computed for the solution of (5.8). If such a decomposition is computed once, it can be reused each time a matrix-vector-product with K has to be computed. Moreover, according to the form of $\mathcal{L}_A(\lambda)$ in (5.12) and Remark 5.1 all matrix-vector-operations required for the computation of Kx can be carried out implicitly on vectors and matrices of size $n \times 1$ and $n \times n$, respectively. Under the reasonable assumptions that $m \ll n$ eigenvalues of $\mathcal{L}_A(\lambda)$ (i.e. of K) are to be found and that $k \ll n$, the costs for the Krylov-Schur algorithm on K will stay within $\mathcal{O}(n^3)$. Therefore, the overall complexity of this approach is $\mathcal{O}(n^3)$. If the EVEN-IRA algorithm is considered for some arbitrary T-even linearization $\mathcal{L}(\lambda)$ for $A(\lambda)$ (for which the results from Section 5.2 do not apply), an LU decomposition of $\mathcal{L}(\lambda)$ in general requires about $\mathcal{O}(k^3 n^3)$ floating point operations.

Remark 5.2. *If the T-even matrix polynomial $A(\lambda) = \sum_{j=0}^{k} A_j \lambda^j$ at hand is of even degree $k = 2d$, we can still almost proceed as before. In fact, we may turn $A(\lambda)$ artificially into a matrix polynomial $\hat{A}(\lambda)$ with $\hat{A}(\lambda) = A_{k+1}\lambda^{k+1} + A(\lambda)$ where the matrix coefficient A_{k+1} is the zero matrix, i.e. $A_{k+1} = 0_{n \times n}$. In this case, $\mathcal{L}_{\hat{A}}(\lambda)$ as in (5.12) will still be a linearization for $A(\lambda)$ [34], although it will not be strong in the sense of Definition 2.13 anymore. Since we are only interested in finite eigenvalues, this causes no troubles at all at this point. Indeed, the linearization $\mathcal{L}_{\hat{A}}(\lambda)$ will still be T-even, so the EVEN-IRA algorithm may be applied to $\mathcal{L}_{\hat{A}}(\lambda)$ as before keeping $A_{k+1} = 0$ in mind.*

5.3.2 Numerical examples

In this section we present two numerical examples on the performance of our implementation of the EVEN-IRA in MATLAB [91] using the results from Section 5.2 for the computation of matrix-vector-products Kx for K as in (5.13). This method is subsequently called *Block Kronecker* EVEN-IRA *algorithm.* Our results are compared to those produced by the `polyeig` command from MATLAB which computes the eigenvalues of a matrix polynomial $A(\lambda)$ from the QZ decomposition of the Frobenius companion pencil $\mathrm{Frob}_1(\lambda)$, cf. (2.13). Both matrix polynomials we consider are T-even and of degree two and four, respectively. In particular, to apply the algorithm, we artificially extend both matrix polynomials by a zero matrix coefficient as described in Remark 5.2. Then the implicitly restarted Krylov-Schur algorithm is applied to the matrix K for some shift $\tau \notin \sigma_f(A)$ where the linearization $\mathcal{L}_A(\lambda)$ is given as in (5.12).

We first consider the example `wiresaw1` from [11]. The matrix polynomial under consideration $Q(\lambda) = M\lambda^2 + C\lambda + K_1$ was constructed to have size 64×64 where

$$M = \frac{1}{2}I_{64}, \qquad K_1 = \mathrm{diag}_{1 \leq j \leq 64}\left[j^2\pi^2(1-\nu^2)/2\right]$$

and $C = [c_{ij}]_{ij} = -C^T$ with

$$c_{ij} = \begin{cases} \frac{4jk}{j^2-k^2}\nu & \text{if } j+k \text{ is odd,} \\ 0 & \text{otherwise.} \end{cases}$$

In our test, the real, nonnegative parameter $0 < \nu < 1$ was chosen as $\nu = 0.5$ and the shift τ as $\tau = 0.2i$. Our goal was to approximate ten eigenvalues of $M(\lambda)$ near the origin. The numerical results are displayed in Figure 5.1. Notice that all eigenvalues of $M(\lambda)$ are purely imaginary but that MATLABs `polyeig` command, which we used

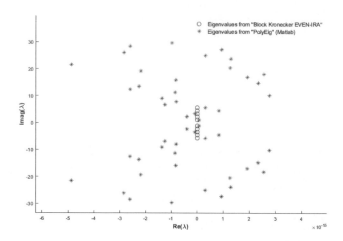

FIGURE 5.1: The Block Kronecker EVEN-IRA algorithm applied
to the example `wiresaw1` from [11].

as a reference, does not anticipate this. In fact, the eigenvalues computed by MATLAB
are not symmetric with respect to the real and imaginary axis (although they are
quite near to the imaginary axis). The situation of only purely imaginary eigenvalues
is certainly captured by EVEN-IRA. Denote the eigenvalues computed from the Block
Kronecker EVEN-IRA algorithm by $\lambda_j, j = 1, \ldots, 10$ and those computed by `polyeig`
by $\mu_i, i = 1, \ldots, 64$. A comparison of the purely imaginary eigenvalues λ_j and the
imaginary parts $\mathrm{imag}(\mu_k)$ of the values μ_k computed by MATLAB yields that

$$\min_{1 \le k \le 64} |\lambda_j - \mathrm{imag}(\mu_k)| < 10^{-10}$$

holds after six restarts. That is, we observe an accordance of the first ten decimal
digits after six restarts.

Our second example is taken from [96] and has already been discussed in [43, Sec. 3].
Here, the matrix polynomial $M(\lambda) = \sum_{j=0}^{4} M_j \lambda^j$ under consideration is of degree
four. The matrix coefficients are build from several Kronecker products as follows: we
set $m = 8$ and $n = m^2 = 64$. Let N denote the $m \times m$ nilpotent Jordan matrix with
ones on the first subdiagonal and define $\tilde{M}_0 = (1/6)(4I_m + N + N^T), \tilde{M}_1 = N - N^T$,
$\tilde{M}_2 = -(2I_m - N - N^T), \tilde{M}_3 = \tilde{M}_1$ and $\tilde{M}_4 = -\tilde{M}_2$. Moreover, we set

$$M_i = c_{i1} I_m \otimes \tilde{M}_i + c_{i2} \tilde{M}_i \otimes I_m$$

with positive constants c_{ij} chosen as $c_{01} = 0.6, c_{02} = 1.3, c_{11} = 1.3, c_{12} = 0.1, c_{21} =
0.1, c_{22} = 1.2, c_{31} = c_{32} = c_{41} = c_{42} = 1.0$. Now the matrix polynomial $M(\lambda) =$

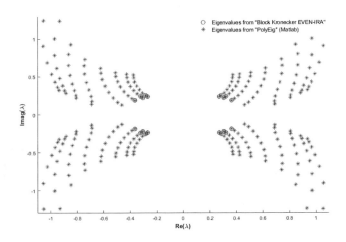

FIGURE 5.2: The Block Kronecker EVEN-IRA algorithm applied
to an example from [96, Sec. 4] (see also [43, Sec. 3]).

$\sum_{j=0}^{4} M_j \lambda^j$ has size 64×64. Our goal is to approximate ten eigenvalues of $M(\lambda)$ in the right half plane near the imaginary axis. To this end, the shift was chosen to be $\tau = 0.2$. The numerical results are displayed in Figure 5.2. This time, in contrast to our first example, MATLAB anticipates the shape of the spectrum. Moreover, notice that, due to the symmetry relations in the spectrum of $A(\lambda)$, we aditionally have found the (matching) ten eigenvalues of $A(\lambda)$ in the left half plane (thus we have found 20 eigenvalue of $A(\lambda)$ in total). Comparing the eigenvalues computed from the block Kronecker EVEN-IRA algorithm and MATLABS `polyeig` function, we again obtain an accordance of the first ten decimal digits in their real and imaginary parts after six restarts.

5.4 Algorithms for symmetric GEPs

In Section 5.3.1 we have discussed how a few eigenvalues of a T-even regular matrix polynomial $A(\lambda) \in M_{n \times n}(\mathbb{R}[\lambda])$ can be found. Our approach was based on the linearization $\mathcal{L}_A(\lambda)$ for $A(\lambda)$ given in (5.12) so that the numerical advantages provided by the EVEN-IRA algorithm apply (since $\mathcal{L}_A(\lambda)$ was T-even) and the economic solution of linear systems with $\mathcal{L}_A(\lambda)$ was possible (due to the results from Section 5.2). In this section we briefly discuss another matrix polynomial structure, i.e. symmetric matrix polynomials (see Table 2.1).

Recall that a regular matrix polynomial

$$A(\lambda) = \sum_{j=0}^{k} A_j \lambda^j = A_k \lambda^k + A_{k-1} \lambda^{k-1} + \cdots + A_1 \lambda + A_0$$

of degree k is called symmetric, if $A(\lambda) = A(\lambda)^T$, i.e. all matrix coefficients A_j are symmetric matrices, $j = 0, \ldots, k$. The ansatz space $\mathbb{SG}_{\eta+1}(A)$ introduced in Section 4.6 provides a great many of symmetric linearizations $\mathcal{L}(\lambda) = \lambda A + B$ for $A(\lambda)$. In this section we will briefly survey a few standard approaches to solve generalized symmetric eigenproblems. In particular, we focus again on Krylov subspace methods that allow the application of the results from Section 5.2.

If all eigenvalues of $\mathcal{L}(\lambda)$ are desired and $\mathcal{L}(\lambda)$ is medium sized, the HR or HZ algorithms can be applied, see [15, 23]. As a first stage, they require the reduction of A and B to diagonal-tridiagonal form as described in [124]. Various refinements of these algorithms have been proposed, e.g. [25, 129]. Moreover, similar methods to the HZ iteration such as the DQR algorithm have been introduced [126]. Other approaches to solve symmetric generalized eigenproblems are based on the QZ algorithm [57], the Jacobi method [2] and the Ehrlich-Aberth algorithm [12, 37]. Both the HZ and DQR algorithms and the Ehrlich-Aberth method compute the desired eigenvalues in $\mathcal{O}(k^2 n^2)$ arithmetic operations. However, all three methods require the reduction of A, B to diagonal-tridiagonal form which usually (i.e. if A, B are dense) has a computational complexity in $\mathcal{O}(k^3 n^3)$. If n is large, the above discussed methods are not appropriate anymore. In this case, the shifted block Lanczos algorithm from [59] or the FEAST algorithm introduced in [107] can be applied. The FEAST algorithm was further developed and improved giving rise to several descendant-algorithms such as the Sakurai-Sugiura method [113] or the CIRR-algorithm [70]. Towards a robust algorithm for the eigenvalue computation of $\mathcal{L}(\lambda)$, Tisseur in [123] proposes to first reduce the pencil $\lambda A + B$ to diagonal-tridiagonal form and then applying the HZ algorithm to compute the eigenvalues (within $\mathcal{O}(k^2 n^2)$ flops). Afterwards, a refinement of the eigenvalues can be accomplished with the Ehrlich-Aberth iteration which requires $\mathcal{O}(kn)$ floating point operations per eigenvalues. In addition, it is proposed to keep track of the structured backward error, cf. [62], and to refine the eigenvalues again if necessary.

In view of the shift-and-invert transformation from (5.1), the symmetric structure of A and B can also be taken into account by applying a pseudo-Lanczos algorithm [6, Sec. 8.6]. In particular, if a shift $\tau \in \mathbb{C}, \tau \notin \sigma_f(\mathcal{L})$, is chosen and $v, w \in \mathbb{C}^{kn}$ are arbitrary, the matrix $\mathcal{L}(\tau)^{-1} A$ satisfies

$$\begin{aligned}
\left(\mathcal{L}(\tau)^{-1} A v\right)^T \left(A w\right) = v^T A^T \mathcal{L}(\tau)^{-T} A w &= v^T A \mathcal{L}(\tau)^{-1} A w \\
&= \left(A v\right)^T \left(\mathcal{L}(\tau)^{-1} A w\right).
\end{aligned} \tag{5.14}$$

In other words, (5.14) reveals that the matrix $\mathcal{L}(\tau)^{-1} A$ is selfadjoint with respect to the indefinite inner product $[v, w] = v^T A w$ (for more information on indefinite inner products consult [55] or the discussion starting in Chapter 8). With this notation at hand, (5.14) is equivalent to $[\mathcal{L}(\tau)^{-1}Av, w] = [v, \mathcal{L}(\tau)^{-1}Aw]$. A modified Lanczos method can now be applied to $\mathcal{L}(\tau)^{-1}A$ based on the indefinite inner product $[v, w] = v^T Aw$. In particular, a Krylov subspace corresponding to $\mathcal{L}(\tau)^{-1}A$ is formed by applying the Arnoldi method incorporating the indefinite inner product for the orthogonalization steps [6, Sec. 8.6]. If a basis $v_1, \ldots, v_m \in \mathbb{C}^{kn \times m}$ is found for the Krylov subspace $\mathcal{K}_m(\mathcal{L}(\tau)^{-1}A, x)$ and $V = [\, v_1 \ \cdots \ v_m \,]$, the matrix $V^T(\mathcal{L}(\tau)^{-1}A)V$ will be symmetric tridiagonal and can be build within the Arnoldi method implicitly. As before, its eigenvalues are used as approximations to the eigenvalues of $\mathcal{L}(\tau)^{-1}A$. The same backshift procedure as in Section 5.1 then yields approximate eigenvalues of $\mathcal{L}(\lambda)$ near the shift τ. The matrix-vector-products with $\mathcal{L}(\tau)^{-1}A$ can once more be carried out efficiently as described in Section 5.2 if $\mathcal{L}(\lambda)$ is a symmetric linearization from $\mathbb{SG}_{\eta+1}(A)$.

As for the standard Arnoldi iteration, a restart procedure can also be incorporated in the pseudo-Lanczos method. For more information see [26, Sec. 3.2]. In conclusion, for symmetric generalized eigenproblems this method exploits the symmetry as described above and therefore reduces the computational complexity. However, due to the indefiniteness of the inner product, the process may be unstable or even (seriously) break down (the latter happens if $[v, w] = 0$ although $v \neq 0, w \neq 0$).

Chapter 6

The connection between $\mathbb{L}_1(A)$, $\mathbb{L}_2(A)$ and $\mathbb{G}_{\eta+1}(A)$

> *The art of doing mathematics consists in finding that special case which contains all the germs of generality.*
>
> David Hilbert (1662 –1943), [109].

Up to now, only the block Kronecker ansatz spaces $\mathbb{G}_{\eta+1}(A)$ and their subspaces $\mathbb{DG}_{\eta+1}(A), \mathbb{BG}_{\eta+1}(A)$ and $\mathbb{SG}_{\eta+1}(A), \eta \geq 0$, have been studied and analyzed. Thereby, they have been discovered as a powerful tool for the construction of new strong linearization families for matrix polynomials $A(\lambda) \in M_{n\times n}(\mathbb{R}[\lambda])$. As it was pointed out before, there is a strong connection between these new vector spaces and the classical ansatz spaces $\mathbb{L}_1(A), \mathbb{L}_2(A)$ and $\mathbb{DL}(A)$ which also provide an inexhaustible source of linearizations. The purpose of this chapter is to elucidate this connection.

In Section 6.1 we derive a new characterization of $\mathbb{L}_1(A)$ and $\mathbb{L}_2(A)$ partially based on the characterization of $\mathbb{G}_{\eta+1}(A)$ from Section 3.1. This at hand, the fundamental relationship between the block Kronecker ansatz spaces and the ansatz spaces $\mathbb{L}_1(A), \mathbb{L}_2(A)$ and $\mathbb{DL}(A)$ is established in Section 6.2. This chapter ends with some conclusions in Section 6.3.

6.1 Block Kronecker ansatz spaces and the spaces $\mathbb{L}_1(A)$ and $\mathbb{L}_2(A)$

Let $A(\lambda) \in M_{n\times n}(\mathbb{R}[\lambda])$ be of degree $k = \epsilon + \eta + 1$. For now, we fix $\eta = 0$ (i.e. $\epsilon = k - 1$) implying that the ansatz equation (3.1) for $\mathbb{G}_1(A)$ takes the form

$$\mathcal{L}(\lambda)\left(\Lambda_{k-1}(\lambda)^T \otimes I_n\right) = \alpha e_1 \otimes A(\lambda). \tag{6.1}$$

This coincides with the ansatz equation for $\mathbb{L}_1(A)$, cf. Definition 2.14 (*i*), for the choice $v = \alpha e_1$. According to Theorem 3.1 every matrix pencil $\widetilde{\mathcal{L}}(\lambda) \in \mathbb{G}_1(A)$ may be expressed as

$$\widetilde{\mathcal{L}}(\lambda) = \left[\begin{array}{c|c} I_n & B_1 \\ \hline 0 & C_1 \end{array}\right] \left[\begin{array}{c} \alpha \Sigma_{0,A}(\lambda) \\ \hline L_{k-1}(\lambda) \otimes I_n \end{array}\right] = \left[\begin{array}{c|c} \alpha I_n & B_1 \\ \hline 0 & C_1 \end{array}\right] \mathcal{F}_{1,0,A}(\lambda).$$

since the post-multiplied matrix $\left[\begin{array}{c|c} I_{(\epsilon+1)n} & \\ \hline B_2 & C_2 \end{array}\right]$ in (3.11) degenerates to I_{kn}. Notice that the first Frobenius companion form for $A(\lambda)$ (recall Example 2.1) shows up here since $\mathcal{F}_{1,0,A}(\lambda) = \text{Frob}_1(\lambda)$. Thus, the space $\mathbb{G}_1(A)$ can actually be obtained from $\text{Frob}_1(\lambda)$ by the premultiplication of special-structured matrices. Moreover, assuming $\alpha = 1$ and multiplying (6.1) from the left with

$$\mathcal{V}_{\text{left}} = \left[\begin{array}{c|c} v \otimes I_n & \dfrac{0_{n \times \epsilon n}}{I_{\epsilon n}} \end{array}\right] \in M_{kn \times kn}(\mathbb{R}), \quad v \in \mathbb{R}^k,$$

gives a pencil $\mathcal{L}(\lambda) := \mathcal{V}_{\text{left}} \widetilde{\mathcal{L}}(\lambda)$ that satisfies $\mathcal{L}(\lambda)(\Lambda_{k-1}(\lambda)^T \otimes I_n) = v \otimes A(\lambda)$. On the other hand it is easily seen, that any matrix pencil of the form

$$\mathcal{L}(\lambda) = \left[\begin{array}{c|c} v \otimes I_n & B_1 \\ & C_1 \end{array}\right] \mathcal{F}_{1,0,A}(\lambda) = \left[\begin{array}{c|c} v \otimes I_n & B_1 \\ & C_1 \end{array}\right] \text{Frob}_1(\lambda) \qquad (6.2)$$

satisfies $\mathcal{L}(\lambda)(\Lambda_{k-1}(\lambda)^T \otimes I_n) = v \otimes A(\lambda)$ as well. Now, verifying that (6.2) is essentially just a reformulation of [89, Thm. 3.5], we have derived an equivalent, but alternative description of $\mathbb{L}_1(A)$.

Theorem 6.1 (Characterization of $\mathbb{L}_1(A)$, [42, Cor. 5]). *Let $A(\lambda) \in M_{n \times n}(\mathbb{R}[\lambda])$ be of degree $k \geq 1$. Then $\mathcal{L}(\lambda)$ satisfies $\mathcal{L}(\lambda)(\Lambda_{k-1}(\lambda)^T \otimes I_n) = v \otimes A(\lambda)$ for $v \in \mathbb{R}^k$ iff*

$$\mathcal{L}(\lambda) = \left[\begin{array}{cc} v \otimes I_n & \mathcal{Z} \end{array}\right] \text{Frob}_1(\lambda) \qquad \mathcal{Z} = \left[\begin{array}{c} B_1 \\ \hline C_1 \end{array}\right] \qquad (6.3)$$

for some arbitrary matrix $\mathcal{Z} \in M_{kn \times (k-1)n}(\mathbb{R})$. Here $\text{Frob}_1(\lambda) = \mathcal{F}_{1,0,A}(\lambda)$ denotes the first Frobenius companion form for $A(\lambda)$ as introduced in Example 2.1. Moreover, for any $\mathcal{L}(\lambda) \in \mathbb{L}_1(A)$, v and $\mathcal{Z}$ are uniquely determined.

Although the original characterization of $\mathbb{L}_1(A)$ is different to the one in Theorem 6.1, the authors of [89] already remarked in their Section 4.2. the possibility of factoring pencils $\mathcal{L}(\lambda) \in \mathbb{L}_1(A)$ as (nonunique) products of constant matrices and the (first) Frobenius companion form. However, the unique characterization in (6.3) yields a very simple linearization condition for pencils in $\mathcal{L}(\lambda) \in \mathbb{L}_1(A)$ for any $A(\lambda) \in M_{n \times n}(\mathbb{R}[\lambda])$.

Corollary 6.1 (Linearization Condition, [42, Cor. 6]). *A matrix pencil $\mathcal{L}(\lambda) \in \mathbb{L}_1(A)$ as in (6.3) is a strong linearization for $A(\lambda) \in M_{n \times n}(\mathbb{R}[\lambda])$ with degree $k \geq 1$ if $[v \otimes I_n \ \mathcal{Z}] \in \text{GL}_{kn}(\mathbb{R})$, that is $\text{rank}([v \otimes I_n \ \mathcal{Z}]) = kn$.*

Proof. As discussed in Example 2.1, $\mathrm{Frob}_1(\lambda) \in M_{kn \times kn}(\mathbb{R}_1[\lambda])$ is always a strong linearization for $A(\lambda) \in M_{n \times n}(\mathbb{R}[\lambda])$ (of degree $k \geq 1$), regardless whether $A(\lambda)$ is regular or singular. Thus, in case $[\, v \otimes I_n \ \mathcal{Z}\,] \in \mathrm{GL}_{kn}(\mathbb{R})$, $\mathcal{L}(\lambda)$ as in (6.3) is strict equivalent to $\mathrm{Frob}_1(\lambda)$ according to Definition 2.12 (ii). Thus, it is a strong linearization for $A(\lambda)$. $\qquad\square$

Notice that, if $\mathcal{L}(\lambda) \in \mathbb{L}_1(A)$ as in (6.3) is a strong linearization for a regular matrix polynomial $A(\lambda)$, then $0 \neq \det([\, v \otimes I_n \ \mathcal{Z}\,]) \det(\mathrm{Frob}_1(\lambda))$ and $\det(\mathrm{Frob}_1(\lambda)) \neq 0$ immediately imply $\det([\, v \otimes I_n \ \mathcal{Z}\,]) \neq 0$. Therefore, $[\, v \otimes I_n \ \mathcal{Z}\,] \in \mathrm{GL}_{kn}(\mathbb{R})$ has to hold and Corollary 6.1 becomes an equivalence for regular $A(\lambda)$.

A characterization of $\mathbb{L}_2(A)$ similar to (6.3) can be derived in an analogous way: in fact, $\mathbb{L}_2(A)$ consists of all matrix pencils $\mathcal{L}(\lambda)$ having the form

$$\mathcal{L}(\lambda) = \mathrm{Frob}_1^{\mathcal{B}}(\lambda) \begin{bmatrix} v^T \otimes I_n \\ \mathcal{Z} \end{bmatrix} = \mathrm{Frob}_2(A) \begin{bmatrix} v^T \otimes I_n \\ \mathcal{Z} \end{bmatrix} \qquad (6.4)$$

for some arbitrary matrix $\mathcal{Z} = [\, B_1 \mid C_1\,] \in M_{(k-1)n \times kn}(\mathbb{R})$. To verify the identity $\mathrm{Frob}_1^{\mathcal{B}}(\lambda) = \mathrm{Frob}_2(\lambda)$ recall once more Example 2.1. The pencils of the form (6.4) with arbitrary $v \in \mathbb{R}^k$ satisfy the (second) classical ansatz equation $(\Lambda_{k-1}(\lambda) \otimes I_n)\mathcal{L}(\lambda) = v^T \otimes A(\lambda)$. Again, (6.4) can be interpreted as a reformulation of [89, Lem. 3.11] and we obtain statements analogous to Theorem 6.1 and Corollary 6.1. An explicit formulation of these results is therefore omitted.

It is important to note that the statement of Corollary 6.1 is essentially equivalent to the well-known Z-rank condition [31, Def. 4.3] (which is explained below). To see this, assume that $\mathcal{L}(\lambda) \in \mathbb{L}_1(A)$ is as in (6.3) for $A(\lambda) \in M_{n \times n}(\mathbb{R}[\lambda])$ of degree $k \geq 2$ and $v \neq 0$. Now let $M \in M_{k \times k}(\mathbb{R})$ be a nonsingular matrix satisfying $Mv = \alpha e_1$ for some nonzero $\alpha \in \mathbb{R}$. Premultiplying $\mathcal{L}(\lambda)$ with $M \otimes I_n$ yields

$$\mathcal{L}^\star(\lambda) := \big(M \otimes I_n \big) \big[\, v \otimes I_n \ \mathcal{Z}\,\big] \mathrm{Frob}_1(\lambda) = \big[\, \alpha e_1 \otimes I_n \mid (M \otimes I_n)\mathcal{Z} \,\big] \mathrm{Frob}_1(\lambda)$$

$$=: \begin{bmatrix} \alpha I_n & B_1^\star \\ 0 & C_1^\star \end{bmatrix} \mathrm{Frob}_1(\lambda),$$

so $\mathcal{L}^\star(\lambda) \in \mathbb{L}_1(A)$ with ansatz vector αe_1. Now $\mathcal{L}^\star(\lambda)$ can be expressed as

$$\mathcal{L}^\star(\lambda) = \begin{bmatrix} \alpha A_k & B_1^\star \\ 0 & C_1^\star \end{bmatrix} \lambda + \begin{bmatrix} \alpha \Sigma_{0,A}(0) + B_1^\star(L_{k-1}(0) \otimes I_n) \\ C_1^\star(L_{k-1}(0) \otimes I_n) \end{bmatrix}.$$

In the form given above $\mathcal{L}^\star(\lambda)$ corresponds to expression (4.3) in [31, Thm. 4.1]. It is said that $\mathcal{L}(\lambda)$ has full Z-rank whenever $C_1^\star$ has full rank for any chosen nonsingular matrix M with the property $Mv = \alpha e_1, \alpha \neq 0$. In particular, the rank of $C_1^\star$ is declared to be the Z-rank of $\mathcal{L}(\lambda)$. Notice that $\mathrm{rank}(C_1^\star)$ is independent of the choice of M as long as $M \in \mathrm{GL}_k(\mathbb{R})$ satisfies $Mv = \alpha e_1$ for some $\alpha \neq 0$ [31, Lem. 4.2]. Moreover, $C_1^\star$ has full rank iff $[\, \alpha e_1 \otimes I_n \ (M \otimes I_n)\mathcal{Z}\,]$ has full rank. Furthermore, since

$$\big[\, \alpha e_1 \otimes I_n \ (M \otimes I_n)\mathcal{Z} \,\big] = (M \otimes I_n)\big[\, v \otimes I_n \ \mathcal{Z} \,\big]$$

and as $M \otimes I_n$ is nonsingular as well, $[\alpha e_1 \otimes I_n \ (M \otimes I_n)\mathcal{Z}]$ has full rank iff $[v \otimes I_n \ \mathcal{Z}]$ is nonsingular. We summarize this observation in the next corollary which is a slightly generalized version of [45, Cor. 3].

Corollary 6.2. *Let $A(\lambda) \in M_{n \times n}(\mathbb{R}[\lambda])$ be of degree $k \geq 1$ and assume $\mathcal{L}(\lambda) \in \mathbb{L}_1(A)$ is given as*

$$\mathcal{L}(\lambda) = \begin{bmatrix} v \otimes I_n \ \mathcal{Z} \end{bmatrix} \mathrm{Frob}_1(\lambda).$$

Then $\mathcal{L}(\lambda)$ has full Z-rank iff $\mathrm{rank}([v \otimes I_n \ \mathcal{Z}]) = kn$. In addition, any matrix pencil $\mathcal{L}(\lambda) \in \mathbb{L}_1(A)$ with full Z-rank is strict equivalent to $\mathrm{Frob}_1(A)$[1].

Moreover, it can be easily checked that the Z-rank-deficiency of $\mathcal{L}(\lambda) \in \mathbb{L}_1(A)$ carries over to the matrix $[v \otimes I_n \ \mathcal{Z}]$, i.e. if the Z-rank of $\mathcal{L}(\lambda)$ is $s < (k-1)n$, so its Z-rank-deficiency is $t = (k-1)n - s$, then it follows that $\mathrm{rank}([v \otimes I_n \ \mathcal{Z}]) = kn - t$. In conclusion, there is in fact no loss of information in considering the rank of $[v \otimes I_n \ \mathcal{Z}]$ for $\mathcal{L}(\lambda) \in \mathbb{L}_1(A)$ as in (6.3) instead of its Z-rank.

Theorem 6.1 and Corollary 6.1 have particularly nice consequences: in fact, many well-known results on $\mathbb{L}_1(A)$, $\mathbb{L}_2(A)$ and $\mathbb{DL}(A)$ admit easily accessible proofs considering the form (6.3) instead of the expressions in [89, Thm. 3.5] or [45]. For instance the dimension of $\mathbb{L}_1(A)$ can immediately be obtained from the characterization (6.3) as $\dim(\mathbb{L}_1(A)) = k + k(k-1)n^2$ (which is exactly the dimension of the vector space containing all matrices of the form $[v \otimes I_n \ \mathcal{Z}]$). In [89, Cor. 3.6] the surjectivity of the mapping $\mathcal{L}(\lambda) \mapsto \mathcal{L}(\lambda)(\Lambda_{k-1}(\lambda)^T \otimes I_n)$ was used for this proof. Furthermore, the strong linearization theorem [89, Thm. 4.3] is easily proven with the characterization from Theorem 6.1 at hand. This will be done in the following chapter. Moreover, the computation of the standard basis of $\mathbb{DL}(A)$ considered in [65] breaks down to the computation of the corresponding matrices $[e_i \otimes I_n \ \mathcal{Z}_i]$ for $i = 1, \ldots, k$. If $[e_i \otimes I_n \ \mathcal{Z}_i]$ is partitioned as in (6.3) it was shown in [45] that C_1 always has to be a block-symmetric matrix. In [44], an extended version of [42], an algorithm for determining the standard basis of $\mathbb{DL}(A)$ is given.

6.2   The central relation between $\mathbb{L}_1(A), \mathbb{L}_2(A)$ and $\mathbb{G}_{\eta+1}(A)$

The central relation between the ansatz spaces $\mathbb{L}_1(A), \mathbb{L}_2(A)$ and $\mathbb{G}_{\eta+1}(A)$ is revealed by the following *fundamental isomorphism theorem.*

Theorem 6.2 (Fundamental Isomorphism Theorem). *Let $A(\lambda) \in M_{n \times n}(\mathbb{R}[\lambda])$ be square and of degree $k = \eta + \epsilon + 1$.*

 (i) *Since $\dim(\mathbb{G}_{\eta+1}(A)) = k(k-1)n^2 + 1$ according to (3.9) it holds that*

$$\mathbb{L}_1(A)\Big|_{\mathrm{span}(e_\ell)} \cong \mathbb{G}_{\eta+1}(A) \cong \mathbb{L}_2(A)\Big|_{\mathrm{span}(e_\ell)}$$

---

[1]As we will show in Example 7.1 (i) the converse is true only for regular matrix polynomials $A(\lambda)$.

for any $\ell \in \{1, 2, \ldots, k\}$ *and any* $\eta \in \{0, 1, \ldots, k-1\}$. *Here* $\mathbb{L}_j(A)|_X$ *denotes the restriction of* $\mathbb{L}_j(A), j = 1, 2$, *to pencils with ansatz vector* $v \in X \subseteq \mathbb{R}^k$. *A canonical isomorphism* Ξ *between* $\mathbb{G}_{\eta+1}(A)$ *and* $\mathbb{L}_1(A)|_{\mathrm{span}(e_\ell)}$ *is given by mapping* $\mathcal{L}(\lambda) \in \mathbb{G}_{\eta+1}(A)$ *as*

$$\mathcal{L}(\lambda) = \left[\begin{array}{c|c} I_{(\eta+1)n} & B_1 \\ \hline 0 & C_1 \end{array} \right] \left[\begin{array}{c|c} \alpha \Sigma_{\eta,A}(\lambda) & L_\eta(\lambda)^T \otimes I_n \\ \hline L_\epsilon(\lambda) \otimes I_n & 0 \end{array} \right] \left[\begin{array}{c|c} I_{(\epsilon+1)n} & 0 \\ \hline B_2 & C_2 \end{array} \right]$$

to the matrix pencil

$$\Xi(\mathcal{L}(\lambda)) = \left[\begin{array}{c|c} \alpha e_\ell \otimes I_n & \begin{array}{c|c} B_1 & B_2^{\mathcal{B}} \\ \hline C_1 & C_2^{\mathcal{B}} \end{array} \end{array} \right] \mathrm{Frob}_1(\lambda) \in \mathbb{L}_1(A)|_{\mathrm{span}(e_\ell)}.$$

A canonical isomorphism between $\mathbb{G}_{\eta+1}(A)$ *and* $\mathbb{L}_2(A)|_{\mathrm{span}(e_\ell)}$ *can be obtained in a similar way.*

(*ii*) *If* $k \geq 2$ *then for* $\eta = 0$ *and* $\eta = k-1$ *it holds that*

$$\mathbb{L}_1(A)|_{\mathrm{span}(e_1)} = \mathbb{G}_1(A) \quad \text{and} \quad \mathbb{L}_2(A)|_{\mathrm{span}(e_1)} = \mathbb{G}_k(A), \text{ respectively.}$$

Proof. The statement in Theorem 6.2 (*i*) follows directly from (3.9) and [89, Cor. 3.6]. The check that Ξ provides an isomorphism between $\mathbb{G}_{\eta+1}(A)$ and $\mathbb{L}_1(A)|_{\mathrm{span}(e_\ell)}$ is straight forward and omitted. The fact of Theorem 6.2 (*ii*) follows from the characterization of $\mathbb{G}_1(A)$ and $\mathbb{G}_k(A)$ in Theorem 3.1 and the characterization of $\mathbb{L}_i(A)|_{\mathrm{span}(e_\ell)}$ in Theorem 6.1 and (6.4) ($i = 1, 2$). $\qquad\Box$

Therefore, the flexibility of $\mathbb{L}_1(A)$ and $\mathbb{L}_2(A)$ that is obtained by taking an ansatz vector $v \in \mathbb{R}^k$ with k (maybe different) components for the construction of some linearization $\mathcal{L}(\lambda) \in \mathbb{L}_1(A)$ is, in the context of the block Kronecker ansatz spaces, reflected by the fact of choosing one out of k different ansatz spaces (i.e. $\eta \in \{0, 1, \ldots, k-1\}$) for $A(\lambda)$. Having chosen η, the construction of $\mathcal{L}(\lambda) \in \mathbb{G}_{\eta+1}(A)$ can then be seen as having equally many degrees of freedom as the construction of a linearization in $\mathbb{L}_1(A)$ or $\mathbb{L}_2(A)$ with ansatz vector $v \in \mathbb{R}^k$ having only one nonzero component[2].

The ansatz space $\mathbb{DL}(A)$ was introduced in [89] as the intersection of $\mathbb{L}_1(A)$ and $\mathbb{L}_2(A)$ (recall Definition 2.14 (*iii*)). As the final result of this section we state the following *intersection lemma* that connects the three general kinds of block Kronecker ansatz spaces from Chapters 3 and 4 and the $\mathbb{DL}(A)$ ansatz space.

Lemma 6.1 (Intersection Lemma, [42, Lem. 4]). *Let* $A(\lambda) \in M_{n \times n}(\mathbb{R}[\lambda])$ *be of degree* $k \geq 1$. *Then*

$$\bigcap_{\eta=0}^{k-1} \mathbb{G}_{\eta+1}(A) = \bigcap_{\eta=0}^{k-1} \mathbb{DG}_{\eta+1}(A) = \bigcap_{\eta=0}^{k-1} \mathbb{BG}_{\eta+1}(A) = \mathbb{DL}(A)|_{\mathrm{span}(e_1)}.$$

---

[2]This reflects the reason why the notation $\mathbb{G}_{\eta+1}(A)$ instead of $\mathbb{G}_\eta(A)$ has been chosen. In this notation we have $\mathbb{G}_1(A) \subseteq \mathbb{L}_1(A)$ and $\mathbb{G}_k(A) \subseteq \mathbb{L}_2(A)$ instead of $\mathbb{G}_0(A) \subseteq \mathbb{L}_1(A)$ etc.

Here, as before, span(e_1) *denotes the one-dimensional subspace of* $\mathbb{R}^k$ *spanned by* e_1.

Proof. Since $\mathbb{G}_1(A) \cap \mathbb{G}_k(A) = \mathbb{D}\mathbb{L}(A)|_{\text{span}(e_1)}$ the lemma follows from the observations in (4.30) and (4.37). □

6.3   Conclusions

In this section we derived a new characterization of the classical ansatz spaces $\mathbb{L}_1(A)$ and $\mathbb{L}_2(A)$ known from [89]. Our characterization uses the first and second Frobenius companion forms as anchor pencils and expresses all pencils from $\mathbb{L}_1(A)$ and $\mathbb{L}_2(A)$ as certain modifications of these companion forms obtained by the pre- and post-multiplication of special-structured matrices. In particular, our characterization gets along without the concept of the row shifted sums and column shifted sums known from [89] which have been introduced for the purpose of characterizing $\mathbb{L}_1(A)$ and $\mathbb{L}_2(A)$. Moreover, our description led to a very simple linearization condition for pencils in $\mathbb{L}_1(A)$ that is equivalent to the well-known Z-rank condition. With the *Fundamental Isomorphism Theorem* we showed how the ansatz spaces $\mathbb{G}_{\eta+1}(A)$ and $\mathbb{L}_1(A)$ can be related by a canonical isomorphism. Finally, we stated the *Intersection Lemma* that reveals the connection between the block Kronecker ansatz spaces $\mathbb{G}_{\eta+1}(A), \mathbb{D}\mathbb{G}_{\eta+1}(A), \mathbb{B}\mathbb{G}_{\eta+1}(A)$ and the double ansatz space $\mathbb{D}\mathbb{L}(A)$.

Chapter 7

Generalized ansatz spaces for orthogonal bases

> *A great deal of my work is just playing with equations and seeing what they give.*
>
> Paul A. M. Dirac (1902 – 1984), [130].

We will now finally leave the framework of block Kronecker ansatz spaces and return to the classical setting originally introduced in [89] with the definition of $\mathbb{L}_1(A), \mathbb{L}_2(A)$ and $\mathbb{DL}(A)$. In particular, the concept established by the ansatz spaces $\mathbb{L}_1(A)$ and $\mathbb{L}_2(A)$ for constructing strong linearizations of $A(\lambda) = \sum_{i=0}^k A_i \lambda^i$ is extended to matrix polynomials which are not expressed in the standard monomial basis. The analytical framework for these nonstandard ansatz spaces will be similar to the one initiated in the previous chapter for $\mathbb{L}_1(A)$ and $\mathbb{L}_2(A)$.

In Section 7.1 the basic framework of orthogonal polynomial bases is presented along with the introduction and characterization of the *generalized ansatz spaces* $\mathbb{M}_1(A), \mathbb{M}_2(A)$ and $\mathbb{DM}(A)$. Those are defined via ansatz equations of the form (2.15) and have already been considered in, e.g., [31, 85, 88, 101]. We will give a simple linearization condition for pencils in these ansatz spaces and state an extended version of the *Strong Linearization Theorem* from [89, Thm. 4.3]. The derivation of simple recovery formulas for eigenvectors from linearizations in these ansatz spaces is the content of Section 7.2. Section 7.3 is dedicated to the study of singular matrix polynomials and their linearizations $\mathcal{L}(\lambda) \in \mathbb{M}_j(A), j = 1, 2$. In this context, a new linearization condition for linearizations of singular matrix polynomials is derived based on their left nullvectors. Moreover, we consider matrix polynomials with all left minimal indices equal to zero and give a complete characterization of their strong linearizations in $\mathbb{M}_1(A)$ based on a certain rank condition. The intersection space $\mathbb{DM}(A) = \mathbb{M}_1(A) \cap \mathbb{M}_2(A)$ is analyzed in Section 7.4 along with an algorithm for the computation of block-symmetric linearizations. Section 7.5 presents the *Eigenvector Exclusion Theorem*, a theorem that makes a statement on linearizations $\mathcal{L}(\lambda)$ in $\mathbb{M}_1(A)$ or $\mathbb{M}_2(A)$ in terms of their left eigenvectors and ansatz vectors. The eigenvalue exclusion theorem from [89] is also reconsidered and proven in a simple algebraic manner for all orthogonal polynomial bases and a certain subclass of matrix polynomials. This chapter ends with some conclusions in Section 7.6.

7.1   The basic framework

This section is devoted to the theoretical foundation of square matrix polynomials not expressed in the standard monomial basis $\{\lambda^j\}_{j=0}^\infty$. In particular, we consider matrix polynomials $A(\lambda) \in M_{n \times n}(\mathbb{R}[\lambda])$ expressed in polynomial bases $\Phi = \{\phi_j(\lambda)\}_{j=0}^\infty$ that follow a three-term recurrence relation. Those polynomial bases are usually called orthogonal, see [118]. Thus, we assume that the relation

$$\alpha_j \phi_{j+1}(\lambda) = (\lambda - \beta_j)\phi_j(\lambda) - \gamma_j \phi_{j-1}(\lambda), \qquad j \geq 0, \tag{7.1}$$

holds for some coefficients $\alpha_j \neq 0, \beta_j, \gamma_j \in \mathbb{R}$, $j \in \mathbb{N}$, with $\phi_{-1}(\lambda) = 0, \phi_0(\lambda) = 1$. Those polynomial bases are always degree-graded, that is $\deg(\phi_j(\lambda)) = j$ holds for any $j \geq 0$. Popular special cases include the monomials, Newton [85] and Chebyshev bases [46] or the Legendre basis.

Whenever this is not further specified, $A(\lambda) \in M_{n \times n}(\mathbb{R}[\lambda])$ will from now on be assumed to be expressed in an orthogonal basis $\Phi = \{\phi_j(\lambda)\}_{j=0}^\infty$ as in (7.2) below with degree $k \geq 2$, that is

$$A(\lambda) = \sum_{j=0}^{k} A_j \phi_j(\lambda) = A_k \phi_k(\lambda) + \cdots + A_1 \phi_1(\lambda) + A_0 \phi_0(\lambda). \tag{7.2}$$

For any basis Φ we define $\Phi_k(\lambda) := [\, \phi_k(\lambda) \;\cdots\; \phi_1(\lambda) \; \phi_0(\lambda)\,] \in M_{1 \times (k+1)}(\mathbb{R}[\lambda])$ for any $k \geq 0$. The following definition is analogous to the definition of $\mathbb{L}_1(A)$ with the monomial basis $1, \lambda, \lambda^2, \dots$ replaced by the orthogonal basis Φ. It can also be found in [101, Sec. 2] or [31, Sec. 7].

Definition 7.1 (Generalized Ansatz Space, [45, Sec. 3]). *Let $A(\lambda) \in M_{n \times n}(\mathbb{R}[\lambda])$ be of degree $k \geq 2$. We define $\mathbb{M}_1(A)$ to be the set of all matrix pencils $\mathcal{L}(\lambda)$ satisfying*

$$\mathcal{L}(\lambda)\Big(\Phi_{k-1}(\lambda)^T \otimes I_n\Big) = v \otimes A(\lambda) \tag{7.3}$$

for some ansatz vector $v \in \mathbb{R}^k$. Equation (7.3) is called the (first) generalized ansatz equation for $A(\lambda)$ corresponding to the polynomial basis Φ.

Recall that for the monomial basis, (7.3) coincides with the ansatz equation for $\mathbb{L}_1(A)$ with $\Phi_{k-1}(\lambda) = [\lambda^{k-1} \;\cdots\; \lambda\; 1] = \Lambda_{k-1}(\lambda)$, see Definition 2.14 (i). Notice further that $\mathbb{M}_1(A)$ simply consists of all scalar multiples $\alpha A(\lambda), \alpha \in \mathbb{R}$, in case one allows $A(\lambda)$ to have degree one. This justifies considering only matrix polynomials of degree $k \geq 2$ (although some results derived subsequently remain true if $k = 1$).

Certainly $\mathbb{M}_1(A)$ is a vector space over $\mathbb{R}$. Moreover, observe that $\mathbb{M}_1(A)$ can certainly be investigated by applying a change-of-basis matrix $S \in \mathrm{GL}_k(\mathbb{R})$ to $\Phi_{k-1}(\lambda)^T$ so that $S\Phi_{k-1}(\lambda)^T = \Lambda_{k-1}(\lambda)^T$. Then, if (7.3) holds, we obtain

$$(\mathcal{L}(\lambda)(S^{-1} \otimes I_n))(S\Phi_{k-1}(\lambda)^T \otimes I_n) = \widetilde{\mathcal{L}}(\lambda)(\Lambda_{k-1}(\lambda)^T \otimes I_n) = v \otimes A(\lambda). \quad (7.4)$$

Thus, $\mathcal{L}(\lambda)$ from $\mathbb{M}_1(A)$ is transformed into a matrix pencil $\widetilde{\mathcal{L}}(\lambda) = \mathcal{L}(\lambda)(S^{-1} \otimes I_n) \in \mathbb{L}_1(A)$. This construction of $\mathbb{M}_1(A)$ was proposed in [31] and yields a possibility to directly apply all results for $\mathbb{L}_1(A)$ to $\widetilde{\mathcal{L}}(\lambda)$. However, our approach is to characterize $\mathbb{M}_1(A)$ intrinsically in the same way as $\mathbb{L}_1(A)$ was characterized in Theorem 6.1. This approach enables us to construct pencils from $\mathbb{M}_1(A)$ without any change of basis.

First of all, recall the fact from Theorem 6.1 that any pencil $\mathcal{L}(\lambda) \in \mathbb{L}_1(A)$ for some matrix polynomial $A(\lambda) \in \mathrm{M}_{n \times n}(\mathbb{R}[\lambda])$ of degree $k \geq 2$ can be expressed as $\mathcal{L}(\lambda) = [\, v \otimes I_n \ B\,]\mathrm{Frob}_1(\lambda)$ for some vector $v \in \mathbb{R}^k$ and the first Frobenius companion form $\mathrm{Frob}_1(\lambda)$ corresponding to $A(\lambda)$. Furthermore, recall that $\mathrm{Frob}_1(\lambda)(\Lambda_{k-1}(\lambda)^T \otimes I_n) = e_1 \otimes A(\lambda)$ holds. This concept for the characterization of $\mathbb{L}_1(A)$ will now be extended to $\mathbb{M}_1(A)$.

To this end, let us assume that $A(\lambda) = \sum_{j=0}^{k} A_j \phi_j(\lambda) \in \mathrm{M}_{n \times n}(\mathbb{R}[\lambda])$ is now given as in (7.2) with degree $k \geq 2$. We first introduce the $n \times kn$ rectangular matrix pencil

$$m_\Phi^A(\lambda) := \left[\ \frac{(\lambda - \beta_{k-1})}{\alpha_{k-1}}A_k + A_{k-1} \quad A_{k-2} - \frac{\gamma_{k-1}}{\alpha_{k-1}}A_k \quad A_{k-3} \quad \cdots \quad A_1 \quad A_0 \ \right].$$

It is easily seen that $m_\Phi^A(\lambda)(\Phi_{k-1}(\lambda)^T \otimes I_n) = A(\lambda)$ holds whenever $A(\lambda)$ has the form (7.2). Moreover, for the $(k-1) \times k$ matrix pencil

$$M_\Phi^\star(\lambda) = \begin{bmatrix} -\alpha_{k-2} & (\lambda - \beta_{k-2}) & -\gamma_{k-2} & & & \\ & -\alpha_{k-3} & (\lambda - \beta_{k-3}) & -\gamma_{k-3} & & \\ & & \ddots & \ddots & \ddots & \\ & & & -\alpha_1 & (\lambda - \beta_1) & -\gamma_1 \\ & & & & -\alpha_0 & (\lambda - \beta_0) \end{bmatrix} \quad (7.5)$$

we have $M_\Phi^\star(\lambda)\Phi_{k-1}(\lambda)^T = 0$. Note that $M_\Phi^\star(\lambda)$ depends only on the chosen basis and, in a sense, encodes the recurrence relation (7.1). Now we define $M_\Phi(\lambda) := M_\Phi^\star(\lambda) \otimes I_n$, so certainly $M_\Phi(\lambda)(\Phi_{k-1}(\lambda)^T \otimes I_n) = 0$. In conclusion, we set

$$F_\Phi^A(\lambda) := \begin{bmatrix} m_\Phi^A(\lambda) \\ M_\Phi(\lambda) \end{bmatrix} \in \mathrm{M}_{kn \times kn}(\mathbb{R}_1[\lambda]). \quad (7.6)$$

It is easily checked that $F_\Phi^A(\lambda)$ is a SBMB pencil as defined in Definition 3.2. Therefore, by the fundamental result from [34, Thm. 3.3], it is a strong linearization for $m_\Phi^A(\lambda)(\Phi_{k-1}(\lambda)^T \otimes I_n) = A(\lambda)$ no matter whether $A(\lambda)$ is regular or singular. In fact, $F_\Phi^A(\lambda)$ can be interpreted as an analogue to $\mathrm{Frob}_1(\lambda)$, cf. Example 2.1, suitable for $A(\lambda)$ expressed in any basis Φ as in (7.2).

For scalar polynomials $A(\lambda) = a(\lambda) \in \mathbb{R}[\lambda]$, matrices (or matrix pencils) of the type (7.6) are called comrade matrices or confederate matrices [7, 90]. In particular, when the Chebyshev basis is considered, the name colleague matrix is also common usage [58]. For matrix polynomials, matrix pencils similar to (7.6) have already appeared in [1, 36] or [27]. Notice that, by construction we have

$$F_{\Phi}^{A}(\lambda)\left(\Phi_{k-1}(\lambda)^{T} \otimes I_{n}\right) = e_{1} \otimes A(\lambda),$$

thus, $F_{\Phi}^{A}(\lambda) \in \mathbb{M}_{1}(A)$ with ansatz vector $e_{1} \in \mathbb{R}^{k}$. In fact, $F_{\Phi}^{A}(\lambda)$ may now be utilized in the same manner as in Section 6.1 as an *anchor pencil* for the construction of $\mathbb{M}_{1}(A)$. To this end, the next theorem gives a concise and succinct characterization of $\mathbb{M}_{1}(A)$ for any matrix polynomial $A(\lambda)$ expressed in some orthogonal polynomial basis. Taking the result from Theorem 6.1 into account, the following result, in fact, comes as no surprise.

Theorem 7.1 (Characterization of $\mathbb{M}_{1}(A)$, [45, Thm. 1]). *Let* $A(\lambda) \in \mathbb{M}_{n \times n}(\mathbb{R}[\lambda])$ *be of degree* $k \geq 2$ *as given in* (7.2). *Then* $\mathcal{L}(\lambda) \in \mathbb{M}_{1}(A)$ *with ansatz vector* $v \in \mathbb{R}^{k}$ *iff*

$$\mathcal{L}(\lambda) = \begin{bmatrix} v \otimes I_{n} & B \end{bmatrix} F_{\Phi}^{A}(\lambda) \tag{7.7}$$

for some matrix $B \in \mathbb{R}^{kn \times (k-1)n}$.

Proof. $\Leftarrow$ It is clear that any pencil $\mathcal{L}(\lambda) = [v \otimes I_{n} \, B] F_{\Phi}^{A}(\lambda)$ satisfies (7.3) since

$$\left(\begin{bmatrix} v \otimes I_{n} & B \end{bmatrix} F_{\Phi}^{A}(\lambda)\right)\left(\Phi_{k-1}(\lambda) \otimes I_{n}\right) = \begin{bmatrix} v \otimes I_{n} & B \end{bmatrix}\left(e_{1} \otimes A(\lambda)\right) = v \otimes A(\lambda).$$

$\Rightarrow$ Now let $\mathcal{L}(\lambda) \in \mathbb{M}_{1}(A)$, thus, $\mathcal{L}(\lambda)(\Phi_{k-1}(\lambda)^{T} \otimes I_{n}) = v \otimes A(\lambda)$ has to hold. As $v \otimes A(\lambda) = \sum_{i=0}^{k}(v \otimes A_{i}\phi_{i}(\lambda))$ it follows that $\mathcal{L}(\lambda)(\Phi_{k-1}(\lambda)^{T} \otimes I_{n})$ has to generate the term $v \otimes A_{k}\phi_{k}(\lambda)$ on the right hand side of (7.3). Since $\phi_{k}(\lambda)$ is not an entry of $\Phi_{k-1}(\lambda)$ and $\phi_{k}(\lambda)$ has degree k, i.e. contains a nonzero term with λ^{k}, we need to operate with $\lambda\phi_{k-1}(\lambda)$ to obtain λ with potency k. One possibility to generate $A_{k}\phi_{k}(\lambda)$ with $\lambda\phi_{k-1}(\lambda)$ (and the other basis polynomials from $\Phi_{k-1}(\lambda)$) is via the recurrence relation (7.1), that is

$$v \otimes A_{k}\phi_{k}(\lambda) = v \otimes \left(\alpha_{k-1}^{-1}\left((\lambda - \beta_{k-1})\phi_{k-1}(\lambda) - \gamma_{k-1}\phi_{k-2}(\lambda)\right)A_{k}\right). \tag{7.8}$$

Note that there are different ways to generate $v \otimes A_{k}\phi_{k}(\lambda)$: for instance, we may replace $\phi_{k-2}(\lambda)$ in (7.8) by its recurrence relation $\alpha_{k-3}\phi_{k-2}(\lambda) = (\lambda - \beta_{j})\phi_{k-3}(\lambda) - \gamma_{j}\phi_{k-4}(\lambda)$ and use (7.1) once again for $\phi_{k-4}(\lambda)$ and so on. However, we can not replace $\Phi_{k-1}(\lambda)$ in (7.8) by its recurrence (7.1) since this would produce a term containing λ^{2} in (7.8). As $\mathcal{L}(\lambda)$ is a matrix pencil this is not possible. Therefore, whatever expression is used for $v \otimes A_{k}\phi_{k}(\lambda)$ in (7.8), it contains a term $v \otimes \lambda\alpha_{k-1}^{-1}\phi_{k-1}(\lambda)$ and we obtain that $\mathcal{L}(\lambda)$ may be expressed as $\mathcal{L}(\lambda) = [v \otimes \alpha_{k-1}^{-1}A_{k} \, \mathcal{L}_{1}]\lambda + [\ell^{\star} \, \mathcal{L}_{0}]$ for some matrices $\ell^{\star} \in \mathbb{M}_{kn \times n}(\mathbb{R})$ and $\mathcal{L}_{1}, \mathcal{L}_{0} \in \mathbb{M}_{kn \times (k-1)n}(\mathbb{R})$.

Now observe that $\mathcal{L}^\star(\lambda) := [\, v \otimes I_n \; \mathcal{L}_1 \,] F_\Phi^A(\lambda) \in \mathbb{M}_1(A)$ can be expressed as $\mathcal{L}^\star(\lambda) = [\, v \otimes \alpha_{k-1}^{-1} A_k \; \mathcal{L}_1 \,]\lambda + [\, v \otimes I_n \; \mathcal{L}_1 \,] F_\Phi^A(0)$ since

$$
F_\Phi^A(\lambda) = F_\Phi^A(0) + \begin{bmatrix} \frac{\lambda}{\alpha_{k-1}} A_k & 0 & \cdots & 0 \\ 0 & & & \\ \vdots & & \lambda I_{(k-1)n} & \\ 0 & & & \end{bmatrix}.
$$

Thus $\Delta\mathcal{L}(\lambda) := \mathcal{L}(\lambda) - \mathcal{L}^\star(\lambda) \in \mathbb{M}_{kn \times kn}(\mathbb{R})$, i.e. it is independent of λ. Moreover, $\Delta\mathcal{L}(\lambda)$ satisfies $\Delta\mathcal{L}(\lambda)(\Phi_{k-1}(\lambda)^T \otimes I_n) = 0$. Since $\phi_0(\lambda), \ldots, \phi_{k-1}(\lambda)$ form a basis of $\mathbb{R}_{k-1}[\lambda]$, this implies $\Delta\mathcal{L} = 0$ and proves that $\mathcal{L}(\lambda) = \mathcal{L}^\star(\lambda)$. $\qquad \square$

In other words, Theorem 7.1 states that $\mathbb{M}_1(A) \subseteq \mathbb{M}_{kn \times kn}(\mathbb{R}[\lambda])$ consists of all matrix pencils $\mathcal{L}(\lambda) \in \mathbb{M}_{kn \times kn}(\mathbb{R}_1[\lambda])$ of the form

$$
\mathcal{L}(\lambda) = \begin{bmatrix} v \otimes I_n \; B \end{bmatrix} F_\Phi^A(\lambda) \quad \text{with } v \in \mathbb{R}^k \text{ and } B \in \mathbb{M}_{kn \times (k-1)n}(\mathbb{R}).
$$

Note that the description of $\mathbb{M}_1(A)$ in Theorem 7.1 reduces to that of $\mathbb{L}_1(A)$ in Theorem 6.1 for the monomial basis $1, \lambda, \ldots, \lambda^{k-1}$: as expected we have $F_\Phi^A(\lambda) = \mathrm{Frob}_1(\lambda)$ whenever $\Phi = \{\lambda^j\}_{j=0}^\infty$ (see Example 2.1).

Beside (3.1) we may also consider the equation $(\Phi_{k-1}(\lambda) \otimes I_n)\mathcal{L}(\lambda) = v^T \otimes A(\lambda)$, i.e. the transposed version of (7.3). As before, all matrix pencils $\mathcal{L}(\lambda)$ satisfying this equation form a vector space over $\mathbb{R}$ which we denote by $\mathbb{M}_2(A)$. For $\Phi_{k-1}(\lambda) = \Lambda_{k-1}(\lambda)$ we obviously have $\mathbb{M}_2(A) = \mathbb{L}_2(A)$, see (6.4). The (second) generalized ansatz space $\mathbb{M}_2(A)$ is characterized analogously to Theorem 7.1 with a similar proof.

Theorem 7.2 (Characterization of $\mathbb{M}_2(A)$, [45, Thm. 2]). *Let $A(\lambda) \in \mathbb{M}_{n \times n}(\mathbb{R}[\lambda])$ be of degree $k \geq 2$ as given in (7.2). Then $\mathcal{L}(\lambda) \in \mathbb{M}_2(A)$ with ansatz vector $v \in \mathbb{R}^k$ iff*

$$
\mathcal{L}(\lambda) = F_\Phi^A(\lambda)^\mathcal{B} \begin{bmatrix} v^T \otimes I_n \\ B^\mathcal{B} \end{bmatrix} \tag{7.9}
$$

for some matrix $B \in \mathbb{M}_{kn \times (k-1)n}(\mathbb{R})$.

Any pencil $\mathcal{L}(\lambda)$ of the form (7.7) or (7.9) can be uniquely identified with the tuple (v, B). This is (for $\mathbb{M}_1(A)$) since the mapping $(v, B) \mapsto [\, v \otimes I_n \; B \,] F_\Phi^A(\lambda)$ is injective. To see this, assume $\mathcal{L}(\lambda) = [\, v_1 \otimes I_n \; B_1 \,] F_\Phi^A(\lambda)$ and $\mathcal{K}(\lambda) = [\, v_2 \otimes I_n \; B_2 \,] F_\Phi^A(\lambda)$ are two pencils from $\mathbb{M}_1(A)$ with $\mathcal{L}(\lambda) = \mathcal{K}(\lambda)$. Then we must have $v_1 = v_2$ as $\mathcal{L}(\lambda)(\Phi_{k-1}(\lambda)^T \otimes I_n) = \mathcal{K}(\lambda)(\Phi_{k-1}(\lambda)^T \otimes I_n)$. Therefore, $0_{kn \times kn} = \mathcal{L}(\lambda) - \mathcal{K}(\lambda) = [\, 0 \otimes I_n \; (B_1 - B_2) \,] F_\Phi^A(\lambda)$. However, according to (7.5),

$$
0_{kn \times kn} = \begin{bmatrix} 0 \otimes I_n \; (B_1 - B_2) \end{bmatrix} F_\Phi^A(\lambda) = \Big(B_1 - B_2\Big) M_\Phi(\lambda)
$$

implies $B_1 - B_2 = 0$ since $M_\Phi(\lambda)^T$ always has full normal rank, i.e. $\mathcal{N}_\ell(M_\Phi(\lambda)) = \{0\}$. Now, since $v_1 = v_2$ and $B_1 = B_2$, the injectivity is proven. For $\mathbb{M}_2(A)$ this is seen in a similar manner. In turn, we obtain the isomorphism[1] $\mathbb{M}_1(A) \cong \mathbb{R}^k \times \mathbb{R}^{kn \times (k-1)n} \cong \mathbb{M}_2(A)$ determining the dimension of $\mathbb{M}_1(A)$ and $\mathbb{M}_2(A)$ as follows:

Corollary 7.1 (Dimension of $\mathbb{M}_1(A)$ and $\mathbb{M}_2(A)$, [45, Cor. 1]). *Let $A(\lambda) \in \mathrm{M}_{n \times n}(\mathbb{R}[\lambda])$ be of degree $k \geq 2$. Then we have*

$$\dim\big(\mathbb{M}_1(A)\big) = \dim\big(\mathbb{M}_2(A)\big) = k(k-1)n^2 + k.$$

Corollary 7.1 essentially reduces to [89, Cor. 3.6] for $\Phi_{k-1}(\lambda) = \Lambda_{k-1}(\lambda)^2$. In particular, $\mathbb{L}_1(A) \cong \mathbb{L}_2(A) \cong \mathbb{M}_1(A) \cong \mathbb{M}_2(A)$ holds for any $A(\lambda) \in \mathrm{M}_{n \times n}(\mathbb{R}[\lambda])$ as the dimensions agree.

Certainly there is a universal linearization condition for pencils $\mathcal{L}(\lambda)$ in $\mathbb{M}_1(A)$ and $\mathbb{M}_2(A)$, too. In particular, as stated below it is similar to Corollary 6.1 and does not depend on the chosen basis at all.

Corollary 7.2 (Linearization Condition, [45, Cor. 2]). *Let $A(\lambda) \in \mathrm{M}_{n \times n}(\mathbb{R}[\lambda])$ be of degree $k \geq 2$. Moreover, let $\mathcal{L}(\lambda) \in \mathbb{M}_1(A)$ of the form (7.7) or $\mathcal{L}(\lambda) \in \mathbb{M}_2(A)$ of the form (7.9) be given. Then the following statements hold:*

  (i) *The matrix pencil $\mathcal{L}(\lambda)$ is a strong linearization for $A(\lambda)$ if $\mathrm{rank}([\, v \otimes I_n \; B \,]) = kn$ holds (regardless whether $A(\lambda)$ is regular or singular). Certainly, this is equivalent to $[\, v \otimes I_n \; B \,] \in \mathrm{GL}_{kn}(\mathbb{R})$.*

  (ii) *If $A(\lambda)$ is a regular matrix polynomial and $\mathcal{L}(\lambda)$ a linearization for $A(\lambda)$, then the rank condition $\mathrm{rank}([\, v \otimes I_n \; B \,]) = kn$ is satisfied.*

As in Theorem 3.2 notice that any pencil $\mathcal{L}(\lambda) \in \mathbb{M}_1(A)$ or $\mathcal{L}(\lambda) \in \mathbb{M}_2(A)$ that does not satisfy the condition in Corollary 7.2 (i) is automatically singular (since we have $\det(\mathcal{L}(\lambda)) = \det([v \otimes I_n \; B \,]) \det(F_\Phi^A(\lambda)) = 0 \cdot \det(F_\Phi^A(\lambda)) = 0$).

Proof. (i) Whenever $\mathrm{rank}([v \otimes I_n \; B \,]) = kn$, $\mathcal{L}(\lambda) = [\, v \otimes I_n \; B \,] F_\Phi^A(\lambda)$ is strict equivalent to $F_\Phi^A(\lambda)$ according to Definition 2.12 (ii) and thus a strong linearization for $A(\lambda)$.

  (ii) If $\mathrm{rank}([v \otimes I_n \; B \,]) < kn$, $\mathcal{L}(\lambda)$ is singular as $\det([\, v \otimes I_n \; B \,]) = 0$ and therefore not a linearization for any regular $A(\lambda)$.   □

Beside Corollary 7.2 there exist other ways to identify strong linearizations in $\mathbb{M}_1(A)$ for $A(\lambda) \in \mathrm{M}_{n \times n}(\mathbb{R}[\lambda])$. As suggested in [88, Prop. 4.9], any $\mathcal{L}(\lambda) \in \mathbb{M}_1(A)$ first can be transformed into an element $\widetilde{\mathcal{L}}(\lambda) \in \mathbb{L}_1(A)$ via a basis change, cf. (7.4). Then, $\widetilde{\mathcal{L}}(\lambda)$ can be expressed with the ansatz vector αe_1, $\alpha \neq 0$, to determine its Z-rank. In particular, Corollary 6.2 applies. Fortunately, Corollary 7.2 applies to the pencils in $\mathbb{M}_1(A)$ expressed as in (7.7) right away without changing the ansatz vector or the polynomial basis. Even if $\mathcal{L}(\lambda) \in \mathbb{M}_1(A)$ is expressed as $\mathcal{L}(\lambda) = X\lambda + Y$,

---

[1]This isomorphism was also observed in the proof of [31, Thm. 4.4] in the context of matrix polynomials in the monomial basis.
[2]Note that Corollary 7.1 remains true if $\deg(A(\lambda)) = 1$.

v and B may easily be recovered to check the rank of $[\,v \otimes I_n\ B\,]$. This is since $\mathcal{L}(\lambda) = [\,v \otimes I_n\ B\,]F_\Phi^A(\lambda)$ can also be written as

$$\mathcal{L}(\lambda) = \Big[\,v \otimes \alpha_{k-1}^{-1}A_k\ B\,\Big]\lambda + \mathcal{L}(0).$$

That is, v and B appear directly in the matrix X. This gives us the following corollary:

Corollary 7.3. *Let* $A(\lambda) = \sum_{j=0}^k A_j\phi_j(\lambda) \in \mathrm{M}_{n\times n}(\mathbb{R})[\lambda]$ *be regular and of degree* $k \geq 2$ *with* $A_k \in \mathrm{GL}_n(\mathbb{R})$. *Moreover, let* $\mathcal{L}(\lambda) = X\lambda + Y \in \mathbb{M}_1(A)$. *Then* $\mathcal{L}(\lambda)$ *is a strong linearization for* $A(\lambda)$ *iff* X *is nonsingular.*

Proof. As above, we write $\mathcal{L}(\lambda) = [\,v \otimes \alpha_{k-1}^{-1}A_k\ B\,]\lambda + \mathcal{L}(0) = X\lambda + Y$. Then

$$X = \Big[\,v \otimes \alpha_{k-1}^{-1}A_k\ B\,\Big] = \Big[\,v \otimes I_n\ B\,\Big]\left[\begin{array}{c|c} \alpha_{k-1}^{-1}A_k & 0 \\ \hline 0 & I_{(k-1)n} \end{array}\right]. \tag{7.10}$$

As A_k was assumed to be nonsingular and $\alpha_{k-1}^{-1} \neq 0$, the second matrix in (7.10) is always nonsingular. Thus, $X \in \mathrm{GL}_{kn}(\mathbb{R})$ holds iff $[\,v \otimes I_n\ B\,]$ is nonsingular. Now the statement follows from Corollary 7.2. □

Notice that, in the context of Corollary 7.3, $\mathcal{L}(\lambda) = X\lambda + Y$ being a strong linearization for some regular $A(\lambda) \in \mathrm{M}_{n\times n}(\mathbb{R}[\lambda])$, $A_k \in \mathrm{GL}_n(\mathbb{R})$, certainly implies X to be nonsingular. Otherwise $\mathcal{L}(\lambda)$ would possess infinite eigenvalues which $A(\lambda)$ does not have. Therefore, the main insight of Corollary 7.3 is essentially, that $\mathcal{L}(\lambda)$ is automatically a strong linearization for $A(\lambda)$ if it has no infinite eigenvalue. In particular, for such $A(\lambda)$, the splitting of $\mathbb{M}_1(A)$ into singular pencils and strong linearizations can also be identified by the pencils $\mathcal{L}(\lambda) \in \mathbb{M}_1(A)$ with and without infinite eigenvalues. In other words, for regular $A(\lambda) \in \mathrm{M}_{n\times n}(\mathbb{R}[\lambda])$, $A_k \in \mathrm{GL}_n(\mathbb{R})$, $\mathcal{L}(\lambda) \in \mathbb{M}_1(A)$ is singular iff $\mathcal{L}(\lambda)$ has infinite eigenvalues.

In [101, Thm. 2.1] the *Strong Linearization Theorem* from [89, Thm. 4.3] was revisited and proven for all generalized ansatz spaces considering any degree-graded basis (reconsider (2.15) or consult [89, Sec. 4.2]). In Section 3.2 we showed that it holds in a similar way for the block Kronecker ansatz space family $\mathbb{G}_{\eta+1}(A)$, $\eta = 0, \ldots, k-1$. We now state this theorem according to our discussion for orthogonal bases. In the form given below it is a generalized version of [45, Thm. 3]. Within our mathematical framework, the theorem admits a very simple and short proof.

Theorem 7.3 (Strong Linearization Theorem). *Assume* $A(\lambda) = \sum_{j=0}^k A_j\phi_j(\lambda) \in \mathrm{M}_{n\times n}(\mathbb{R})[\lambda]$ *to be regular and of degree* $k \geq 2$. *In addition, let* $\mathcal{L}(\lambda) \in \mathbb{M}_1(A)$ *be as in (7.7). Then the following statements are equivalent:*

(i) $\mathcal{L}(\lambda)$ *is a strong linearization for* $A(\lambda)$.

(ii) $\mathcal{L}(\lambda)$ *is a linearization for* $A(\lambda)$.

(iii) $\mathcal{L}(\lambda)$ *is a regular matrix pencil.*

(iv) $\operatorname{rank}([\, v \otimes I_n \ B\,]) = kn$.

(v) $\mathcal{L}(\lambda)$ is strict equivalent to $F_\Phi^A(\lambda)$.

If $\mathcal{L}(\lambda) = X\lambda + Y$ and $A_k \in \mathrm{GL}_n(\mathbb{R})$, then (i) to (v) are additionally equivalent to

(vi) The matrix $X \in \mathrm{M}_{kn \times kn}(\mathbb{R})$ is nonsingular.

(vii) $\mathcal{L}(\lambda)$ has no eigenvalue at infinity.

Proof. Similarly to the proof of Theorem 3.4, $(iv) \Rightarrow (v) \Rightarrow (i) \Rightarrow (ii) \Rightarrow (iii)$ is clear. For $(iii) \Rightarrow (iv)$ suppose that $\mathcal{L}(\lambda) = [\, v \otimes I_n \ B\,]F_\Phi^A(\lambda)$ is regular. Then, in particular

$$\det(\mathcal{L}(\lambda)) = \det\left(\left[\, v \otimes I_n \ B\,\right]\right) \det\left(F_\Phi^A(\lambda)\right) \neq 0$$

since $A(\lambda)$ was assumed to be regular and $F_\Phi^A(\lambda)$ is a strong linearization for $A(\lambda)$. Certainly this implies $\det([\, v \otimes I_n \ B\,]) \neq 0$ and $[\, v \otimes I_n \ B\,]$ is nonsingular. That is, $\operatorname{rank}([\, v \otimes I_n \ B\,]) = kn$, so $(iii) \Rightarrow (iv)$. The equivalence $(vi) \Leftrightarrow (vii)$ is clear since the nonsingularity of X is equivalent to the nonexistence of infinite eigenvalues of $\mathcal{L}(\lambda)$. Under the condition $\infty \notin \sigma(A)$ (which is in turn equivalent to $A_k \in \mathrm{GL}_n(\mathbb{R})$) the equivalence of (iv) and (i) follows from Corollary 7.3 and the proof is complete. $\square$

Certainly Theorem 7.3 is valid in a similar fashion for matrix pencils $\mathcal{L}(\lambda) \in \mathbb{M}_2(A)$. Moreover, since almost every matrix of the form $[\, v \otimes I_n \ B\,]$ has full rank, we obtain the following genericity statement similarly to [89, Thm. 4.7].

Corollary 7.4 (Genericity of Linearizations, [45, Cor. 4]). *For any $A(\lambda) \in \mathrm{M}_{n \times n}(\mathbb{R}[\lambda])$ of degree $k \geq 1$ being regular or singular, almost every matrix pencil in $\mathbb{M}_1(A)$ and $\mathbb{M}_2(A)$ is a strong linearization for $A(\lambda)$.*

7.2   The recovery of eigen- and nullspaces

In this section we show how eigenvectors for regular $A(\lambda) \in \mathrm{M}_{n \times n}(\mathbb{R}[\lambda])$ as in (7.2) may be recovered from eigenvectors of linearizations in $\mathbb{M}_1(A)$. The main ideas behind this derivation follow mainly the approach in [89] and [31, Sec. 5]. However, Theorem 7.1 simplifies things and shows that we can actually restrict the study of (right) eigenvectors to $F_\Phi^A(\lambda)$. This yields a new kind of linearization condition for pencils in $\mathbb{M}_1(A)$ and $\mathbb{M}_2(A)$, respectively. We begin with a proposition relating the right eigenvectors of any $\mathcal{L}(\lambda) \in \mathbb{M}_1(A)$ to those of $F_\Phi^A(\lambda)$.

Proposition 7.1 ([45, Prop. 1]). *Let $A(\lambda) \in \mathrm{M}_{n \times n}(\mathbb{R}[\lambda])$ be regular of degree $k \geq 2$ and $\mathcal{L}(\lambda) \in \mathbb{M}_1(A)$ as in (7.7). Then the following statements hold:*

(i) *Every right eigenvector of $F_\Phi^A(\lambda)$ (for any eigenvalue) is a right eigenvector of $\mathcal{L}(\lambda)$.*

(*ii*) Let $\mathcal{L}(\lambda)$ be a (strong) linearization for $A(\lambda)$. Then every right eigenvector of $\mathcal{L}(\lambda)$ (for any eigenvalue) is a right eigenvector of $F_\Phi^A(\lambda)$.

Proof. (*i*) For $\alpha \in \sigma_f(A)$ this is clear since $\mathrm{null}(F_\Phi^A(\alpha)) \subseteq \mathrm{null}(\mathcal{L}(\alpha))$ always holds, see (7.7), that is $\mathrm{Eig}_r(F_\Phi^A, \alpha) \subseteq \mathrm{Eig}_r(\mathcal{L}, \alpha)$. For $\infty \in \sigma(A)$ the situation in the same considering the reversals of both matrix polynomials. To this end, notice that the reversal of $\mathcal{L}(\lambda)$ as in (7.7) is given as $\mathrm{rev}\mathcal{L}(\lambda) = [\, v \otimes I_n \; B\,]\mathrm{rev}F_\Phi^A(\lambda)$.
(*ii*) Whenever $\mathcal{L}(\lambda) \in \mathbb{M}_1(A)$ is a (strong) linearization for $A(\lambda)$, we obtain from (*i*) that $\mathrm{null}(F_\Phi^A(\alpha)) = \mathrm{null}(\mathcal{L}(\alpha))$ has to hold. Thus every right eigenvector of $\mathcal{L}(\lambda)$ is a right eigenvector of $F_\Phi^A(\lambda)$. The situation for $\infty \in \sigma(A)$ is again handled similar. □

Obviously, Proposition 7.1 extends to singular matrix polynomials. That is, if $A(\lambda) \in \mathrm{M}_{n\times n}(\mathbb{R}[\lambda])$ is singular and $\mathcal{L}(\lambda) \in \mathbb{M}_1(A)$ is as in (7.7), then $q(\lambda) \in \mathcal{N}_r(F_\Phi^A(\lambda))$ with $q(\lambda) \in \mathbb{R}(\lambda)^{kn}$ implies $q(\lambda) \in \mathcal{N}_r(\mathcal{L}(\lambda))$. Moreover, in case $[\, v \otimes I_n \; B\,]$ is nonsingular every $q(\lambda) \in \mathcal{N}_r(\mathcal{L}(\lambda))$ is also an element of $\mathcal{N}_r(F_\Phi^A(\lambda))$. However, for singular $A(\lambda)$, the condition $[\, v \otimes I_n \; B\,] \in \mathrm{GL}_{kn}(\mathbb{R})$ is in general not necessary for a matrix pencil $\mathcal{L}(\lambda) = [\, v \otimes I_n \; B\,]F_\Phi^A(\lambda)$ to be a strong linearization for $A(\lambda)$. This situation will be dealt with in Section 7.3.

From Proposition 7.1 we directly obtain the following linearization condition for regular matrix polynomials $A(\lambda)$ that relates the eigenvectors of any $\mathcal{L}(\lambda) \in \mathbb{M}_1(A)$ to those of $F_\Phi^A(\lambda)$:

Proposition 7.2 ([45, Prop. 2]). *Let $A(\lambda) \in \mathrm{M}_{n\times n}(\mathbb{R}[\lambda])$ be regular and let $\mathcal{L}(\lambda) \in \mathbb{M}_1(A)$. Then $\mathcal{L}(\lambda)$ is a strong linearization for $A(\lambda)$ iff every right eigenvector of $\mathcal{L}(\lambda)$ is a right eigenvector of $F_\Phi^A(\lambda)$ for any (finite or infinite) eigenvalue.*

Proof. $\Rightarrow$ Let $\mathcal{L}(\lambda) \in \mathbb{M}_1(A)$ be a strong linearization for $A(\lambda) \in \mathrm{M}_{n\times n}(\mathbb{R}[\lambda])$ of degree $k \geq 2$. Suppose $\alpha \in \sigma_f(A) = \sigma_f(\mathcal{L})$ is an eigenvalue of $\mathcal{L}(\lambda)$ and $u \in \mathbb{C}^{kn}$ is a corresponding eigenvector, i.e. $\mathcal{L}(\alpha)u = 0$. Then, since

$$\mathcal{L}(\alpha) = \big[\, v \otimes I_n \; B\,\big]F_\Phi^A(\alpha)$$

assuming that $F_\Phi^A(\alpha)u \neq 0$ we obtain that $F_\Phi^A(\alpha)u \in \mathrm{null}([\, v \otimes I_n \; B\,])$. Therefore, $[\, v \otimes I_n \; B\,]$ is singular and $\mathcal{L}(\lambda)$ is no linearization for $A(\lambda)$, a contradiction. The proof for $\alpha = \infty$ is essentially the same considering the reversals of $\mathcal{L}(\lambda)$ and $F_\Phi^A(\lambda)$.
$\Leftarrow$ On the other hand, assume that $[\, v \otimes I_n \; B\,]$ is singular, i.e. $\mathcal{L}(\lambda)$ is not a linearization for $A(\lambda)$. Then there exists some $w \in \mathbb{C}^{kn}$ so that $[\, v \otimes I_n \; B\,]w = 0$. Now for any $\beta \in \mathbb{C}$ that is not an eigenvalue of $A(\lambda)$, $F_\Phi^A(\beta)$ is nonsingular. Therefore we may solve $F_\Phi^A(\beta)z = w$ for $z \in \mathbb{C}^{kn}$ and thus z is a right eigenvector of $\mathcal{L}(\beta)$ that is not an eigenvector of $F_\Phi^A(\beta)$. This completes the proof. □

Note that Proposition 7.1 states a linearization condition for any pencil $\mathcal{L}(\lambda)$ in $\mathbb{M}_1(A)$ in terms of (a comparison of) the right eigenvectors of $\mathcal{L}(\lambda)$ and $F_\Phi^A(\lambda)$.[3]

---

[3]Another linearization condition based upon left eigenvectors is derived in Section 7.5.

In particular, taking Proposition 7.2 into account, any $\mathcal{L}(\lambda) \in \mathbb{M}_1(A)$ is a strong linearization for $A(\lambda) \in M_{n \times n}(\mathbb{R}[\lambda])$ being regular iff

$$\text{Eig}_r(\mathcal{L}, \alpha) = \text{Eig}_r(F_{\Phi}^A, \alpha) \qquad \forall \, \alpha \in \sigma(A). \tag{7.11}$$

Certainly, similar statements to Proposition 7.1 and 7.2 hold for $\mathbb{M}_2(A)$ and the left eigenspaces $\text{Eig}_\ell(\mathcal{L}, \alpha)$ and $\text{Eig}_\ell((F_{\Phi}^A)^B, \alpha)$, $\alpha \in \sigma(A)$. Moreover, notice that, for regular $A(\lambda)$, is suffices to find $\text{Eig}_r(F_{\Phi}^A, \alpha)$ for $\alpha \in \sigma(A)$ in order to determine $\text{Eig}_r(\mathcal{L}, \alpha)$ for any strong linearization $\mathcal{L}(\lambda) \in \mathbb{M}_1(A)$ since, due to (7.11), both spaces coincide.

The following proposition shows how eigenvectors of regular $A(\lambda)$ can be recovered from those of $F_{\Phi}^A(\lambda)$. This has already been observed in a slightly different form in [31, Sec. 7] and can be proven by straight forward calculations (see also [89, Thm. 3.8, Thm. 3.14, Thm. 4.4]). Notice that, taking Proposition 7.1, Proposition 7.2 and, in particular, (7.11) into account, we obtain the complete right (left, respectively) eigenvector recovery for regular linearizations in $\mathbb{M}_1(A)$ ($\mathbb{M}_2(A)$, respectively).

Proposition 7.3 (Eigenvector Recovery, [45, Prop. 3]). *Let $A(\lambda) \in M_{n \times n}(\mathbb{R}[\lambda])$ be regular and of degree $k \geq 2$. Then the following statements hold:*

   (i) *Let $\alpha \in \sigma_f(A)$. Then we have $u \in \text{Eig}_r(A, \alpha)$ iff $\Phi_{k-1}(\alpha)^T \otimes u \in \text{Eig}_r(F_{\Phi}^A, \alpha)$ and $u \in \text{Eig}_\ell(A, \alpha)$ iff $\Phi_{k-1}(\alpha)^T \otimes u \in \text{Eig}_\ell((F_{\Phi}^A)^B, \alpha)$. Moreover, every right (left) eigenvector w of $F_{\Phi}^A(\lambda)$ ($F_{\Phi}^A(\lambda)^B$) for α has the form $w = \Phi_{k-1}(\alpha)^T \otimes u$ for some right (left) eigenvector u of $A(\lambda)$ corresponding to α.*

  (ii) *Let α be infinity. Then we have $u \in \text{Eig}_r(A, \infty)$ iff $e_1 \otimes u \in \text{Eig}_r(F_{\Phi}^A, \infty)$ and $u \in \text{Eig}_\ell(A, \infty)$ iff $e_1 \otimes u \in \text{Eig}_\ell((F_{\Phi}^A)^B, \infty)$. Moreover, every right (left) eigenvector w of $F_{\Phi}^A(\lambda)$ ($F_{\Phi}^A(\lambda)^B$) for α has the form $w = e_1 \otimes u$ for some right (left) eigenvector u of $A(\lambda)$ corresponding to α.*

7.2.1   The recovery of minimal indices and minimal bases

If $A(\lambda) \in M_{n \times n}(\mathbb{R}[\lambda])$ is singular and $\mathcal{L}(\lambda) \in \mathbb{M}_1(A)$ is a linearization for $A(\lambda)$, a statement analogous to Proposition 7.3 holds (see Theorem 7.4 *(ii)* below). Additionally, we have to draw attention to the recovery of the minimal indices of $A(\lambda)$ from those of $\mathcal{L}(\lambda)$. The situation for recovering minimal indices and minimal bases was already considered in [31] and is stated here only for completeness.

Theorem 7.4 (Recovery of right Minimal Indices, [31, Thm. 7.2]). *Let a singular matrix polynomial $A(\lambda) \in M_{n \times n}(\mathbb{R}[\lambda])$ of degree $k \geq 2$ be given. Moreover, let $\mathcal{L}(\lambda) \in \mathbb{M}_1(A)$ be a linearization for $A(\lambda)$ with ansatz vector $0 \neq v \in \mathbb{R}^k$. Then the following statements hold:*

   (i) *If $\epsilon_1 \leq \cdots \leq \epsilon_p$ are the right minimal indices of $A(\lambda)$, then*

$$(k-1) + \epsilon_1 \leq (k-1) + \epsilon_2 \leq \cdots \leq (k-1) + \epsilon_p$$

are the right minimal indices of $\mathcal{L}(\lambda)$.

(ii) *Every minimal basis of* $\mathcal{N}_r(\mathcal{L}(\lambda))$ *has the form* $\Phi_{k-1}(\lambda)^T \otimes x_1(\lambda), \ldots, \Phi_{k-1}(\lambda)^T \otimes$ $x_p(\lambda)$ *for a minimal basis* $x_1(\lambda), \ldots, x_p(\lambda)$ *of* $\mathcal{N}_r(A(\lambda))$.

In particular, Theorem 7.4 states that the right minimal indices of $F_\Phi^A(\lambda)$ are those of $A(\lambda)$ shifted by $k-1$ (choose $[\,v \otimes I_n\ B\,] = I_{kn}$). However, neither (i) nor (ii) from Theorem 7.4 are sufficient for $\mathcal{L}(\lambda) \in \mathbb{M}_1(A)$ to be a linearization for $A(\lambda)$. A comprehensive example on this situation is provided in Section 7.3. Concerning the left minimal indices of any linearization $\mathcal{L}(\lambda) = [\,v \otimes I_n\ B\,]F_\Phi^A(\lambda) \in \mathbb{M}_1(A)$ the situation is more complicated as $[\,v \otimes I_n\ B\,]$ need not have full rank for $\mathcal{L}(\lambda)$ to be a linearization. However, if $[\,v \otimes I_n\ B\,]$ has full rank, the situation simplifies significantly.

Theorem 7.5 (Recovery of left Minimal Indices, [31, Thm. 7.3]). *Let a singular matrix polynomial* $A(\lambda) \in \mathrm{M}_{n \times n}(\mathbb{R}[\lambda])$ *of degree* $k \geq 2$ *be given. Moreover, let* $\mathcal{L}(\lambda) = [\,v \otimes I_n\ B\,]F_\Phi^A(\lambda) \in \mathbb{M}_1(A)$ *be a linearization for* $A(\lambda)$ *so that* $[\,v \otimes I_n\ B\,]$ *is nonsingular. Then the following statements hold:*

(i) *The left minimal indices of* $\mathcal{L}(\lambda)$ *and* $A(\lambda)$ *coincide.*

(ii) *Every minimal basis of* $\mathcal{N}_\ell(A(\lambda))$ *has the form* $(v^T \otimes I_n)x_1(\lambda), \ldots, (v^T \otimes I_n)x_r(\lambda)$ *for a minimal basis* $x_1(\lambda), \ldots, x_r(\lambda)$ *of* $\mathcal{N}_\ell(\mathcal{L}(\lambda))$.

Surely, analogous results hold for linearizations $\mathcal{L}(\lambda) \in \mathbb{M}_2(A)$: the left minimal indices of $\mathcal{L}(\lambda)$ are those of $A(\lambda)$ shifted by $k-1$ whereas the right minimal indices of $\mathcal{L}(\lambda) = F_\Phi^A(\lambda)^{\mathcal{B}}[\,v \otimes I_n\ B\,]^{\mathcal{B}}$ coincide with those of $A(\lambda)$ if $[\,v \otimes I_n\ B\,]^{\mathcal{B}}$ has full rank [31]. The recovering of minimal bases is analogous to Theorem 7.4 (ii) and Theorem 7.5 (ii). For more information on the recovering of minimal indices and minimal bases we would like to point the reader to [31, Sec. 7].

7.3 A note on singular matrix polynomials

According to Theorem 7.3, when $A(\lambda) \in \mathrm{M}_{n \times n}(\mathbb{R}[\lambda])$ is regular, any linearization from $\mathbb{L}_1(A)$ (or $\mathbb{M}_1(A)$) is necessarily a strong linearization. In [31, Ex. 3] is was shown that the equivalence of strong linearizations and linearizations does in general not hold in $\mathbb{L}_1(A)$ for singular $A(\lambda)$. In particular, this situation carries over to $\mathbb{M}_1(A)$. In this section we consider singular matrix polynomials $A(\lambda)$ and give a sufficient condition on when the equivalence of being a linearization and a strong linearization holds. The main result of this section is the following theorem which extends [31, Lem. 5.5] by complementing it to an equivalence statement. Moreover, it is extended to orthogonal bases.

Theorem 7.6 ([45, Thm. 4]). *Let* $A(\lambda) \in \mathrm{M}_{n \times n}(\mathbb{R}[\lambda])$ *be singular and of degree* $k \geq 2$ *and assume* $\mathcal{L}(\lambda) \in \mathbb{M}_1(A)$ *is as in (7.7) with ansatz vector* $0 \neq v \in \mathbb{R}^k$. *Then* $\mathrm{rank}([\,v \otimes I_n\ B\,]) = kn$ *iff for every nonzero* $u(\lambda) \in \mathcal{N}_\ell(\mathcal{L}(\lambda))$ *it holds that*

$$u(\lambda)^T(v \otimes I_n) \neq 0.$$

Proof. $\Rightarrow$ Assume $\mathcal{L}(\lambda) \in \mathbb{M}_1(A)$ in (7.7) satisfies rank($[\,v \otimes I_n\ B\,]$) $= kn$ and let $u(\lambda) \in \mathcal{N}_\ell(\mathcal{L}(\lambda))$, $u(\lambda) \neq 0$. Defining $w(\lambda) \in \mathbb{R}(\lambda)^{kn}$ as

$$w(\lambda)^T = \left[\, w_1(\lambda)\ w_2(\lambda)\ \cdots\ w_{kn}(\lambda)\,\right] := u(\lambda)^T \left[\, v \otimes I_n\ B\,\right]$$

and assuming that $u(\lambda)^T(v \otimes I_n) = 0$, we obviously obtain $[\,w_1(\lambda)\ \cdots\ w_n(\lambda)\,] = 0$. Moreover, since $u(\lambda) \in \mathcal{N}_\ell(\mathcal{L}(\lambda))$, we have

$$u(\lambda)^T \mathcal{L}(\lambda) = u(\lambda)^T \left[\, v \otimes I_n\ B\,\right] F_\Phi^A(\lambda) = w(\lambda)^T F_\Phi^A(\lambda) = 0. \tag{7.12}$$

Using the fact $[\,w_1(\lambda)\ \cdots\ w_n(\lambda)\,] = 0$, equation (7.12) and the block-Hessenberg structure of $F_\Phi^A(\lambda)$ imply $\alpha_{k-2}[w_{n+1}(\lambda)\ \cdots\ w_{2n}(\lambda)\,] = [\,0\ \cdots\ 0\,]$, cf. (7.5) and (7.6). Therefore, $w_{n+1}(\lambda) = \cdots = w_{2n}(\lambda) = 0$ since $\alpha_{k-2} \neq 0$. In particular, in (7.12) we actually have

$$\left[\, 0\ 0\ \cdots\ 0\ 0\ w_{2n+1}(\lambda)\ \cdots\ w_{kn}(\lambda)\,\right] F_\Phi^A(\lambda) = \left[\, 0\ \cdots\ 0\,\right]. \tag{7.13}$$

From (7.13) the same observation yields $\alpha_{k-3}[w_{2n+1}(\lambda)\ \cdots\ w_{3n}(\lambda)\,] = [\,0\ \cdots\ 0\,]$ implying $w_{2n+1}(\lambda) = \cdots = w_{3n}(\lambda) = 0$ since $\alpha_{k-3} \neq 0$. Continuing this procedure up to α_0 we obtain $w(\lambda) \equiv [\,0\ \cdots\ 0\,]$. In other words, $u(\lambda)^T[\,v \otimes I_n\ B\,] = 0$. This implies $[\,v \otimes I_n\ B\,]$ to be singular. Since we assumed $[\,v \otimes I_n\ B\,]$ to have full rank, $u(\lambda)^T(v \otimes I_n) = 0$ in turn implies $u(\lambda) \equiv 0$, a contradiction. Thus, $u(\lambda)^T(v \otimes I_n) = 0$ cannot hold whenever $[\,v \otimes I_n\ B\,]$ is nonsingular and we have $u(\lambda)^T(v \otimes I_n) \neq 0$ for every nonzero $u(\lambda) \in \mathcal{N}_\ell(\mathcal{L}(\lambda))$.

$\Leftarrow$ Now suppose $u(\lambda)^T(v \otimes I_n) \neq 0$ holds for all nonzero $u(\lambda) \in \mathcal{N}_\ell(\mathcal{L}(\lambda))$. Assuming rank($[\,v \otimes I_n\ B\,]$) $< kn$ implies the existence of at least one vector $0 \neq q \in \mathbb{R}^{kn}$ with $q^T[\,v \otimes I_n\ B\,] = 0$. Since $\mathcal{L}(\lambda) = [\,v \otimes I_n\ B\,]F_\Phi^A(\lambda)$ we have $q^T \mathcal{L}(\lambda) = 0$, so obviously $q \in \mathcal{N}_\ell(\mathcal{L}(\lambda))$. In particular, q satisfies $q^T(v \otimes I_n) = 0$ which is a contradiction for we assumed $u(\lambda)^T(v \otimes I_n) \neq 0$. Thus we must have rank($[\,v \otimes I_n\ B\,]$) $= kn$ and the proof is complete. $\qquad\square$

We obtain an immediate corollary:

Corollary 7.5 (Linearization Condition, [45, Cor. 5]). *Let a singular matrix polynomial $A(\lambda) \in \mathrm{M}_{n \times n}(\mathbb{R}[\lambda])$ of degree $k \geq 2$ be given and assume $\mathcal{L}(\lambda) \in \mathbb{M}_1(A)$ with ansatz vector $v \in \mathbb{R}^k$. If $u(\lambda)^T(v \otimes I_n) \neq 0$ holds for all nonzero $u(\lambda) \in \mathcal{N}_\ell(\mathcal{L}(\lambda))$ then $\mathcal{L}(\lambda)$ is a strong linearization for $A(\lambda)$.*

Proof. This is a direct consequence of Theorem 7.6 since $u(\lambda)^T(v \otimes I_n) \neq 0$ for all nonzero $u(\lambda) \in \mathcal{N}_\ell(\mathcal{L}(\lambda))$ implies for $\mathcal{L}(\lambda) = [\,v \otimes I_n\ B\,]F_\Phi^A(\lambda)$ that rank($[\,v \otimes I_n\ B\,]$) $= kn$ holds. In turn this implies $\mathcal{L}(\lambda)$ to be a strong linearization for $A(\lambda)$ according to Corollary 7.2 (i). $\qquad\square$

We now state a modified version of the "Strong Linearization Theorem" adapted for singular matrix polynomials.

Theorem 7.7 (Strong Linearization Theorem, [45, Thm. 5]). *Let $A(\lambda) \in M_{n \times n}(\mathbb{R}[\lambda])$ be singular and of degree $k \geq 2$. Moreover, let $\mathcal{L}(\lambda) \in \mathbb{M}_1(A)$ be as in (7.7) with ansatz vector $0 \neq v \in \mathbb{R}^k$. Additionally assume that $u(\lambda)^T(v \otimes I_n) \neq 0$ holds for all $0 \neq u(\lambda) \in \mathcal{N}_\ell(\mathcal{L}(\lambda))$. Then the following statements are equivalent:*

(i) rank$([\, v \otimes I_n \; B\,]) = kn$.

(ii) $\mathcal{L}(\lambda)$ *is a strong linearization for $A(\lambda)$.*

(iii) $\mathcal{L}(\lambda)$ *is a linearization for $A(\lambda)$.*

Proof. It is clear that $(i) \Rightarrow (ii) \Rightarrow (iii)$ holds even without the assumption $u(\lambda)^T(v \otimes I_n) \neq 0$ for all $0 \neq u(\lambda) \in \mathcal{N}_\ell(\mathcal{L}(\lambda))$ and that $(iii) \Rightarrow (i)$ follows from Theorem 7.6 taking into account the assumption that $u(\lambda)^T(v \otimes I_n) \neq 0$ holds for all $u(\lambda) \in \mathcal{N}_\ell(\mathcal{L}(\lambda))$. □

As already discussed, if $\mathcal{L}(\lambda) = [\, v \otimes I_n \; B\,] F_\Phi^A(\lambda) \in \mathbb{M}_1(A)$ is a (strong) linearization for some singular $A(\lambda) \in M_{n \times n}(\mathbb{R}[\lambda])$ of degree $k \geq 2$, then rank$([\, v \otimes I_n \; B\,]) = kn$ does not necessarily follow. In particular, the condition from Corollary 7.2 (i) turns out to be neither necessary for linearizations nor for strong linearizations. This is illustrated by the following example where we confine ourselves to $\mathbb{L}_1(A)$ for simplicity. Recall that the abbreviation SCF refers to the Smith canonical form, cf. Theorem 2.1. Further examples are given in [31, Ex. 2, Ex. 3].

Example 7.1. *Consider the matrix polynomials $A(\lambda), B(\lambda), C(\lambda) \in M_{2 \times 2}(\mathbb{R}[\lambda])$ of degree $k = 2$ and normal rank one given below. For each matrix polynomial we construct a matrix pencil $\mathcal{L}(\lambda) = [\, v \otimes I_2 \; B\,] \text{Frob}_1(\lambda) \in \mathbb{L}_1(\cdot)$ with the property that $\text{nrank}(\mathcal{L}(\lambda)) = \text{nrank}(\text{Frob}_1(\lambda)) = 3$. Notice that this condition is certainly necessary for $\mathcal{L}(\lambda)$ to be a (strong) linearization. Moreover, although in each case additionally rank$([\, v \otimes I_2 \; B\,]) = 3$ holds, all three pencils behave quite differently.*

(i) *Consider $A(\lambda) = \begin{bmatrix} \lambda^2 & \lambda \\ \lambda^2 & \lambda \end{bmatrix}$, its Frobenius companion form $\text{Frob}_1(\lambda)$ and*

$$\mathcal{L}(\lambda) = \begin{bmatrix} 1 & 0 & 0 & 0 \\ 0 & 1 & 0 & 1 \\ 0 & 0 & -1 & 0 \\ 0 & 0 & 0 & 0 \end{bmatrix} \text{Frob}_1(\lambda) = \begin{bmatrix} \lambda & 1 & 0 & 0 \\ \lambda & 0 & 0 & \lambda \\ 1 & 0 & -\lambda & 0 \\ 0 & 0 & 0 & 0 \end{bmatrix} \in \mathbb{L}_1(A).$$

As $[\, -1 \; 1 \; 0 \; 0\,]\text{Frob}_1(\lambda) = 0$, it is easily checked that

$$P\text{Frob}_1(\lambda) = \mathcal{L}(\lambda) \quad \text{for } P = \begin{bmatrix} 1 & 0 & 0 & 0 \\ 0 & 1 & 0 & 1 \\ 0 & 0 & -1 & 0 \\ -1 & 1 & 0 & 0 \end{bmatrix}.$$

As P is nonsingular, we have $\text{Frob}_1(\lambda) = P^{-1}\mathcal{L}(\lambda)$ and $\mathcal{L}(\lambda)$ is strict equivalent to $\text{Frob}_1(\lambda)$. Thus it follows that $\mathcal{L}(\lambda)$ is a strong linearization for $A(\lambda)$. We will show in Theorem 7.8 that this had to be the case.

(ii) Consider $B(\lambda) = \begin{bmatrix} \lambda^2 & \lambda \\ \lambda & 1 \end{bmatrix}$, its Frobenius companion form $\mathrm{Frob}_1(\lambda)$ and

$$\mathcal{L}(\lambda) = \begin{bmatrix} 1 & 0 & 0 & 0 \\ 0 & 1 & 0 & 0 \\ 0 & 0 & -1 & 0 \\ 0 & 0 & 0 & 0 \end{bmatrix} \mathrm{Frob}_1(\lambda) = \begin{bmatrix} \lambda & 1 & 0 & 0 \\ 1 & 0 & 0 & 1 \\ 1 & 0 & -\lambda & 0 \\ 0 & 0 & 0 & 0 \end{bmatrix} \in \mathbb{L}_1(B).$$

The SCFs of $B(\lambda)$ and $\mathrm{Frob}_1(\lambda)$ are given as $S_B = \mathrm{diag}[\, 1\,, 0]$ and $S_{\mathrm{Frob}} = I_3 \oplus [0]$, respectively. Moreover, the SCF of $\mathcal{L}(\lambda)$ is again $S_{\mathcal{L}} = I_3 \oplus [0]$, so $\mathcal{L}(\lambda)$ is a linearization for $B(\lambda)$. However, the SCFs of $\mathrm{rev}(\mathrm{Frob}_1(\lambda))$ and $\mathrm{rev}(\mathcal{L}(\lambda))$ are $S_{\mathrm{rev}(\mathrm{Frob}_1)} = I_3 \oplus [0]$ and $S_{\mathrm{rev}(\mathcal{L})} = \mathrm{diag}[\, 1\,, 1\,, \lambda\,, 0]$. Thus $S_{\mathrm{rev}(\mathrm{Frob}_1)} \neq S_{\mathrm{rev}(\mathcal{L})}$. Indeed, we have $\mathrm{rank}(\mathrm{rev}(\mathcal{L}(0))) = 2 < 3 = \mathrm{nrank}(\mathrm{rev}(\mathcal{L}(\lambda)))$. Therefore, $\mathcal{L}(\lambda)$ is a linearization for $B(\lambda)$ but not strong. This example is taken from [31, Ex. 3].

Thus, we see that $\mathrm{nrank}(\mathcal{L}(\lambda)) = \mathrm{nrank}(\mathrm{Frob}_1(\lambda))$ is not sufficient for $\mathcal{L}(\lambda)$ being a strong linearization. In particular, it is not even sufficient for $\mathcal{L}(\lambda)$ being a linearization as the third example shows:

(iii) Consider $C(\lambda) = \begin{bmatrix} -\lambda^2 & \lambda^2 \\ \lambda - 1 & 1 - \lambda \end{bmatrix}$, its Frobenius companion form $\mathrm{Frob}_1(\lambda)$ and

$$\mathcal{L}(\lambda) = \begin{bmatrix} 1 & 0 & 0 & 0 \\ 0 & 1 & 0 & 0 \\ 0 & 0 & -1 & 0 \\ 0 & 0 & 0 & 0 \end{bmatrix} \mathrm{Frob}_1(\lambda) = \begin{bmatrix} -\lambda & \lambda & 0 & 0 \\ 1 & -1 & -1 & 1 \\ 1 & 0 & \lambda & 0 \\ 0 & 0 & 0 & 0 \end{bmatrix} \in \mathbb{L}_1(C).$$

The SCF of $C(\lambda)$ and $\mathrm{Frob}_1(\lambda)$ are $S_C(\lambda) = \mathrm{diag}[\, 1\,, 0]$ and $S_{\mathrm{Frob}}(\lambda) = I_3 \oplus [0]$, respectively. In particular, $\sigma_f(C) = \emptyset$ and there is no $\lambda_0 \in \mathbb{C}$ for which $\mathrm{rank}(C(\lambda_0)) < 1$ or $\mathrm{rank}(\mathrm{Frob}_1(\lambda_0)) < 3$ holds. However, the SCF of $\mathcal{L}(\lambda)$ is $S_{\mathcal{L}} = \mathrm{diag}[\, 1\,, 1\,, \lambda\,, 0]$. Indeed $\mathrm{rank}(\mathcal{L}(0)) = 2 < 3$ and $\mathcal{L}(\lambda)$ can not even be a linearization for $C(\lambda)$.

We will now, at least for one special situation, analyse to what extend we may predict if $\mathcal{L}(\lambda) \in \mathbb{M}_1(A)$ is a (strong) linearization for some (singular) $A(\lambda) \in M_{n \times n}(\mathbb{R}[\lambda])$ without deeper computations. The fundamental result is the following Theorem 7.8 stating that linearizations for matrix polynomials with only left minimal indices equal to zero are "simple" to identify.

Theorem 7.8. Let $A(\lambda) \in M_{n \times n}(\mathbb{R}[\lambda])$ be singular of degree $k \geq 2$ and assume all left minimal indices of $A(\lambda)$ are zero. Then $\mathcal{L}(\lambda) \in \mathbb{M}_1(A)$ with ansatz vector $0 \neq v \in \mathbb{R}^k$ is a strong linearization for $A(\lambda)$ iff

$$\mathrm{nrank}(\mathcal{L}(\lambda)) = (k-1)n + \mathrm{nrank}(A(\lambda)).$$

Proof. $\Rightarrow$ Assume that $\mathcal{L}(\lambda) = [\, v{\otimes}I_n \; B \,]F_\Phi^A(\lambda) \in \mathbb{M}_1(A)$ and notice that $\mathrm{nrank}(\mathcal{L}(\lambda))$ $\leq t := (k-1)n + \mathrm{nrank}(A(\lambda))$ holds as $\mathrm{nrank}(F_\Phi^A(\lambda)) = t = \mathrm{rank}(I_{(k-1)n} \oplus A(\lambda))$ (recall Definition 2.13 (i)). In particular, we will never have a linearization in case of strict inequality and $\mathrm{nrank}(\mathcal{L}(\lambda)) = t$ is necessary for $\mathcal{L}(\lambda)$ to be a linearization.

$\Leftarrow$ Now assume that $\mathrm{nrank}(F_\Phi^A(\lambda)) = \mathrm{nrank}(\mathcal{L}(\lambda)) = t$. As $\dim(\mathcal{N}_\ell(F_\Phi^A(\lambda))) = kn - t$ and the left minimal indices of $F_\Phi^A(\lambda)$ are also all equal to zero according to Theorem 7.5 (i), there are $kn - t$ vectors $u_1, \ldots, u_{kn-t} \in \mathbb{R}^{kn}$ with $u_j^T F_\Phi^A(\lambda) = 0$ for all $j = 1, \ldots, kn-t$. Let $M_1 = [\, w_1 \; \cdots \; w_t \; u_1 \; \cdots \; u_{kn-t} \,] \in \mathbb{M}_{kn \times kn}(\mathbb{R})$ be nonsingular and observe that the last $kn - t$ rows of $\widetilde{F}_\Phi^A(\lambda) := M_1^T F_\Phi^A(\lambda)$ vanish completely. Moreover, $\widetilde{F}_\Phi^A(\lambda)$ and $F_\Phi^A(\lambda)$ are strict equivalent due to the nonsingularity of M_1. In particular, $\mathrm{nrank}(\widetilde{F}_\Phi^A(\lambda)) = t$ and $F_\Phi^A(\lambda) = M_1^{-T} \widetilde{F}_\Phi^A(\lambda)$. Now we consider $\mathcal{L}(\lambda)$ which has also normal rank t. Let $P \in \mathbb{M}_{kn \times kn}(\mathbb{R})$ be some permutation matrix so that the first t rows of $P\mathcal{L}(\lambda)$ are linearly independent as vector polynomials over $\mathbb{R}(\lambda)$. Then

$$P\mathcal{L}(\lambda) = P\big[\, v \otimes I_n \; B \,\big]F_\Phi^A(\lambda) = P\big[\, v \otimes I_n \; B \,\big]M_1^{-T} \widetilde{F}_\Phi^A(\lambda) =: \widehat{M} \widetilde{F}_\Phi^A(\lambda)$$

with $\widehat{M} = P[\, v \otimes I_n \; B \,]M_1^{-T}$. Let $\widehat{M} = [\, \widehat{m}_1 \; \cdots \; \widehat{m}_{kn} \,] = [\widehat{m}_{ij}]_{ij}$. As the last $kn - t$ rows of $\widetilde{F}_\Phi^A(\lambda)$ are entirely zero we may, without any restriction, replace the last $kn - t$ columns of $\widehat{M}$ by zero columns. Doing that modification, the matrix pencil $\widehat{M} \widetilde{F}_\Phi^A(\lambda)$ does not change at all. Thus, with $\overline{M} := [\, \widehat{m}_1 \; \cdots \; \widehat{m}_t \; 0 \; \cdots \; 0 \,]$ we have $P\mathcal{L}(\lambda) = \overline{M} \widetilde{F}_\Phi^A(\lambda)$. Moreover, as the first t rows of $P\mathcal{L}(\lambda)$ were assume to be linearly independent, it is easy to see the the $t \times t$ leading principal submatrix $\overline{M}_t$ of $\overline{M}$ must be nonsingular, i.e. $\overline{M}_t = [\widehat{m}_{ij}]_{i,j=1}^t \in \mathrm{GL}_t(\mathbb{R})$. Since we do not change $\overline{M} \widetilde{F}_\Phi^A(\lambda)$ by adding nonzero entries to the last $kn - t$ (zero) columns of $\overline{M}$ we construct

$$\widetilde{M} := \left[\begin{array}{ccc|c} \widehat{m}_1 & \cdots & \widehat{m}_t & \dfrac{0_{t \times (kn-t)}}{I_{kn-t}} \end{array}\right] = \overline{M} + \sum_{i=t+1}^{kn} e_i e_i^T \in \mathbb{M}_{kn \times kn}(\mathbb{R}).$$

Notice that $\widetilde{M}$ is nonsingular and that $\widetilde{M} \widetilde{F}_\Phi^A(\lambda) = \overline{M} \widetilde{F}_\Phi^A(\lambda) = P\mathcal{L}(\lambda)$. In particular, $\mathcal{L}(\lambda) = P^{-1} \widetilde{M} \widetilde{F}_\Phi^A(\lambda)$. Thus, we see that

$$\mathcal{L}(\lambda) = P^{-1} \widetilde{M} \widetilde{F}_\Phi^A(\lambda) = P^{-1} \widetilde{M} M_1^T F_\Phi^A(\lambda).$$

As $P^{-1} \widetilde{M} M_1^T$ is a product of three nonsingular matrices, it is certainly nonsingular, too. Thus $\mathcal{L}(\lambda)$ is in fact strict equivalent to $F_\Phi^A(\lambda)$. In particular, it is thus a strong linearization for $A(\lambda)$. This completes the proof. $\qquad\square$

We immediately obtain the following corollary which can be interpreted as an analogous *Strong Linearization Theorem* on $\mathbb{M}_1(A)$ for matrix polynomials $A(\lambda) \in \mathbb{M}_{n \times n}(\mathbb{R}[\lambda])$ with all left minimal indices equal to zero.

Corollary 7.6. *Let* $A(\lambda) \in \mathbb{M}_{n \times n}(\mathbb{R}[\lambda])$ *be singular of degree* $k \geq 2$ *with all left minimal indices equal to zero. Then, for* $\mathcal{L}(\lambda) \in \mathbb{M}_1(A)$ *with ansatz vector* $0 \neq v \in \mathbb{R}^k$ *the following statements are equivalent:*

(i) $\mathcal{L}(\lambda)$ *is a strong linearization for* $A(\lambda)$.

(ii) $\mathcal{L}(\lambda)$ is a linearization for $A(\lambda)$.

(iii) $\mathrm{nrank}(\mathcal{L}(\lambda)) = (k-1)n + \mathrm{nrank}(A(\lambda))$.

Proof. Certainly (i) $\Rightarrow$ (ii) is clear. Therefore, assume that $\mathcal{L}(\lambda)$ is a linearization for $A(\lambda)$. Then, according to Definition 2.13 (i) there exist unimodular matrices $U(\lambda), V(\lambda) \in \mathrm{M}_{kn \times kn}(\mathbb{R}[\lambda])$ with

$$U(\lambda)\mathcal{L}(\lambda)V(\lambda) = \begin{bmatrix} I_{(k-1)n} & \\ & A(\lambda) \end{bmatrix}.$$

Therefore, $\mathrm{nrank}(\mathcal{L}(\lambda)) = (k-1)n + \mathrm{nrank}(A(\lambda))$. If in turn ($iii$) holds, $\mathcal{L}(\lambda)$ is a strong linearization for $A(\lambda)$ according to Theorem 7.8 and the proof is complete. $\square$

7.4 Double generalized ansatz spaces

In this section, we characterize matrix pencils that are contained in both generalized ansatz spaces $\mathbb{M}_1(A)$ and $\mathbb{M}_2(A)$ for $A(\lambda) \in \mathrm{M}_{n \times n}(\mathbb{R}[\lambda])$ of degree $k \geq 2$. Certainly, if some matrix pencil $\mathcal{L}(\lambda)$ satisfies (7.3), $\mathcal{L}(\lambda)^{\mathcal{B}}$ satisfies $(\Phi_{k-1}(\lambda) \otimes I_n)\mathcal{L}(\lambda) = v^T \otimes A(\lambda)$ and vice versa. This is easily seen by block-transposing the ansatz equation (7.3). Consequently, if $\mathcal{L}(\lambda) = \mathcal{L}(\lambda)^{\mathcal{B}}$, $\mathcal{L}(\lambda) \in \mathbb{M}_1(A) \cap \mathbb{M}_2(A)$ holds. Thus, the vector space $\mathbb{DM}(A) := \mathbb{M}_1(A) \cap \mathbb{M}_2(A)$, called *double generalized ansatz space* in the following, contains all block-symmetric pencils from $\mathbb{M}_1(A)$ and $\mathbb{M}_2(A)$. Similarly, in the monomial case, the double ansatz space $\mathbb{DL}(A) = \mathbb{L}_1(A) \cap \mathbb{L}_2(A)$ contains all block-symmetric pencils from $\mathbb{L}_1(A)$, see [65]. We now give a rather surprising auxiliary statement on block-skew-symmetric pencils in $\mathbb{M}_1(A)$ that will, later on, be beneficial for the characterization of $\mathbb{DM}(A)$.

Proposition 7.4 ([45, Prop. 4]). *Let $A(\lambda) \in \mathrm{M}_{n \times n}(\mathbb{R}[\lambda])$ be of degree $k \geq 2$ and let $\mathcal{L}(\lambda) \in \mathbb{M}_1(A)$ be block-skew-symmetric. Then $\mathcal{L}(\lambda)$ satisfying (7.3) with $v = [\,0 \ v_2 \ v_3 \ \cdots \ v_k\,]^T \in \mathbb{R}^k$ implies $\mathcal{L}(\lambda) \equiv 0$.*

Proof. Assuming that $\mathcal{L}(\lambda)$ is block-skew-symmetric, i.e. $\mathcal{L}(\lambda) = -\mathcal{L}(\lambda)^{\mathcal{B}}$, we have

$$\mathcal{L}(\lambda) = [\,v \otimes I_n \ B\,]F_\Phi^A(\lambda) = [\,v \otimes \alpha_{k-1}^{-1}A_k \ B\,]\lambda + [\,v \otimes I_n \ B\,]F_\Phi^A(0) \qquad (7.14)$$

$$= -\begin{bmatrix} v^T \otimes \alpha_{k-1}^{-1}A_k \\ B^{\mathcal{B}} \end{bmatrix}\lambda - F_\Phi^A(0)^{\mathcal{B}}\begin{bmatrix} v^T \otimes I_n \\ B^{\mathcal{B}} \end{bmatrix}.$$

Considered as a $k \times k$ matrix with $n \times n$ blocks, the blocks along the main diagonal of $\mathcal{L}(\lambda)$ need certainly be zero due to the block-skew-symmetry of $\mathcal{L}(\lambda)$. Now assume that $v = [\,0 \ v_2 \ v_3 \ \cdots \ v_k\,]^T$ and consider $\mathcal{L}(\lambda)$ in the form (7.14). For $\mathcal{L}(\lambda) = \lambda X + Y$ with $X = [\,v \otimes \alpha_{k-1}^{-1}A_k \ B\,]$ and $Y = [\,v \otimes I_n \ B\,]F_\Phi^A(0)$, both X and Y have to be block-skew-symmetric. The block-skew-symmetry of X a priori implies B to have the form

$$B = \begin{bmatrix} Z \\ B^\star \end{bmatrix} = \begin{bmatrix} Z_2 & \cdots & Z_k \\ & B^\star & \end{bmatrix} := \begin{bmatrix} [\,v_2 \ v_3 \ \cdots \ v_k\,] \otimes (-\alpha_{k-1}^{-1}A_k) \\ B^\star \end{bmatrix}$$

with $Z \in M_{n \times (k-1)n}(\mathbb{R})$ and a block-skew-symmetric $(k-1)n \times (k-1)n$ matrix $B^\star$. With $B^\star = [B_{i,j}^\star]_{i,j=1}^{k-1}$ and $B_{i,j}^\star \in M_{n \times n}(\mathbb{R})$, the block-skew-symmetry yields $B_{j,j}^\star = 0_{n \times n}$ for all $j = 1, \ldots, k-1$. Moreover, all $jn \times jn$ leading principal submatrices of X and Y (which are denoted $[X]_{jn}$ and $[Y]_{jn}$ in the following) have to be block-skew-symmetric themselves $(j = 1, \ldots, k)$. As $v_1 = 0$ we obviously have $[X]_n = 0_{n \times n}$ according to (7.14). For $[Y]_n$ we obtain $[Y]_n = -\alpha_{k-2}Z_2 = 0_{n \times n}$ so, due to the form of Z_2, this implies $v_2 = 0$. In turn, $v_1 = v_2 = 0$ implies $[X]_{2n} = 0$ and we get for $[Y]_{2n}$ a matrix of the form

$$[Y]_{2n} = \begin{bmatrix} 0_{n \times n} & -\alpha_{k-3}Z_3 \\ 0_{n \times n} & -\alpha_{k-3}B_{12}^\star \end{bmatrix} \in M_{2n \times 2n}(\mathbb{R}).$$

As $[Y]_{2n}$ needs to be block-skew-symmetric, this implies $v_3 = 0$ and $B_{12}^\star = 0$. From this, we obtain $[X]_{3n} = 0$. The proof now continues in a similar manner by induction: assuming $[X]_{jn} = 0$ for some $j \geq 3$ (which implies $v_1 = \cdots = v_j = 0$) we obtain for $[Y]_{jn}$ a matrix of the form

$$[Y]_{jn} = \begin{bmatrix} 0_{n \times n} & \cdots & \cdots & 0_{n \times n} & -\alpha_{k-j}Z_{j+1} \\ \vdots & & & \vdots & -\alpha_{k-j}B_{1,j}^\star \\ \vdots & & & \vdots & \cdots \\ 0_{n \times n} & \cdots & \cdots & 0_{n \times n} & -\alpha_{k-j}B_{j-1,j}^\star \end{bmatrix} \in M_{jn \times jn}(\mathbb{R})$$

which has to be block-skew-symmetric. Therefore, we must have $Z_{j+1} = 0$ (which is the case iff $v_{j+1} = 0$) and $B_{1,j}^\star = \cdots = B_{j-1,j}^\star = 0$. With $[X]_{jn} = 0$ we thus obtain $[X]_{(j+1)n} = 0$ (as we also have $B_{j,j}^\star = 0$). In the end, this inductive procedure yields $[X]_{kn} = 0$ and $[Y]_{kn} = 0$, thus $\mathcal{L}(\lambda) = 0$. This completes the proof. $\square$

Using Proposition 7.4 we are able to formulate a simple proof of the following theorem.

Theorem 7.9 (Block-Symmetry of $\mathbb{DM}(A)$, [45, Thm. 6]). *Let $A(\lambda) \in M_{n \times n}(\mathbb{R}[\lambda])$ be of degree $k \geq 2$. Then any matrix pencil $\mathcal{L}(\lambda) \in \mathbb{DM}(A)$ is block-symmetric.*

Proof. Let $\mathcal{L}(\lambda) \in \mathbb{DM}(A)$. Then $\mathcal{L}(\lambda)$ can be expressed as

$$\mathcal{L}(\lambda) = \begin{bmatrix} v \otimes I_n & B_1 \end{bmatrix} F_\Phi^A(\lambda) = (F_\Phi^A(\lambda))^{\mathcal{B}} \begin{bmatrix} w^T \otimes I_n \\ B_2^{\mathcal{B}} \end{bmatrix}$$

as an element of $\mathbb{M}_1(A)$ and $\mathbb{M}_2(A)$ respectively (with $B_1, B_2 \in M_{kn \times (k-1)n}(\mathbb{R})$ and $v, w \in \mathbb{R}^k$). Regarding $\mathcal{L}(\lambda)$ in the form $\mathcal{L}(\lambda) = X\lambda + Y$ this shows that we have $[X]_n = v_1 \alpha_{k-1}^{-1} A_k = w_1 \alpha_{k-1}^{-1} A_k$ for the $n \times n$ leading principal submatrix $[X]_n$ of X. Thus it follows that $v_1 = w_1$. Now note that $\mathcal{L}(\lambda)$ (seen as an element of $\mathbb{M}_2(A)$) via block-transposition becomes an element of $\mathbb{M}_1(A)$. Therefore, since $\mathbb{M}_1(A)$ is a vector space,

$$\tilde{\mathcal{L}}(\lambda) := \mathcal{L}(\lambda) - \mathcal{L}(\lambda)^{\mathcal{B}} = \begin{bmatrix} (v - w) \otimes I_n & (B_1 - B_2) \end{bmatrix} F_\Phi^A(\lambda)$$

$$=: \begin{bmatrix} \tilde{v} \otimes I_n & \tilde{B} \end{bmatrix} F_\Phi^A(\lambda)$$

is a block-skew-symmetric pencil in $\mathbb{M}_1(A)$. Since $\tilde{v} = [\,0 \ \tilde{v}_2 \ \tilde{v}_3 \ \cdots \ \tilde{v}_k\,]^T$, applying Proposition 7.4 to $\tilde{\mathcal{L}}(\lambda)$ we obtain $\tilde{\mathcal{L}}(\lambda) = 0$ and so $\mathcal{L}(\lambda) = \mathcal{L}(\lambda)^{\mathcal{B}}$. $\qquad \square$

Following the previous proof we obtain the next result on the explicit form of pencils in the double generalized ansatz space.

Corollary 7.7 ([45, Cor. 6]). *Let $A(\lambda) \in \mathbb{M}_{n \times n}(\mathbb{R}[\lambda])$ be of degree $k \geq 2$ and let*

$$\mathcal{L}(\lambda) = \Big[\, v \otimes I_n \ B_1 \,\Big] F_\Phi^A(\lambda) = F_\Phi^A(\lambda)^{\mathcal{B}} \begin{bmatrix} w^T \otimes I_n \\ B_2 \end{bmatrix} \in \mathbb{DM}(A).$$

Then $v = w$ and $B_1 = B_2^{\mathcal{B}}$.

Do not overlook that pencils in $\mathbb{DM}(A)$ not only have to have equal left and right ansatz vectors. Corollary 7.7 makes a stronger statement. In fact, the matrices B_1 and B_2 are additionally related to each other as in a way that $B_1 = B_2^{\mathcal{B}}$. Moreover, recalling that any block-symmetric matrix pencil $\mathcal{L}(\lambda) = \mathcal{L}(\lambda)^{\mathcal{B}}$ from $\mathbb{M}_1(A)$ is in $\mathbb{DM}(A)$ we obtain $\mathbb{DM}(A) = \{\mathcal{L}(\lambda) \in \mathbb{M}_1(A) \mid \mathcal{L}(\lambda) = \mathcal{L}(\lambda)^{\mathcal{B}}\}$.

Clearly, also all pencils in $\mathbb{DL}(A)$ for matrix polynomials $A(\lambda)$ in the monomial basis are block-symmetric. This has first been proven in [65, 80]. In particular, we may restrict the subsequent study of $\mathbb{DM}(A)$ to regular matrix polynomials due to the nonexistence of linearizations in $\mathbb{DM}(A)$ in case $A(\lambda)$ is singular. This is the content of the following theorem whose proof follows exactly the same reasoning as that of [31, Thm. 6.1].

Theorem 7.10 ([45, Thm. 7]). *Let $A(\lambda) \in \mathbb{M}_{n \times n}(\mathbb{R}[\lambda])$ be singular and of degree $k \geq 2$. Then none of the pencils in $\mathbb{DM}(A)$ is a linearization for $A(\lambda)$.*

Proof. Let $\mathcal{L}(\lambda) \in \mathbb{DM}(A)$ be a linearization for $A(\lambda)$ and first assume that $\operatorname{rank}([\,v \otimes I_n \ B\,]) = kn$. According to Theorem 7.4 (i), seeing $\mathcal{L}(\lambda)$ as an element of $\mathbb{M}_1(A)$, the right minimal indices of $\mathcal{L}(\lambda)$ are

$$(k - 1) + \epsilon_1 \leq (k - 1) + \epsilon_2 \leq \cdots \leq (k - 1) + \epsilon_p$$

if the right minimal indices of $A(\lambda)$ are $\epsilon_1 \leq \epsilon_2 \leq \cdots \leq \epsilon_p$. This leads to a contradiction interpreting $\mathcal{L}(\lambda)$ as an element of $\mathbb{M}_2(A)$ (recall the discussion subsequent to Theorem 7.5). Thus $\operatorname{rank}([\,v \otimes I_n \ B\,]) = s < kn$ has to hold. But then there are vectors $y_1, \ldots y_p \in \mathbb{R}^{kn}$ $(p = kn - s)$ with $y_i^T[\,v \otimes I_n \ B\,] = 0$ and therefore $y_i \in \mathcal{N}_\ell(\mathcal{L}(\lambda))$ for all $i = 1, \ldots, p$. Thus $\mathcal{L}(\lambda)$ has at least p left minimal indices equal to zero. This again contradicts the discussion following Theorem 7.5 for $\mathcal{L}(\lambda)$ seen as an element of $\mathbb{M}_2(A)$. Therefore, $\mathcal{L}(\lambda)$ will never be a strong linearization for $A(\lambda)$. $\qquad \square$

In [31] it is shown that if $A(\lambda)$ is a singular matrix polynomial of degree $k \geq 2$, then none of the pencils in $\mathbb{DL}(A)$ is a linearization of $A(\lambda)$.

7.4.1   Construction of block-symmetric pencils

This section is dedicated to the construction of elements $\mathcal{L}(\lambda) \in \mathbb{DM}(A)$ for regular $A(\lambda) \in M_{n \times n}(\mathbb{R}[\lambda])$. According to Theorem 7.9 each such pencil is block-symmetric. In fact, it turns out that the characterizations (7.7) of $\mathbb{M}_1(A)$ and (7.9) of $\mathbb{M}_2(A)$ together constitute a simple procedure to construct those pencils.

As before, assume $A(\lambda)$ to have the form (7.2) with degree $k \geq 2$. Moreover, let $\mathcal{L}(\lambda) = [\, v \otimes I_n \ B\,]F_{\Phi}^{A}(\lambda) \in \mathbb{M}_1(A)$ for some ansatz vector $v \in \mathbb{R}^k$ and some matrix $B \in M_{kn \times (k-1)n}(\mathbb{R})$. In particular, $\mathcal{L}(\lambda)$ can be expressed as $\mathcal{L}(\lambda) = [\, v \otimes \alpha_{k-1}^{-1}A_k \ B\,]\lambda + [\, v \otimes I_n \ B\,]F_{\Phi}^{A}(0)$ as was frequently used in the previous section. Similar to the block-skew-symmetric case (recall the proof of Proposition 7.4), $[\, v \otimes \alpha_{k-1}^{-1}A_k \ B\,]$ being block-symmetric implies

$$B = \begin{bmatrix} Z \\ B^{\star} \end{bmatrix} = \begin{bmatrix} Z_2 & \cdots & Z_k \\ & B^{\star} & \end{bmatrix} := \begin{bmatrix} [v_2 \ v_3 \ \cdots \ v_k] \otimes (\alpha_{k-1}^{-1}A_k) \\ B^{\star} \end{bmatrix} \tag{7.15}$$

with some $(k-1)n \times (k-1)n$ block-symmetric matrix $B^{\star}$. Here, our goal is to determine the blocks of $B^{\star}$ in a simple way with low computational costs. Therefore, considering $B^{\star}$ as a $(k-1) \times (k-1)$ block matrix with $n \times n$ blocks, it suffices to compute only the blocks of the lower triangular part of $B^{\star}$ and to only consider $[\, v \otimes I_n \ B\,]F_{\Phi}^{A}(0) = \mathcal{L}(0)$ for the remaining derivations (since $B^{\star}$ being block-symmetric implies $[\, v \otimes \alpha_{k-1}^{-1}A_k \ B\,]$ to be block-symmetric if B is constructed as in (7.15)). Notice that, in terms of the matrix $B = [B_{ij}]_{ij}$ with $i = 1, \ldots, k$ and $j = 1, \ldots, k-1$, it holds that $B_{s,t} = B_{t+1,s-1}$ for $t < s$ and $s \geq 2$. The block-symmetry of $\mathcal{L}(0)$ certainly requires

$$(e_i^T \otimes I_n)\mathcal{L}(0)(e_1 \otimes I_n) = (e_1^T \otimes I_n)\mathcal{L}(0)(e_i \otimes I_n) \tag{7.16}$$

for all $i = 1, \ldots, k$. Since the first block column $v \otimes I_n$ and the first block row Z of $[\, v \otimes I_n \ B\,]$ are already known, equation (7.16) reads for $2 \leq i \leq k$

$$v_i\left(-\frac{\beta_{k-1}}{\alpha_{k-1}}A_k + A_{k-1}\right) - \alpha_{k-2}B_{i,1}$$

$$= v_1 A_{k-i} - \frac{\left(v_{i-1}\gamma_{k-i+1} + v_i\beta_{k-i} + v_{i+1}\alpha_{k-i-1}\right)}{\alpha_{k-1}}A_k$$

whereby we set $v_{k+1} = \alpha_{-1} = 0$ for $i = k$.[4] This equation can directly be verified by computing the left and right hand side of (7.16). Moreover, this formula can easily be solved for the matrix $B_{i,1}$ and yields

$$B_{i,1} = \frac{\left(v_{i-1}\gamma_{k-i+1} + v_i(\beta_{k-i} - \beta_{k-1}) + v_{i+1}\alpha_{k-i-1}\right)}{\alpha_{k-1}\alpha_{k-2}}A_k$$

$$+ \frac{(v_i A_{k-1} - v_1 A_{k-i})}{\alpha_{k-2}}. \tag{7.17}$$

---

[4]Terms involving α_{-1} will show up for $i = k$ in subsequent formulas, too. We will always assume $\alpha_{-1} = 0$.

Equation (7.17) shows that $B_{i,1}$ can in fact be expressed and computed as a particular linear combination of A_k, A_{k-1} and A_{k-i}. In this way, the blocks $B_{i,1}$ can be determined for all $i = 2, \ldots, k$ one by one. Recall that the $n \times n$ blocks $B_{2,1}, B_{3,1}, \ldots, B_{k,1}$ constitute the first block column of $B^\star$. Thus, since $B^\star = (B^\star)^{\mathcal{B}}$ has to hold, this completely and uniquely determines the first block column and the first block row of $B^\star$. In the same way, considering $(e_i^T \otimes I_n)\mathcal{L}(0)(e_2 \otimes I_n) = (e_2^T \otimes I_n)\mathcal{L}(0)(e_i \otimes I_n)$ for $i = 3, \ldots, k$ gives the equation

$$v_i\left(A_{k-2} - \frac{\gamma_{k-1}}{\alpha_{k-1}}A_k\right) - \beta_{k-2}B_{i,1} - \alpha_{k-3}B_{i,2}$$
$$= v_2 A_{k-i} - (\gamma_{k-i+1}B_{2,i-2} + \beta_{k-i}B_{2,i-1} + \alpha_{k-i-1}B_{2,i}). \tag{7.18}$$

Notice that we have computed the first block column and block row of $B^\star$ so far but this equation involves $n \times n$ blocks from other positions in B. However, it follows from the block-symmetry of $B^\star$ that $B_{2,i-2} = B_{i-1,1}$, $B_{2,i-1} = B_{i,1}$ and $B_{2,i} = B_{i+1,1}$. Thus, in fact everything is known in (7.18) but $B_{i,2}$ and we obtain an explicit expression for $B_{i,2}$:

$$B_{i,2} = \frac{(\gamma_{k-i+1}B_{i-1,1} + (\beta_{k-i} - \beta_{k-2})B_{i,1} + \alpha_{k-i-1}B_{i+1,1})}{\alpha_{k-3}}$$
$$+ \frac{(v_i A_{k-2} - v_2 A_{k-i})}{\alpha_{k-3}} - v_i\frac{\gamma_{k-1}}{\alpha_{k-3}\alpha_{k-1}}A_k. \tag{7.19}$$

Recall that, due to the block-symmetry of $B^\star$, it suffices to consider (7.19) only for $i \geq 3$. Therefore, the blocks $B_{3,2}, \ldots, B_{k,2}$ may be computed via (7.19). Following the same pattern, the equation $(e_i^T \otimes I_n)\mathcal{L}(0)(e_j \otimes I_n) = (e_j^T \otimes I_n)\mathcal{L}(0)(e_i \otimes I_n)$ yields in its most general form for $j \geq 3$ and $i \geq j$ an explicit expression for $B_{i,j}$:

$$B_{i,j} = \frac{\gamma_{k-i+1}B_{i-1,j-1} + (\beta_{k-i} - \beta_{k-j})B_{i,j-1} + \alpha_{k-i-1}B_{i+1,j-1} - \gamma_{k-j+1}B_{i,j-2}}{\alpha_{k-j-1}}$$
$$+ \frac{(v_i A_{k-j} - v_j A_{k-i})}{\alpha_{k-j-1}}. \tag{7.20}$$

Thus, the remaining blocks of B can now be computed by (7.20). That is, for $j = 3$ we compute $B_{i,3}$ for $i = 4, \ldots, k$; for $j = 4$ we determine $B_{i,4}$ for $i = 5, \ldots, k$ etc. In conclusion, we may interpret the blockwise computation of B (in particular $B^\star$) according to (7.20) as some kind of updated recurrence relation. Moreover, the derivation shows that (7.17), (7.19) and (7.20) are sufficient and necessary for $\mathcal{L}(\lambda) \in \mathbb{DM}(A)$ being block-symmetric and having ansatz vector $v \in \mathbb{R}^k$.

Notice that the algorithmic approach for constructing block-symmetric pencils in $\mathbb{DM}(A)$ presented in this section does not require a single matrix-matrix-multiplication. Instead only scalar-matrix-multiplications are needed. The complexity of this procedure is thus easily determined as $\mathcal{O}(k^2 n^2)$. This equals the complexity of the construction algorithm presented in [101, Sec. 7]. Although there are structural similarities between both algorithms, they rise from quite different viewpoints. To the authors knowledge, there are no other algorithms known for this problem.

Before we consider a particular example, we summarize the procedure for the computation of block-symmetric pencils in $\mathbb{M}_1(A)$: For any matrix polynomial $A(\lambda) \in M_{n \times n}(\mathbb{R}[\lambda])$ expressed in some orthogonal basis Φ as in (7.2) and of degree $k \geq 2$ choose any ansatz vector $v \in \mathbb{R}^k$ and compute

$$B = \begin{bmatrix} Z \\ B^\star \end{bmatrix} \in M_{kn \times (k-1)n}(\mathbb{R}) \qquad B = [B_{i,j}], B_{i,j} \in M_{n \times n}(\mathbb{R})$$

according to (7.17), (7.19) and (7.20). Set $Z = \begin{bmatrix} v_2 & v_3 & \cdots & v_k \end{bmatrix} \otimes \alpha_{k-1}^{-1} A_k$. Then $\mathcal{L}(\lambda) = \begin{bmatrix} v \otimes I_n & B \end{bmatrix} F_\Phi^A(\lambda) \in \mathbb{DM}(A)$ is block-symmetric with ansatz vector v. The following example shows the results of this algorithm for the Chebyshev polynomial basis.

Example 7.2 ([45, Ex. 1]). *The Chebyshev polynomials of first kind follow the recurrence relation*

$$\phi_{j+1}(\lambda) = 2\lambda\phi_j(\lambda) - \phi_{j-1}(\lambda) \qquad j \geq 1$$

with $\phi_1(\lambda) = \lambda$ and $\phi_0(\lambda) = 1$. Now let a matrix polynomial $A(\lambda) = \sum_{j=0}^{3} A_k \phi_j(\lambda)$ of degree $\deg(A(\lambda)) = 3$ be given in the Chebyshev basis. Using the algorithm for the construction of block-symmetric pencils in $\mathbb{DM}(A)$ we may easily compute the block-symmetric pencils that correspond to the standard unit vectors $v = e_1, e_2, e_3 \in \mathbb{R}^3$. Considering the ansatz vector $e_1 \in \mathbb{R}^3$ we compute

$$\begin{bmatrix} e_1 \otimes I_n & B_1 \end{bmatrix} = \begin{bmatrix} I_n & 0_n & 0_n \\ 0_n & 2(A_3 - A_1) & -2A_0 \\ 0_n & -2A_0 & A_3 - A_1 \end{bmatrix}$$

which gives the matrix pencil $\mathcal{L}(\lambda) = \begin{bmatrix} e_1 \otimes I_n & B_1 \end{bmatrix} F_\Phi^A(\lambda)$ as

$$\mathcal{L}(\lambda) = \begin{bmatrix} 2A_3 & 0 & 0 \\ 0 & 2(A_1 - A_3) & -2A_0 \\ 0 & -2A_0 & A_1 - A_3 \end{bmatrix} \lambda + \begin{bmatrix} A_2 & A_1 - A_3 & A_0 \\ A_1 - A_3 & 2A_0 & A_1 - A_3 \\ A_0 & A_1 - A_3 & A_0 \end{bmatrix}.$$

The computation of $\begin{bmatrix} e_2 \otimes I_n & B_2 \end{bmatrix}$ yields $\begin{bmatrix} e_2 \otimes I_n & B_2 \end{bmatrix} = \begin{bmatrix} 0_n & 2A_3 & 0_n \\ I_n & 2A_2 & 2A_3 \\ 0_n & 2A_3 & A_2 - A_0 \end{bmatrix}$ so the corresponding matrix pencil $\mathcal{K}(\lambda) = \begin{bmatrix} e_2 \otimes I_n & B_2 \end{bmatrix} F_\Phi^A(\lambda)$ is easily determined as

$$\mathcal{K}(\lambda) = \begin{bmatrix} 0 & 2A_3 & 0 \\ 2A_3 & 2A_2 & 2A_3 \\ 0 & 2A_3 & A_2 - A_0 \end{bmatrix} \lambda + \begin{bmatrix} -A_3 & 0 & -A_3 \\ 0 & A_1 - 3A_3 & A_0 - A_2 \\ -A_3 & A_0 - A_2 & -A_3 \end{bmatrix}.$$

Finally, for $\begin{bmatrix} e_3 \otimes I_n & B_3 \end{bmatrix}$ we obtain $\begin{bmatrix} e_3 \otimes I_n & B_3 \end{bmatrix} = \begin{bmatrix} 0_n & 0_n & 2A_3 \\ 0_n & 4A_3 & 2A_2 \\ I_n & 2A_2 & A_3 + A_1 \end{bmatrix}$ so the corresponding matrix pencil $\mathcal{G}(\lambda) = \begin{bmatrix} e_3 \otimes I_n & B_3 \end{bmatrix} F_\Phi^A(\lambda)$ is given by

$$\mathcal{G}(\lambda) = \begin{bmatrix} 0 & 0 & 2A_3 \\ 0 & 4A_3 & 2A_2 \\ 2A_3 & 2A_2 & A_1 + A_3 \end{bmatrix} \lambda + \begin{bmatrix} 0 & -2A_3 & 0 \\ -2A_3 & -2A_2 & -2A_3 \\ 0 & -2A_3 & A_0 - A_2 \end{bmatrix}.$$

Notice that such pencils need not be (strong) linearizations for $A(\lambda)$. For instance, $\mathcal{L}(\lambda) = [\, e_3 \otimes I_n \; B_3 \,] F_\Phi^A(\lambda)$ can only be a (strong) linearization for $A(\lambda)$ iff A_3 is nonsingular (due to the anti-lower-block-triangular form of $[\, e_3 \otimes I_n \; B_3 \,]$).

Fortunately, now we obtain the following corollary without real effort.

Corollary 7.8 (Dimension of $\mathbb{DM}(A)$, [45, Cor. 7]). *For any matrix polynomial $A(\lambda) \in \mathrm{M}_{n \times n}(\mathbb{R}[\lambda])$ of degree $k \geq 2$ we have* $\dim(\mathbb{DM}(A)) = k$.

Proof. First observe that $\dim(\mathbb{DM}(A)) \geq k$ certainly holds because $\mathcal{B}_1(\lambda)$, ..., $\mathcal{B}_k(\lambda) \in \mathbb{DM}(A)$ with $\mathcal{B}_j(\lambda) = [\, e_j \otimes I_n \; B_j \,] F_\Phi^A(\lambda)$ are obviously linear independent. Now observe that any $\mathcal{L}(\lambda) = \sum_{i=1}^k \alpha_i \mathcal{B}_i(\lambda)$ for arbitrary coefficients $\alpha_i \in \mathbb{R}$ is block-symmetric with ansatz vector $v = \sum_{i=1}^k \alpha_i e_i$. Therefore, whenever any $\mathcal{L}^\star(\lambda) \in \mathbb{DM}(A)$ has the same ansatz vector $v \in \mathbb{R}^k$, $\mathcal{L}^\star(\lambda)$ necessarily coincides with $\mathcal{L}(\lambda)$ due to the uniqueness of the expressions (7.17) - (7.20). Thus $\dim(\mathbb{DM}(A)) \leq k$ and Corollary 7.8 follows. $\qquad\square$

7.5 The eigenvector exclusion theorem

In this section we present a new linearization condition for pencils in $\mathbb{M}_1(A)$ and $\mathbb{M}_2(A)$ that we call *Eigenvector Exclusion Theorem*. It states the linearization property of any $\mathcal{L}(\lambda) \in \mathbb{M}_1(A)$ for any matrix polynomial $A(\lambda) \in \mathrm{M}_{n \times n}(\mathbb{R}[\lambda])$ of degree $k \geq 2$ in terms of a condition relating the left eigenvectors of $\mathcal{L}(\lambda)$ to its ansatz vector $v \in \mathbb{R}^k$. In particular, due to the form of $\mathcal{L}(\lambda) = [\, v \otimes I_n \; B \,] F_\Phi^A(\lambda) \in \mathbb{M}_1(A)$, a statement on the nonsingularity of $[\, v \otimes I_n \; B \,]$ can be derived from the behavior of the left eigenvectors of $\mathcal{L}(\lambda)$ multiplied solely by $v \otimes I_n$. Notice that the following theorem generalizes and extends [31, Lem. 5.5, Lem. 5.6]. Moreover, it will be central for the proofs in Section 7.5.1 on linearizations in $\mathbb{DM}(A)$. In addition, do not overlook the similarity to Theorem 7.6.

Theorem 7.11 (Eigenvector Exclusion Theorem, [45, Thm. 8]). *Let a regular matrix polynomial $A(\lambda) \in \mathrm{M}_{n \times n}(\mathbb{R}[\lambda])$ of degree $k \geq 2$ be given. Moreover, assume $\mathcal{L}(\lambda) \in \mathbb{M}_1(A)$ with ansatz vector $v \in \mathbb{R}^k$. Then $\mathcal{L}(\lambda)$ is a strong linearization for $A(\lambda)$ iff*

$$u^T \!\left(v \otimes I_n \right) \neq 0 \tag{7.21}$$

holds for any left eigenvector $u \neq 0$ of $\mathcal{L}(\lambda)$ for every eigenvalue α of $A(\lambda)$.

Proof. We confine ourselves to a sketch of the proof since it is similar to that of Theorem 7.6. Assume that $\mathcal{L}(\lambda) \in \mathbb{M}_1(A)$ as given in (7.7) is a strong linearization for $A(\lambda) \in \mathrm{M}_{n \times n}(\mathbb{R}[\lambda])$ of degree $k \geq 2$. Moreover, let $0 \neq u \in \mathrm{Eig}_\ell(\mathcal{L}, \alpha)$ for some $\alpha \in \sigma_f(A) = \sigma_f(\mathcal{L})$. Thus, in particular it holds that $u^T \mathcal{L}(\alpha) = 0$. Then, defining $w \in \mathbb{C}^{kn}$ as

$$w^T = [\, w_1 \; w_2 \; \cdots \; w_{kn} \,] := u^T [\, v \otimes I_n \; B \,]$$

and assuming that $u^T(v \otimes I_n) = 0$, we obviously obtain $[\, w_1 \cdots w_n \,] = 0$. Moreover, since $u \in \text{Eig}_\ell(\mathcal{L}, \alpha)$, we have $u^T \mathcal{L}(\alpha) = u^T[\, v \otimes I_n \ B\,] F^A_\Phi(\alpha) = w^T F^A_\Phi(\alpha) = 0$. Now a similar argumentation as in the proof of Theorem 7.6 gives that $w \equiv 0$. So, in particular $u^T[\, v \otimes I_n \ B\,] = 0$ and $[\, v \otimes I_n \ B\,]$ is singular. However, this is a contradiction since we assumed $\mathcal{L}(\lambda)$ to be a strong linearization for $A(\lambda)$ (i.e. $[\, v \otimes I_n \ B\,]$ has full rank, see Theorem 7.3). On the other hand, whenever $[\, v \otimes I_n \ B\,]$ is singular, there is a vector $0 \neq q \in \mathbb{R}^{kn}$ such that $q^T[\, v \otimes I_n \ B\,] = 0$, so $q \in \text{Eig}_\ell(\mathcal{L}, \alpha)$ for *any* $\alpha \in \mathbb{C}$. Now clearly $u^T(v \otimes I_n) = 0$ with $u = q$ holds. The proof follows the same arguments when $\alpha = \infty$ using $\text{rev}_1 \mathcal{L}(0)$ instead of $\mathcal{L}(\alpha)$ noting that $\text{rev}_1 \mathcal{L}(\lambda) = [\, v \otimes I_n \ B\,]\text{rev}_1 F^A_\Phi(\lambda)$. $\qquad \square$

Now Theorem 7.11 enables us to give a statement on the recovery of left eigenvectors for regular matrix polynomials $A(\lambda) \in \text{M}_{n \times n}(\mathbb{R}[\lambda])$ of degree $k \geq 2$ from linearizations $\mathcal{L}(\lambda) \in \mathbb{M}_1(A)$. To this end, assume that $\mathcal{L}(\lambda)$ is a strong linearization for $A(\lambda)$, so (7.21) holds for any $0 \neq u \in \text{Eig}_\ell(\mathcal{L}, \alpha)$ for any eigenvalue $\alpha \in \sigma(A)$. Then from (3.1) we obtain

$$0 = u^T \mathcal{L}(\alpha)(\Phi_k(\alpha) \otimes I_n) = u^T(v \otimes I_n)A(\alpha)$$

and therefore, since $u^T(v \otimes I_n) \neq 0$, $u^T(v \otimes I_n)$ is a left eigenvector for $A(\lambda)$ with corresponding eigenvalue α. In other words, Theorem 7.11 states that for strong linearizations $\mathcal{L}(\lambda) \in \mathbb{M}_1(A)$ the mapping $u \mapsto (v^T \otimes I_n)u$ of left eigenvectors of $\mathcal{L}(\lambda)$ to left eigenvectors of $A(\lambda)$ is injective for every eigenvalue α of $A(\lambda)$.

In fact, if $u_1^T(v \otimes I_n) = u_2^T(v \otimes I_n)$ for two distinct vectors $u_1, u_2 \in \text{Eig}_\ell(\mathcal{L}, \alpha)$, then $0 \neq w := u_1 - u_2 \in \text{Eig}_\ell(\mathcal{L}, \alpha)$ and $w^T(v \otimes I_n) = 0$. According to Theorem 7.11 this contradicts the linearization property of $\mathcal{L}(\lambda)$. Moreover, the mapping is surjective since a basis of $\text{Eig}_\ell(\mathcal{L}, \alpha)$ is mapped to a basis of $\text{Eig}_\ell(A, \alpha)$ for any $\alpha \in \sigma(A) = \sigma(\mathcal{L})$. To see this, notice that if $u_1, \dots, u_s \subseteq \mathbb{R}^{kn}$ are linearly independent, then so are $u_j^T(v \otimes I_n), j = 1, \dots, s$. In particular, whenever $\mathcal{L}(\lambda) \in \mathbb{M}_1(A)$ is a strong linearization for $A(\lambda)$ we have $\dim(\text{Eig}_\ell(\mathcal{L}, \alpha)) = \dim(\text{Eig}_\ell(A, \alpha))$ and thus obtain a bijection between the left eigenvectors of $\mathcal{L}(\lambda)$ and the left eigenvectors of $A(\lambda)$ for every eigenvalue of $A(\lambda)$.

Corollary 7.9 (Left Eigenvector Recovery, [45, Cor. 8]). *Let $A(\lambda) \in \text{M}_{n \times n}(\mathbb{R}[\lambda])$ be regular of degree $k \geq 2$ and let $\mathcal{L}(\lambda) \in \mathbb{M}_1(A)$ with ansatz vector $v \in \mathbb{R}^k$ be a strong linearization for $A(\lambda)$. Then any left eigenvector $w \in \text{Eig}_\ell(A, \alpha), \alpha \in \sigma(A)$, has the form $w = u^T(v \otimes I_n)$ for some left eigenvector $u \in \text{Eig}_\ell(\mathcal{L}, \alpha)$. Moreover, if $u_1, \dots u_s \subseteq \mathbb{C}^{kn}$ is a basis of $\text{Eig}_\ell(\mathcal{L}, \alpha)$ then $w_1 = (v^T \otimes I_n)u_1, \dots, w_s = (v^T \otimes I_n)u_s \subseteq \mathbb{C}^n$ is a basis of $\text{Eig}_\ell(A, \alpha)$.*

Certainly, a statement similar to Theorem 7.11 holds for pencils from $\mathbb{M}_2(A)$. In particular, whenever $A(\lambda) \in \text{M}_{n \times n}(\mathbb{R}[\lambda])$ is regular of degree $k \geq 2$, an analogous proof to that of Theorem 7.11 shows that a pencil $\mathcal{L}(\lambda) \in \mathbb{M}_2(A)$ with ansatz vector $v \in \mathbb{R}^k$ is a strong linearization for $A(\lambda)$ iff $(v^T \otimes I_n)u \neq 0$ holds for any right eigenvector $u \in \text{Eig}_r(\mathcal{L}, \alpha)$ for every eigenvalue α of $A(\lambda)$. Obviously this is the same

condition as (7.21) using right instead of left eigenvectors. Of course Corollary 7.9
holds in a similar way for $\mathbb{M}_2(A)$, too.

We conclude this section with a simple observation:

Corollary 7.10. *Let $A(\lambda) \in M_{n \times n}(\mathbb{R}[\lambda])$ be regular and of degree $k \geq 2$. Moreover,
assume $\mathcal{L}(\lambda) \in \mathbb{DM}(A)$ with ansatz vector $v \in \mathbb{R}^k$. Then the following statements
hold:*

   (*i*) *The matrix pencil $\mathcal{L}(\lambda)$ is a strong linearization for $A(\lambda)$ if $u^T(v \otimes I_n) \neq 0$
holds for either every right or every left eigenvector of $\mathcal{L}(\lambda)$ corresponding to
any eigenvalue $\alpha \in \sigma(A)$.*

  (*ii*) *If $u^T(v \otimes I_n) \neq 0$ holds for every right eigenvector of $\mathcal{L}(\lambda)$ corresponding to any
eigenvalue $\alpha \in \sigma(A)$ the same follows for every left eigenvector of $\mathcal{L}(\lambda)$ and
vice versa.*

7.5.1   The eigenvalue exclusion theorem

Let $A(\lambda) \in M_{n \times n}(\mathbb{R}[\lambda])$ be regular of degree $k \geq 2$. According to Corollary 7.8
we have $\dim(\mathbb{DM}(A)) = k$. In particular, any $\mathcal{L}(\lambda) \in \mathbb{DM}(A)$ with ansatz vector
$v \in \mathbb{R}^k$ is uniquely determined by v and the formulas in Section 7.4.1 apply for the
construction of $B \in M_{kn \times (k-1)n}(\mathbb{R})$ so that $\mathcal{L}(\lambda) = [\, v \otimes I_n \; B \,] F_\Phi^A(\lambda)$. Therefore, due
to (7.17) - (7.20), it is impossible to "design" B so that $\mathcal{L}(\lambda) \in \mathbb{DM}(A)$ becomes a
strong linearization once v has been fixed. With this in mind, it comes as no surprise
that there has to be a relationship between $A(\lambda)$ and $v \in \mathbb{R}^k$ that determines whether
$\mathcal{L}(\lambda) = [\, v \otimes I_n \; B \,] F_\Phi^A(\lambda) \in \mathbb{DM}(A)$ is a strong linearization for $A(\lambda)$ or not. This
relation was first discovered and investigated for $\mathbb{DL}(A)$ in [89, Sec. 6]. In particular,
for $\mathbb{DL}(A)$ the following theorem holds.

Theorem 7.12 (Eigenvalue Exclusion Theorem, [89, Thm. 6.7])**.** *Let a regular matrix
polynomial $A(\lambda) \in M_{n \times n}(\mathbb{R}[\lambda])$ of degree $k \geq 2$ be given and assume that $\mathcal{L}(\lambda) \in
\mathbb{DL}(A)$ with nonzero ansatz vector $v \in \mathbb{R}^k$. Then $\mathcal{L}(\lambda)$ is a strong linearization for
$A(\lambda)$ iff no root of the polynomial*

$$\Lambda_{k-1}(\lambda)v = \sum_{i=1}^{k} v_i \lambda^{k-i} = v_1 \lambda^{k-1} + v_2 \lambda^{k-2} + \cdots + v_{k-1}\lambda + v_k \qquad (7.22)$$

*coincides with an eigenvalue of $A(\lambda)$. Hereby, in case $\infty \in \sigma(A)$, the polynomial
$\Lambda_{k-1}(\lambda)v$ is declared to have the root ∞ if $v_1 = 0$.*

Theorem 7.12 became famous as the *Eigenvalue Exclusion Theorem* since it does
not allow the roots of $\Lambda_{k-1}(\lambda)v$ to be eigenvalues of $A(\lambda)$ if $\mathcal{L}(\lambda) \in \mathbb{DL}(A)$ with ansatz
vector $v \in \mathbb{R}^k$ is designated to be a strong linearization for $A(\lambda)$. As it was proven in
[101, Sec. 4], Theorem 7.12 remains valid in a similar fashion for nonstandard ansatz
spaces as in (2.15) corresponding to arbitrary polynomial bases $\tau_0(\lambda), \ldots, \tau_{k-1}(\lambda)$ of

$\mathbb{R}_{k-1}[\lambda]$. In this context, $\Lambda_{k-1}(\lambda)$ in (7.22) simply has to be replaced by $\mathcal{T}_{k-1}(\lambda)$, cf. page 18. In particular, Theorem 7.12 applies in its adapted form to orthogonal bases. The proofs of Theorem 7.12 in [89, Sec. 6] and its extended version [101, Thm. 4.1] are quite involved and rather long and technical. In fact, in [89] the authors gave an explicit formula for the determinant of pencils $\mathcal{L}(\lambda) \in \mathbb{DL}(A)$ whereas in [101] the authors deal with bivariate polynomials and generalized Bezoutian matrices, cf. [77]. In this section we show that is is possible to formulate a proof of Theorem 7.12 adapted to $\mathbb{DM}(A)$ based on the characterizations from Section 7.1 and, in particular, the results from Section 7.5, in a comparatively simple fashion. With these tools at hand, the necessity part of the *Generalized Eigenvalue Exclusion Theorem* is easily proven.

Theorem 7.13 ([45, Thm. 9]). *Let $A(\lambda) \in \mathrm{M}_{n\times n}(\mathbb{R}[\lambda])$ be regular of degree $k \geq 2$ and assume $\mathcal{L}(\lambda) \in \mathbb{DM}(A)$ with ansatz vector $0 \neq v \in \mathbb{R}^k$. Then, if $\mathcal{L}(\lambda)$ is a strong linearization for $A(\lambda)$, no root of the polynomial*

$$\Phi_{k-1}(\lambda)v = \sum_{j=1}^{k} \phi_{k-j}(\lambda)v_j = \phi_{k-1}(\lambda)v_1 + \phi_{k-2}(\lambda)v_2 + \cdots + \phi_0(\lambda)v_k$$

coincides with an eigenvalue of $A(\lambda)$. Moreover, if $\alpha = \infty$ is an eigenvalue of $A(\lambda)$, then $v_1 \neq 0$ holds.

Proof. Let $\mathcal{L}(\lambda) \in \mathbb{DM}(A)$ be a strong linearization for $A(\lambda) \in \mathrm{M}_{n\times n}(\mathbb{R}[\lambda])$ with ansatz vector $v \in \mathbb{R}^k$. According to Theorem 7.11 we know that (7.21) holds for any $u \in \mathrm{Eig}_\ell(\mathcal{L}, \alpha)$ for every eigenvalue $\alpha \in \sigma(A)$. Moreover, since $\mathcal{L}(\lambda) \in \mathrm{M}_2(A)$, we know from Proposition 7.3 that any $0 \neq u \in \mathrm{Eig}_\ell(\mathcal{L}, \alpha)$ has the form $u = \Phi_{k-1}(\alpha)^T \otimes w$ for some $0 \neq w \in \mathrm{Eig}_\ell(A, \alpha)$ (or $e_1 \otimes w$ in the case $\alpha = \infty$, $e_1 \in \mathbb{R}^k$). Therefore, according to Theorem 7.11, $\mathcal{L}(\lambda)$ is a linearization for $A(\lambda)$ iff

$$0 \neq \left(\Phi_{k-1}(\alpha) \otimes w^T\right)\left(v \otimes I_n\right) = \Phi_k(\alpha)v \otimes w^T$$

for all $\alpha \in \sigma(A)$ and $w \in \mathrm{Eig}_\ell(A, \alpha)$. Since $w \neq 0$ this holds iff $\Phi_{k-1}(\alpha)v \neq 0$. If $\alpha = \infty$ is an eigenvalue of $A(\lambda)$ we obtain according to Theorem 7.13 $(e_1^T \otimes w^T)(v \otimes I_n) = v_1 \otimes w^T \neq 0$, so $v_1 \neq 0$. This completes the proof. $\square$

Notice that the proof of Theorem 7.13 would have worked using $\mathrm{M}_1(A)$ and $\mathrm{M}_2(A)$ in reversed roles.

Although we were able to give a simple proof of Theorem 7.13, it seems that our characterizations of $\mathrm{M}_1(A), \mathrm{M}_2(A)$ and $\mathbb{DM}(A)$ do not offer a simple approach to prove the reverse statement, i.e. the equivalence in Theorem 7.13. This problem objectively seems to be harder. However, our characterizations admit indeed a basic algebraic proof for the opposite statement if certain generic conditions are imposed on $A(\lambda)$. Notice that the concept introduced in Definition 7.2 is not standard terminology.

Definition 7.2. *A matrix polynomial $A(\lambda) = \sum_{j=0}^{k} A_j \phi_j(\lambda) \in \mathrm{M}_{n\times n}(\mathbb{R}[\lambda])$ of degree $k \geq 2$ shall be called simple, if $A_k \in \mathrm{GL}_n(\mathbb{R})$ is nonsingular and all structural indices of $A(\lambda)$ are equal to one.*

It is readily checked that a matrix polynomial $A(\lambda) \in M_{n \times n}(\mathbb{R}[\lambda])$ is simple iff $A_k \in GL_n(\mathbb{R})$ is nonsingular and all eigenvalues $\alpha \in \sigma_f(A)$ are semisimple according to [119, Def. 1]. In particular, Definition 7.2 is the straight-forward generalization of a *simple matrix pencil* given in [76, Sec. 2.1]. Certainly, almost all $A(\lambda) \in M_{n \times n}(\mathbb{R}[\lambda])$ satisfy these conditions. The simplicity of any $A(\lambda)$ can be characterized via $F_\Phi^A(\lambda)$ as in Proposition 7.5 below.

Proposition 7.5. *Let* $A(\lambda) \in M_{n \times n}(\mathbb{R}[\lambda])$ *be a matrix polynomial as in* (7.2) *of degree* $k \geq 2$. *Moreover, let* $F_\Phi^A(\lambda)$ *be the strong linearization of* $A(\lambda)$ *introduced in* (7.6) *given as*

$$F_\Phi^A(\lambda) = \lambda \begin{bmatrix} \frac{1}{\alpha_{k-1}} A_k & 0 & \cdots & 0 \\ 0 & & & \\ \vdots & & I_{(k-1)n} & \\ 0 & & & \end{bmatrix} + F_\Phi^A(0) \tag{7.23}$$

$$=: \lambda X + Y.$$

Then $A(\lambda)$ *is simple iff* YX^{-1} *is diagonalizable (i.e. semisimple).*

Proof. $\Rightarrow$ First assume $A(\lambda) \in M_{n \times n}(\mathbb{R}[\lambda])$ of degree $k \geq 2$ is simple. Then, since A_k is nonsingular, the matrix $X \in M_{kn \times kn}(\mathbb{R})$ in (7.23) is nonsingular. Therefore, if $\alpha \in \sigma_f(A) = \sigma(A)$ and $z \in \mathrm{Eig}_\ell(F_\Phi^A, \alpha)$ we have

$$z^T F_\Phi^A(\alpha) = 0 \quad \Leftrightarrow \quad \alpha z^T X = -z^T Y \quad \Leftrightarrow \quad \alpha z^T = -z^T Y X^{-1}. \tag{7.24}$$

Therefore, $z \in \mathrm{Eig}_\ell(F_\Phi^A, \alpha)$ iff z is a left eigenvector of $-YX^{-1} \in M_{kn \times kn}(\mathbb{R})$ for α. Since the structural indices of $A(\lambda)$ for each eigenvalue of $A(\lambda)$ are all equal to one, we have geomult$(A, \alpha) =$ algmult(A, α) for all $\alpha \in \sigma(A)$ (recall Remark 2.1). Moreover, as $F_\Phi^A(\lambda)$ is a strong linearization for $A(\lambda)$, this equally holds for $F_\Phi^A(\lambda)$. Making use of Theorem 2.2 (Index Sum Theorem) we have $\sum_{\alpha \in \sigma(A)}$ algmult$(F_\Phi^A, \alpha) = kn$ so in turn we obtain $\sum_{\alpha \in \sigma(A)}$ geomult$(F_\Phi^A, \alpha) = kn$ once more due to the simplicity of $A(\lambda)$. Taking (7.24) into account we have that, whenever $\alpha \in \sigma(A)$ has algebraic multiplicity s_α then $\alpha \in \sigma(-YX^{-1})$ has algebraic multiplicity s_α. Moreover, $\mathrm{Eig}_\ell(F_\Phi^A, \alpha) = \mathrm{Eig}_\ell(-YX^{-1}, \alpha)$ for all eigenvalues α of $A(\lambda)$. Since eigenvectors of $-YX^{-1}$ for different eigenvalues are linearly independent, we conclude that $-YX^{-1}$ has kn linearly independent eigenvectors. That is, there exists a basis of $\mathbb{C}^{kn}$ consisting entirely of eigenvectors of $-YX^{-1}$. Therefore, $-YX^{-1}$ is semisimple. Then, certainly YX^{-1} is semisimple, too.

$\Leftarrow$ Assuming now that YX^{-1} is semisimple, we obtain that A_k has to be nonsingular (i.e. X^{-1} exists) and that $-YX^{-1}$ is semisimple. In particular, we have for all $\alpha \in \sigma(-YX^{-1})$ that geomult$(-YX^{-1}, \alpha) =$ algmult$(-YX^{-1}, \alpha)$ holds. Taking again the relation (7.24) into account, we find that geomult$(F_\Phi^A, \alpha) =$ algmult(F_Φ^A, α) holds for all $\alpha \in \sigma(F_\Phi^A) = \sigma(A)$. However, this can only hold (by definition) if the structural indices for each eigenvalue of $A(\lambda)$ are all equal to one. This completes the proof. $\qquad\square$

We are now ready to prove the reverse statement to Theorem 7.13 for simple matrix polynomials in a basic algebraic way. Nevertheless, notice that the statement holds in general for all regular $A(\lambda) \in M_{n \times n}(\mathbb{R}[\lambda])$ [101, Thm. 4.1].

Theorem 7.14. *Let $A(\lambda) \in M_{n \times n}(\mathbb{R}[\lambda])$ be regular and simple of degree $k \geq 2$. Moreover, assume $\mathcal{L}(\lambda) \in \mathbb{DM}(A)$ with ansatz vector $v \in \mathbb{R}^k$. Then, if no root of the polynomial*

$$\Phi_{k-1}(\lambda)v = \phi_{k-1}(\lambda)v_1 + \phi_{k-2}(\lambda)v_2 + \cdots + \phi_0(\lambda)v_k$$

coincides with an eigenvalue of $A(\lambda)$, $\mathcal{L}(\lambda)$ is a strong linearization for $A(\lambda)$.

Proof. Let $A(\lambda) \in M_{n \times n}(\mathbb{R}[\lambda])$ of degree $k \geq 2$ be simple. Moreover, assume that $\mathcal{L}(\lambda) = [\, v \otimes I_n \;\; B \,]F_\Phi^A(\lambda) \in \mathbb{DM}(A)$ is given with ansatz vector $v \in \mathbb{R}^k$ and that $\Phi_{k-1}(\alpha)v \neq 0$ holds for all eigenvalues α of $A(\lambda)$. Our goal will be to use the simplicity of $A(\lambda)$ to show that range$([\, v \otimes I_n \;\; B \,]) = \mathbb{C}^{kn}$ (with $[\, v \otimes I_n \;\; B \,]$ considered as a matrix from $M_{kn \times kn}(\mathbb{C})$) which implies the nonsingularity of $[\, v \otimes I_n \;\; B \,]$. To this end, let $\alpha \in \sigma(A)$ be arbitrary.

According to Proposition 7.5 there exists a basis of $\mathbb{C}^{kn}$ consisting entirely of eigenvectors of $F_\Phi^A(\lambda)$. First notice that $\mathcal{L}(\lambda)$ can equally be expressed in the form (7.7) and (7.9). For any basis $u_1, \ldots, u_s$ of $\text{Eig}_\ell(A, \alpha)$ the vectors $\tilde{u}_1 := \Phi_{k-1}(\alpha)^T \otimes u_1, \ldots, \tilde{u}_s := \Phi_{k-1}(\alpha)^T \otimes u_s$ form a basis of $\text{Eig}_\ell((F_\Phi^A)^\mathcal{B}, \alpha)$. In fact, if $u_1, \ldots, u_s$ are linearly independent so are $\tilde{u}_1, \ldots, \tilde{u}_s$ and we have

$$\tilde{u}_j^T F_\Phi^A(\alpha)^\mathcal{B} = 0 \quad \Rightarrow \quad \tilde{u}_j^T \mathcal{L}(\alpha) = \tilde{u}_j^T F_\Phi^A(\alpha)^\mathcal{B} \begin{bmatrix} v^T \otimes I_n \\ B^\mathcal{B} \end{bmatrix} = 0 \quad \forall j = 1, \ldots, s$$

where we used the expression of $\mathcal{L}(\lambda)$ as an element of $\mathbb{M}_2(A)$.

Now consider $\mathcal{L}(\lambda)$ as an element of $\mathbb{M}_1(A)$. According to our assumption we have $\Phi_{k-1}(\alpha)v \neq 0$, so in particular

$$\tilde{u}_j^T(v \otimes I_n) = (\Phi_{k-1}(\alpha) \otimes u_j^T)(v \otimes I_n) = \Phi_{k-1}(\alpha)v \otimes u_j^T \neq 0.$$

Thus, $\hat{u}_j^T := \tilde{u}_j^T[\, v \otimes I_n \;\; B \,] \neq 0$ for all $j = 1, \ldots, s$. Moreover, $\hat{u}_1, \ldots, \hat{u}_s$ are still linearly independent. This follows immediately since, if there would exists coefficients $\beta_1, \ldots, \beta_s$ so that $\sum_{j=1}^s \beta_j \hat{u}_j = 0$, then

$$\beta_1 \hat{u}_1^T + \cdots + \beta_s \hat{u}_s^T = \left(\beta \tilde{u}_1^T + \cdots + \beta_s \tilde{u}_s^T \right)[\, v \otimes I_n \;\; B \,] = 0$$

and $\sum_{j=1}^s \beta_j \tilde{u}_j = \Phi_{k-1}(\alpha)^T \otimes (\sum_{j=1}^s u_j) =: \Phi_{k-1}(\alpha)^T \otimes u$. Notice that u is still a nonzero left eigenvector of $A(\lambda)$ for α and that $(\Phi_{k-1}(\alpha) \otimes u^T)[\, v \otimes I_n \;\; B \,] = 0$ would imply $(\Phi_{k-1}(\alpha) \otimes u^T)(v \otimes I_n) = \Phi_{k-1}(\alpha)v \otimes u^T = 0$. Therefore, in particular, $\Phi_{k-1}(\alpha)v = 0$ which contradicts our assumption. Thus $\hat{u}_1, \ldots, \hat{u}_s$ are linearly independent.

Since $\tilde{u}_j^T \mathcal{L}(\alpha) = \hat{u}_j^T F_\Phi^A(\alpha) = 0$ holds for all $j = 1, \ldots, s$, the vectors $\hat{u}_1, \ldots, \hat{u}_s$ form a basis of $\text{Eig}_\ell(F_\Phi^A, \alpha)$ which is also s-dimensional. Since α was arbitrary, we obtain bases for all eigenspaces $\text{Eig}_\ell(F_\Phi^A, \alpha), \alpha \in \sigma(A)$, in this way. In particular, since $A(\lambda)$ was assumed to be simple, if we collect all such bases for all eigenspaces of $F_\Phi^A(\lambda)$, we obtain exactly kn vectors and have a basis of $\mathbb{C}^{kn}$. In conclusion, this means that $\{z \in \mathbb{C}^{kn} \mid z^T = w^T[\, v \otimes I_n \; B\,]$ for some $w \in \mathbb{C}^{kn}\}$ equals $\mathbb{C}^{kn}$. Thus, the matrix $[\, v \otimes I_n \; B\,]$ has full rank and is therefore nonsingular. This implies $\mathcal{L}(\lambda)$ to be a strong linearization for $A(\lambda)$ and completes the proof. $\square$

7.6   Conclusions

In this chapter we considered the ansatz spaces $\mathbb{M}_1(A)$ and $\mathbb{M}_2(A)$ of the form (2.15) for orthogonal polynomial bases $\Phi = \{\phi_j(\lambda)\}_{j=0}^\infty$. We gave a comprehensive characterization of the matrix pencils in $\mathbb{M}_1(A)$ and $\mathbb{M}_2(A)$ that substantially deviates from the one given in [89] for $\mathbb{L}_1(A)$ and $\mathbb{L}_2(A)$. Based on this new presentation, we were able to derive several (old and new) statements in a simple algebraic way and to give short proofs for various results. Among those, we stated a simplified linearization condition for pencils in $\mathbb{M}_1(A)$ and $\mathbb{M}_2(A)$ and showed how the eigenvector recovery can be done quite easily from the corresponding anchor pencil. For completeness, we stated some results on the recovery of minimal indices and minimal bases that already appeared in [31]. For singular matrix polynomials $A(\lambda)$ we derived a new condition for the strict equivalence of a pencil $\mathcal{L}(\lambda) \in \mathbb{M}_1(A)$ to $F_\Phi^A(\lambda)$ based on the left nullvectors of $\mathcal{L}(\lambda)$. Along with several examples we presented a necessary and sufficient linearization condition for pencils in $\mathbb{M}_1(A)$ whenever the matrix polynomial $A(\lambda)$ at hand has only left minimal indices equal to zero. Motivated by the study of $\mathbb{DL}(A)$ we analyzed the double generalized ansatz space $\mathbb{DM}(A)$ comprehensively and showed that this vector space contains only block-symmetric matrix pencils. Moreover, we gave an $\mathcal{O}(k^2 n^2)$ algorithm for their computation. Finally, we have proven the *Eigenvector Exclusion Theorem* that reveals a connection between the linearization property of pencils in $\mathbb{M}_1(A)$, their left eigenvectors and their ansatz vectors. With this result, we were able to restate the *Strong Linearization Theorem* for generalized ansatz spaces from [101]. For the sufficiency result we confined ourselves to simple matrix polynomials for which we showed a particular diagonalization result related to the anchor pencil $F_\Phi^A(\lambda)$.

It should be mentioned that the basic results presented in this chapter directly extend to other degree-graded polynomial bases. However, since the essential ideas developed here actually remain the same for other polynomial bases (while the notation becomes more cumbersome) we omit a detailed discussion. All necessary information related to this situation can be found in [45, Sec. 9]. Moreover, the article [28] shows how this ansatz space framework (for the monomial basis) can be extended to rectangular matrix polynomials.

Chapter 8

Sesquilinear forms

As already mentioned in Chapter 5, the generalized eigenproblem $(\lambda X + Y)v = 0$ corresponding to a linearization $\mathcal{L}(\lambda) = \lambda X + Y \in \mathrm{M}_{kn \times kn}(\mathbb{R}[\lambda])$ for some matrix polynomial $A(\lambda) \in \mathrm{M}_{n \times n}(\mathbb{R}[\lambda])$ of degree $k \geq 2$ can always be transformed into a standard eigenvalue problem of the form $\lambda v = -X^{-1}Y$ if X is nonsingular. In this and the following chapters, we consider matrices of the form $X^{-1}Y$ that arise, among other things, from transforming specially-structured generalized eigenvalue problems $(\lambda X + Y)v = 0$ into standard eigenproblems $\lambda v = -X^{-1}Y$. To achieve the most general results, we will be dealing with complex matrices $X, Y \in \mathrm{M}_{n \times n}(\mathbb{C})$ from now on. In particular, we consider matrices $X^{-1}Y$ where X and Y are both either Hermitian (i.e. $X^H = X, Y^H = Y$), skew-Hermitian (i.e. $X^H = -X, Y^H = -Y$) or a Hermitian/skew-Hermitian pair (that is, $X^H = \pm X, Y^H = \mp Y$).

Notice that, if the matrices X, Y are both Hermitian or skew-Hermitian, the matrix $X^{-1}Y$ is unlikely to preserve that structure. For instance, if $X^H = X$ and $Y^H = Y$ holds, the matrix $X^{-1}Y$ will in general not be Hermitian. Nevertheless, matrices of the form $X^{-1}Y$ arise frequently in applications for certain choices of X and can indeed be seen as *structured* in a particular way:

(*i*) Assume that the matrix X is given by

$$X := J_{2n} = \begin{bmatrix} 0_{n \times n} & I_n \\ -I_n & 0_{n \times n} \end{bmatrix} \in \mathrm{M}_{2n \times 2n}(\mathbb{R}). \tag{8.1}$$

If $Y \in \mathrm{M}_{2n \times 2n}(\mathbb{C})$ is Hermitian, a matrix $A \in \mathrm{M}_{2n \times 2n}(\mathbb{C})$ of the form $A = J_{2n}^{-1}Y = J_{2n}^T Y$ is called *skew-Hamiltonian*. It can be regarded as particularly structured, as a skew-Hamiltonian matrix A satisfies (by definition) $(J_{2n}A)^H = J_{2n}A$. On the other hand, if Y is skew-Hermitian and $A = J_{2n}^T Y$, then A is called *Hamiltonian* [79]. Those matrices satisfy the identity $(J_{2n}A)^H = -J_{2n}A$.

(ii) Now assume that the matrix X is given by

$$X := R_{2n} = \begin{bmatrix} 0_{n \times n} & R_n \\ R_n & 0_{n \times n} \end{bmatrix} \in M_{2n \times 2n}(\mathbb{R}), \qquad R_n = \begin{bmatrix} & & 1 \\ & \cdot^{\cdot^{\cdot}} & \\ 1 & & \end{bmatrix}.$$

If $Y \in M_{2n \times 2n}(\mathbb{C})$ is Hermitian, a matrix $A \in M_{2n \times 2n}(\mathbb{C})$ of the form $A = R_{2n}^{-1} Y = R_{2n}^{T} Y = R_{2n} Y$ is called *perhermitian* [81]. It can also be regarded as particularly structured, as a perhermitian matrix A satisfies (again by definition) $(R_{2n} A)^H = R_{2n} A$. On the other hand, if Y is skew-Hermitian and $A = R_{2n} A$, then A is called *perskew-Hermitian*. Those matrices satisfy the identity $(R_{2n} A)^H = -R_{2n} A$.

In matrix analysis and numerical linear algebra, it is in general desirable to handle structured matrices in a structure-preserving way. For instance, if X and Y are both Hermitian and X is positive definite, then there exist a Cholesky factorization $X = LL^H$ of X with a lower triangular matrix L (see [57, Thm. 4.2.7]). In this case, the GEP $(\lambda X + Y)v = 0$ can be transformed into a standard eigenproblem in two ways: either by considering $\lambda v = -X^{-1}Y$ or by regarding $\lambda v = -L^{-1}YL^{-H}$. Notice that the matrix $-L^{-1}YL^{-H}$ is still Hermitian and has the same eigenvalues as $\mathcal{L}(\lambda) = \lambda X + Y$. The Hermitian eigenproblem corresponding to $-L^{-1}YL^{-H}$ can thus be efficiently solved via any appropriate method. It is immediate that transforming $(\lambda X + Y)v = 0$ into the standard Hermitian eigenvalue problem $\lambda v = -L^{-1}YL^{-H}$ is usually more adequate than considering the unstructured problem $\lambda v = -X^{-1}Y$. In fact, as the Cholesky factorization can be computed in a backward stable manner (optionally with complete pivoting), $-L^{-1}YL^{-H}$ can be computed in a numerically reliable way, [29, 61]. Another striking fact that the transformation $\lambda v = -L^{-1}YL^{-H}$ should be preferred over the problem $\lambda v = -X^{-1}Y$ is, that the Hermitian matrix $-L^{-1}YL^{-H}$ can be brought to diagonal form by a unitary similarity transformation [67]. Recall that, in general, unitary similarity transformations of (skew-)Hermitian matrices preserve this structure.

Now consider again a matrix $A := X^{-1}Y$ with $X = J_{2n}$ given as in (i) above and Y being either Hermitian or skew-Hermitian. Recall that A is structured in a way that $(J_{2n}A)^H = \pm J_{2n}A$. Analogously to the situation of unitary diagonalization of (skew-)Hermitian matrices, the question arises whether there exists similarity transformations $\tilde{A} = U^{-1}AU$ of A such that the new matrix $\tilde{A}$ is still structured as A, that is $(J_{2n}\tilde{A})^H = \pm J_{2n}\tilde{A}$ holds, and has a simpler form compared to A (e.g. is diagonal). In view of structure-preservation, it can be checked by a direct calculation that for any matrix $U \in M_{2n \times 2n}(\mathbb{C})$ that satisfies $U^H J_{2n} U = J_{2n}$, the matrix $\tilde{A} = U^{-1}AU$ indeed satisfies $(J_{2n}\tilde{A})^H = \pm J_{2n}\tilde{A}$ if $(J_{2n}A)^H = \pm J_{2n}A$ holds. Thus, A and $\tilde{A}$ belong to the same structure class, i.e. are both (skew-)Hamiltonian. In this and the following chapter we will answer the question under what conditions there exists a matrix $U \in M_{2n \times 2n}(\mathbb{C})$ with $U^H J_{2n} U = J_{2n}$ such that $\tilde{A} = U^{-1}AU$ is diagonal.

Instead of considering only Hamiltonian and skew-Hamiltonian matrices A as above, we will analyse this situation in a more general setting by considering arbitrary Hermitian or skew-Hermitian (complex) generalized permutation matrices B in place of J_{2n}. In our context, a complex generalized permutation matrix B is a matrix with exactly one nonzero entry per row and column. Certainly, J_{2n} and R_{2n} both belong to the class of (complex) generalized permutation matrices. The structures considered above can thus be characterized by the equations $(BA)^H = \pm BA$ and $U^H BU = B$. The matrix equations characterising these structures actually come from the consideration of sesquilinear forms $\mathbb{C}^n$. We will study such forms in detail in this chapter with particular focus on matrices that are structured as above.

A sesquilinear form on $\mathbb{C}^n$ is a mapping $[\cdot, \cdot] : \mathbb{C}^n \times \mathbb{C}^n \to \mathbb{C}$ that is linear in its second and semilinear in its first argument, i.e. it holds that

$$[u, \lambda v + \mu w] = \lambda[u, v] + \mu[u, w] \quad \text{and} \quad [\lambda u + \mu v, w] = \overline{\lambda}[u, w] + \overline{\mu}[v, w] \qquad (8.2)$$

for all $u, v, w \in \mathbb{C}^n$ and all $\lambda, \mu \in \mathbb{C}$. The sesquilinear form most often considered on $\mathbb{C}^n$ is certainly the Euclidean form

$$[x, y] = (x, y) := x^H y, \quad x, y \in \mathbb{C}^n. \qquad (8.3)$$

It provides the background to geometric concepts such as angles or orthogonality. The Euclidean sesquilinear form satisfies the additional property that $(x, y) = \overline{(y, x)}$ holds for all $x, y \in \mathbb{C}^n$. A sesquilinear form with this property is usually referred to as Hermitian. Furthermore, sesquilinear forms satisfying $[x, y] = -\overline{[y, x]}$ for all $x, y \in \mathbb{C}^n$ are called skew-Hermitian. It immediately follows that $[x, x] \in \mathbb{R}$ holds for any Hermitian form while $[x, x] \in i\mathbb{R}$ is true for any skew-Hermitian form. Moreover, for the Euclidean sesquilinear form (8.3) it holds that (x, x) is always nonnegative (with $(x, x) = 0$ iff $x = 0$). This property is called positive definiteness. A Hermitian and positive definite sesquilinear form on $\mathbb{C}^n$ is called a scalar product (on $\mathbb{C}^n$). The vector space $\mathbb{C}^n$ equipped with such a form thus is a (finite-dimensional) Hilbert space [33].

In the following sections we are interested in Hermitian and skew-Hermitian sesquilinear forms which are indefinite in the sense that $[x, x]$ may take positive as well as negative real/purely imaginary values. Concepts such as orthogonality or adjoint matrices may be defined for indefinite sesquilinear forms as for the Euclidean scalar product. In particular, matrices satisfying $(BA)^H = \pm BA$ and $U^H BU = B$ considered previously are referred to as the Jordan and Lie algebras and the Lie group corresponding to the sesquilinear form $[x, y] = x^H By$ [82]. For the Euclidean scalar product, these are exactly the Hermitian and skew-Hermitian matrices as well as the unitary ones.

The vector space $\mathbb{C}^n$ equipped with an indefinite sesquilinear form is also often called an indefinite inner product space. There are important properties that indefinite forms and their definite companions (such as the Euclidean scalar product) share although not all desirable properties known from scalar products are maintained. The research field discussing vector spaces with those (particularly Hermitian) indefinite sesquilinear forms is called *Indefinite Linear Algebra*, a name that was invented by the authors of [55]. In particular, the monograph [55] provides a nice survey on their theoretical aspects and applications. In the following chapters, our central object of investigation is the development of canonical decompositions of matrices that posses special properties with respect to indefinite forms. Thereby we will particularly focus on diagonalization.

The main concepts needed for the following sections are introduced in Section 8.1. Among others, the idea of normal matrices in indefinite inner product spaces is introduced. In Section 8.2 the three fundamental classes associated with sesquilinear forms - automorphisms, selfadjoint and skewadjoint matrices - are presented along with some of their basic properties. Most of the material in these two sections is standard and can also be found in, e.g., [13, 55, 82, 83]. Finally, in Section 8.3 we introduce and analyze a new class of sesquilinear forms that plays a key role in the sequel. Namely, we consider sesquilinear forms induced by generalized permutation matrices B and provide several results concerning the structure-preserving diagonalization of matrices which are (by themselves) structured in a certain way with respect to the sesquilinear form $[x, y] = x^H B y$.

8.1   Introduction and basic definitions

From here on square matrices $A \in \mathrm{M}_{n \times n}(\mathbb{C})$ are considered. Recall that the spectrum $\sigma(A)$ of A, its eigenvalues, eigenvectors as well as their algebraic and geometric multiplicities are defined for A in terms of the corresponding matrix pencil $\mathcal{L}(\lambda) := \lambda I_n - A$. Thus, all concepts introduced in Section 2 apply for A via $\mathcal{L}(\lambda)$. Keep in mind that a single matrix A never has infinite eigenvalues as $I_n \in \mathrm{GL}_n(\mathbb{C})$. As we are only dealing with right eigenspaces from now on, we slightly simplify our nomenclature and define $E(A, \mu) := \mathrm{Eig}_r(\mathcal{L}, \mu) = \mathrm{null}(\mathcal{L}(\mu))$ for all $\mu \in \sigma(A)$. In particular, $E(A, \mu) = \{0\}$ whenever $\mu \notin \sigma(A)$. If $S \subset \mathbb{C}^n$ is some subset and $\mu \in \mathbb{C}$ we set $E(A, \mu, S) := E(A, \mu) \cap S$. A matrix $A \in \mathrm{M}_{n \times n}(\mathbb{C})$ is called diagonalizable if there exists a complete set of (linear independent) eigenvectors $r_1, \ldots, r_n \subseteq \mathbb{C}^n$ of A forming a basis of $\mathbb{C}^n$. In this case, $R^{-1}AR$ is a diagonal matrix for $R = [\, r_1 \; \cdots \; r_n \,] \in \mathrm{GL}_n(\mathbb{C})$. The term nondefective is often used synonymic to diagonalizable. For any two vector spaces $S \subseteq T \subseteq \mathbb{C}^n$, T/S denotes the quotient vector space of T modulo S, that is, the vector space of equivalence classes induced by the equivalence relation $x \sim y$ iff $x - y \in S$.

To begin, we restate the definition of sesquilinear forms from (8.2). Notice that Definition 8.1 slightly deviates from the definition of sesquilinear forms given in [55, Sec. 2.1].

Definition 8.1 (Sesquilinear Form). *A sesquilinear form $[\cdot, \cdot]$ on $\mathbb{C}^n$ is a mapping $[\cdot, \cdot] : \mathbb{C}^n \times \mathbb{C}^n \to \mathbb{C}$ so that for all $u, v, w \in \mathbb{C}^n$ and all $\alpha, \beta \in \mathbb{C}$ the following relations (i) and (ii) hold*

$$(i) \; [\alpha u + \beta v, w] = \overline{\alpha}[u, w] + \overline{\beta}[v, w] \qquad (ii) \; [u, \alpha v + \beta w] = \alpha[u, v] + \beta[u, w].$$

If $[\cdot, \cdot]$ is some sesquilinear form and $x := \alpha e_j, y := \beta e_k \in \mathbb{C}^n$ with $\alpha, \beta \in \mathbb{C}$ are two vectors that are multiples of the jth and kth unit vectors, then $[x, y] = \overline{\alpha}\beta[e_j, e_k]$. Thus any sesquilinear form is uniquely determined by the images of the standard unit vectors $[e_j, e_k], j, k = 1, \ldots, n$. In particular, $[\cdot, \cdot]$ on $\mathbb{C}^n$ can be expressed as

$$[x, y] = x^H B y \qquad (8.4)$$

for the particular matrix $B = [b_{jk}]_{jk} \in M_{n \times n}(\mathbb{C})$ with $b_{jk} = [e_j, e_k], j, k = 1, \ldots, n$. A form $[\cdot, \cdot]$ as in (8.4) is called Hermitian if $[x, y] = \overline{[y, x]}$ holds for all $x, y \in \mathbb{C}^n$. It is easy to see that $[\cdot, \cdot]$ is Hermitian iff $B \in M_{n \times n}(\mathbb{C})$ is Hermitian, i.e. $B = B^H$ [55]. The form $[\cdot, \cdot]$ is called skew-Hermitian if $[x, y] = -\overline{[y, x]}$ holds for all $x, y \in \mathbb{C}^n$. This is the case iff $B = -B^H$. Skew-Hermitian forms are also commonly considered over $\mathbb{R}^n$ with a skew-symmetric matrix $B = -B^T$ in which case (8.4) reads $[x, y] = x^T B y$. Those forms are often called alternating and can equivalently be characterized by the fact that $[x, x] = 0$ holds for all $x \in \mathbb{R}^n$. Note that this alternative characterization is invalid for sesquilinear forms over $\mathbb{C}^n$. Unless $B = B^H$ ($B = -B^H$) and all real (purely imaginary) eigenvalues of B have the same sign, the sesquilinear form (8.4) is called indefinite. The following definition introduces two important classes of sesquilinear forms that will be central for our discussion in the sequel.

Definition 8.2 (Classification of Sesquilinear Forms, [82, Def. 3.1, Def. 4.1]).

(i) *The form $[x, y] = x^H B y$ on $\mathbb{C}^n$ is called orthosymmetric if $B^H = \beta B$ for some scalar $\beta \in \mathbb{C}$ with $|\beta| = 1$.*

(ii) *The form $[x, y] = x^H B y$ on $\mathbb{C}^n$ is called unitary if $B = \beta U$ for some unitary matrix $U \in M_{n \times n}(\mathbb{C})$, $U^H = U^{-1}$, and some scalar $\beta \in \mathbb{R}, \beta > 0$.*

Throughout our discussion we will only deal with sesquilinear forms $[x, y] = x^H B y$ that are orthosymmetric or unitary. This obviously includes the cases $B = B^H$ and $B = -B^H$ on which we will particularly focus. In fact, forms arising from applications often belong to one of the two classes introduced in Definition 8.2. Such examples are discussed in Sections 9.4.1 to 9.4.3.

Any two vectors $x, y \in \mathbb{C}^n$ are called orthogonal with respect to $[x, y] = x^H B y$ if $[x, y] = 0$. The assumption that $[\cdot, \cdot]$ is orthosymmetric ensures that orthogonality is a symmetric relation, i.e. $[x, y] = 0$ iff $[y, x] = 0$ [82, Thm. 3.2(b)]. In addition, for any subset $S \subseteq \mathbb{C}^n$, the orthogonal companion $S^{[\perp]}$ with respect to an orthosymmetric form $[\cdot, \cdot]$ is defined to be the subspace of all vectors $u \in \mathbb{C}^n$ that satisfy $[u, s] = 0$ for all $s \in S$. If $[t, s] = 0$ holds for all $t \in T \subset \mathbb{C}^n, s \in S \subset \mathbb{C}^n$ we write $S \perp T$. For indefinite sesquilinear forms, a distinction as in Definition 8.3 below on the nature of subspaces is necessary. This situation does not occur for definite scalar products.

Definition 8.3 (Isotropic and Nondegenerate Subspace, [55, Sec. 2]). *Let $[x, y] = x^H B y$ be some sesquilinear form.*

(i) *A subspace $S \subseteq \mathbb{C}^n$ of dimension $\dim(S) = k \geq 1$ is called isotropic with respect to $[\cdot, \cdot]$ iff $\text{rank}(V^H B V) = 0$ for any basis $v_1, \ldots, v_k$ of S and $V = [v_1 \ \cdots \ v_k]$.*

(ii) *A subspace $S \subseteq \mathbb{C}^n$ of dimension $\dim(S) = k \geq 1$ is called nondegenerate with respect to $[\cdot, \cdot]$ iff $V^H BV$ is nonsingular, i.e. $\operatorname{rank}(V^H BV) = k$, for any basis $v_1, \ldots, v_k$ of S and $V = [v_1 \cdots v_k]$. Otherwise, S is called degenerate.*

Notice that $\operatorname{rank}(V^H BV)$ as considered above is independent of the chosen basis, i.e. isotropic and (non)degenerate subspaces are well-defined by Definition 8.3. Equivalently to Definition 8.3 (i), a subspace $S \subseteq \mathbb{C}^n$ is isotropic iff $[v, w] = 0$ holds for all $v, w \in S$. In addition, a third definition of isotropic subspaces $S \subseteq \mathbb{C}^n$ is that the property $S \subseteq S^{[\perp]}$ holds. It is easily seen that all three concepts are equivalent.

A nondegenerate subspace $S \subseteq \mathbb{C}^n$ can alternatively be described by the property that, for any $u \in S$, $[u, v] = 0$ for all $v \in S$ implies that $u = 0$. This characterization is equivalent to Definition 8.3 (ii). The sesquilinear form $[x, y] = x^H By$ itself is called nondegenerate, if $S = \mathbb{C}^n$ is nondegenerate. Notice that, according to Definition 8.3 (ii), $[x, y] = x^H By$ is nondegenerate iff B is nonsingular. All forms considered in the sequel will implicitly be understood as nondegenerate.

The result from Proposition 8.1 below is central for the upcoming discussion and can be found in, e.g., [67, Sec. 4.5] (for the case $A = A^H$). The statement for $A = -A^H$ is easily verified by noting that $A = A^H$ is Hermitian iff iA is skew-Hermitian and vice versa.

Proposition 8.1 (Sylvesters Law of Inertia). *Let $A \in M_{n \times n}(\mathbb{C})$ and assume that either $A = A^H$ or $A = -A^H$ holds. Then there exists a nonsingular matrix $U \in GL_n(\mathbb{C})$ so that*

$$U^H A U = \begin{bmatrix} -\alpha I_p & & \\ & \alpha I_q & \\ & & 0_{r \times r} \end{bmatrix}$$

where $\alpha = 1$ if A is Hermitian and $\alpha = i$ otherwise. Hereby, p coincides with the number of negative real/purely imaginary eigenvalues of A, q coincides with the number of positive real/purely imaginary eigenvalues of A and r is the algebraic multiplicity of zero as an eigenvalue of A.

The triple (p, q, r) from Proposition 8.1 is usually referred to as the inertia of A. Two Hermitian or skew-Hermitian matrices $A, B \in M_{n \times n}(\mathbb{C})$ with the same inertia are called congruent. Following directly from Proposition 8.1 we have:

Proposition 8.2. *Let $A, B \in M_{n \times n}(\mathbb{C})$ be two matrices which are either both Hermitian or skew-Hermitian. Then there exists a nonsingular matrix $S \in M_{n \times n}(\mathbb{C})$ so that $S^H A S = B$ iff A and B have the same inertia.*

From now on we call a diagonal matrix $D \in M_{n \times n}(\mathbb{C})$ a real/complex *signature matrix*, if all diagonal entries of D are either $+1$ and -1 in some combination or, in the complex case, $+i$ and $-i$. Notice that the term "complex signature matrix" does not seem to be standard.

It is well-known that any subspace $S \subseteq \mathbb{C}^n$ can always be decomposed as the direct and orthogonal sum $S = S_1 \oplus S_2$ of a nondegenerate subspace $S_1 \subseteq S$ and an isotropic subspace $S_2 \subseteq S$ [13, Lem. 5.1]. We will call S_1 maximal nondegenerate in S if there does not exist a nondegenerate subspace $\tilde{S}_1 \subseteq S$ that properly contains S_1. Maximal isotropic subspaces in S are defined analogously. We will prove the existence of a decomposition of S into a maximal nondegenerate and a maximal isotropic subspace in Lemma 8.1 based on Definition 8.3.

Lemma 8.1. *Let $[x, y] = x^H B y$ be some nondegenerate Hermitian or skew Hermitian sesquilinear form on $\mathbb{C}^n$. Moreover, let S be some nonzero subspace of $\mathbb{C}^n$. Then there exists a decomposition $S = S_1 \oplus S_2$ of S as the direct and orthogonal sum of a maximal nondegenerate subspace S_1 in S and a maximal isotropic subspace S_2 in S. The subspace S_2 is unique and equals $S \cap S^{[\perp]}$ in any such decomposition.*

Proof. First assume that $[x, y] = x^H B y$ is Hermitian. Let $0 \neq S \subseteq \mathbb{C}^n$ be some subspace and let the columns of $R = [\, r_1 \; \cdots \; r_k \,]$ be a basis of S. If rank$(R^H B R) = 0$, then S is isotropic, we set $S_1 = \{0\}$ and $S_2 = S$. On the other hand, if rank$(R^H B R) = \ell > 0$, then $R^H B R$ is again real Hermitian and, according to Proposition 8.2, there always exists a matrix $U \in \mathrm{GL}_k(\mathbb{C})$ so that

$$U^H\big(R^H B R\big) U = \left[\begin{array}{c|c} K & 0 \\ \hline 0 & 0_{k-\ell} \end{array}\right] \in \mathrm{M}_{k\times k}(\mathbb{C}), \tag{8.5}$$

where K is a real signature matrix, i.e. $K = \mathrm{diag}[\pm 1, \ldots, \pm 1]$. Then, since

$$\big[\, Ru_1 \; \cdots \; Ru_\ell \,\big]^H B \big[\, Ru_1 \; \cdots \; Ru_\ell \,\big] = K$$

has full rank ℓ, $S_1 := \mathrm{span}(Ru_1, \ldots, Ru_\ell)$ is nondegenerate according to Definition 8.3 (ii). Further, $S_2 := \mathrm{span}(Ru_{\ell+1}, \ldots, Ru_k)$ is a direct complement of S_1 is S, i.e. $S = S_1 \oplus S_2$. As can be seen from (8.5) and $[\, Ru_{\ell+1} \; \cdots \; Ru_k \,]^H B [\, Ru_{\ell+1} \; \cdots \; Ru_k \,] = 0$, S_2 is isotropic and $S_1 \perp S_2$. Certainly, S_1 is maximal nondegenerate in S (since there cannot exist a nondegenerate subspace $\tilde{S} \subseteq S$ with $\dim(\tilde{S}) > \ell$) and S_2 is maximal isotropic in S. In addition we have $S/S_2 \simeq S_1$ and it is easily checked that $S_2 = S \cap S^{[\perp]}$. Set $RU =: V = [\, v_1 \; \cdots \; v_k \,]$ and notice that $v_1, \ldots, v_k$ is just another basis of S (different from $r_1, \ldots, r_k$). Thus, S can be decomposed as the direct sum $S_1 \oplus S_2$ of a maximal nondegenerate subspace (span$(v_1, \ldots, v_\ell)$) and a maximal isotropic subspace span$(v_{\ell+1}, \ldots, v_k)$ (in S). Next, we discuss the question of uniqueness.

To this end, assume that we have two different expressions as in (8.5), that is $w_1, \ldots, w_k$ is yet another basis of S beside $v_1, \ldots, v_k$ with $W = [\, w_1 \; \cdots \; w_k \,]$ satisfying $W^H B W = L \oplus 0_{k-\ell}$ for some signature matrix $L \in \mathrm{GL}_\ell(\mathbb{C})$. Then the inertia of K and L must be equal according to Proposition 8.2 and there exists some matrix $Q \in \mathrm{GL}_k(\mathbb{C})$ so that $W = VQ$ (since W and V both span S) giving

$$W^H B W = \begin{bmatrix} Q_{11} & Q_{12} \\ Q_{21} & Q_{22} \end{bmatrix}^H (V^H B V) \begin{bmatrix} Q_{11} & Q_{12} \\ Q_{21} & Q_{22} \end{bmatrix} = Q^H \left[\begin{array}{c|c} K & 0 \\ \hline 0 & 0_{k-\ell} \end{array}\right] Q = \left[\begin{array}{c|c} L & 0 \\ \hline 0 & 0_{k-\ell} \end{array}\right]$$

with $Q_{11} \in M_{\ell \times \ell}(\mathbb{C})$. The last equality directly implies $Q_{12} = 0$. Thus, since $Q \in GL_k(\mathbb{C})$ we obtain $Q_{11} \in GL_\ell(\mathbb{C})$ and $Q_{22} \in GL_{k-\ell}(\mathbb{C})$. In particular,

$$\begin{bmatrix} w_1 \cdots w_\ell \end{bmatrix} = \begin{bmatrix} v_1 \cdots v_\ell \end{bmatrix} Q_{11} + \begin{bmatrix} v_{\ell+1} \cdots v_k \end{bmatrix} Q_{21} \tag{8.6}$$

and $\begin{bmatrix} w_{\ell+1} \cdots w_k \end{bmatrix} = \begin{bmatrix} v_{\ell+1} \cdots v_k \end{bmatrix} Q_{22}$. Therefore, $\mathrm{span}(w_{\ell+1}, \ldots, w_k) = S_2$. However, unless $Q_{21} = 0$, $\mathrm{span}(w_1, \ldots, w_\ell) \neq S_1$. Thus S_2 is unique (i.e. the same in any such decomposition), but S_1 is not. Following the above discussion, a similar result holds in case B is skew-Hermitian using Proposition 8.1 and Proposition 8.2 on skew-Hermitian matrices. □

A priori there is no canonical choice for S_1 (recall (8.6) for $Q_{21} \neq 0$). However, concerning the eigenspaces of some classes of matrices, there is one as we will see later on (i.e. Corollary 9.1). Those special matrices are defined in Definition 8.4 (*ii*) below.

Definition 8.4 (Adjoint, Normal Matrix, [82, Sec. 2], [55]). *Let $[x, y] = x^H B y$ be some nondegenerate, orthosymmetric sesquilinear form.*

(*i*) *Let $A \in M_{n \times n}(\mathbb{C})$. The adjoint $A^\star$ of A is the uniquely defined matrix that satisfies $[Au, v] = [u, A^\star v]$ for all $u, v \in \mathbb{C}^n$. It can be explicitly expressed as $A^\star = B^{-1} A^H B$.*

(*ii*) *A matrix $A \in M_{n \times n}(\mathbb{C})$ is called normal with respect to $[x, y] = x^H B y$, or simply B-normal, iff $AA^\star = A^\star A$ holds.*

For orthosymmetric, nondegenerate sesquilinear forms $[x, y] = x^H B y$ the unique matrix $\hat{A}$ satisfying $[u, Av] = [\hat{A}u, v]$ for all $u, v \in \mathbb{C}^n$ is also equal to $A^\star$, that is $\hat{A} = A^\star$ [82]. More information on B-normal matrices in case $B = B^H$ is provided in [55, Sec. 8]. In the following section we consider special classes of B-normal matrices.

8.2 Special classes of B-normal matrices

Let $[x, y] = x^H B y$ be some nondegenerate and orthosymmetric sesquilinear form on $\mathbb{C}^n$. There are three classes of B-normal matrices that deserve special attention.

Definition 8.5 ([82, Sec. 2]). *Let $[x, y] = x^H B y$ be some nondegenerate, orthosymmetric sesquilinear form. For $M \in M_{n \times n}(\mathbb{C})$ let $M^\star = B^{-1} M^H B$ be the adjoint of M.*

(*i*) *The matrices $G \in M_{n \times n}(\mathbb{C})$ satisfying $[Gu, Gv] = [u, v]$ for all $u, v \in \mathbb{C}^n$ are called automorphisms for $[x, y] = x^H B y$. The set of automorphisms (corresponding to $[x, y] = x^H B y$) is denoted by $\mathbb{G}(B)$ in the following.*

(*ii*) *A matrix $J \in M_{n \times n}(\mathbb{C})$ satisfying $J^\star = J$ is called selfadjoint. The set of all selfadjoint matrices is denoted $\mathbb{J}(B)$ in the following.*

(iii) *A matrix* $L \in M_{n \times n}(\mathbb{C})$ *satisfying* $L^\star = -L$ *is called skewadjoint. The set of all skewadjoint matrices is denoted* $\mathbb{L}(B)$ *in the following.*

For any automorphism $G \in \mathbb{G}(B)$, notice that $[Gu, Gv] = [u, G^\star Gv] = [u, v]$ for all $u, v \in \mathbb{C}^n$. This implies that $G^\star G$ needs to be the identity matrix, i.e. $G^\star G = I_n$. Since $G^\star = B^{-1} G^H B$, the identity $G^\star G = I_n$ is thus equivalent to $G^H BG = B$. Moreover, $G^\star G = I_n$ implies $G^\star = G^{-1}$ (as the inverse of a matrix is uniquely determined if it exists). In addition, notice that the equation $G^H BG = B$ implies $|\det(G)| = 1$ for any $G \in \mathbb{G}(B)$ regardless of the form of B (here $|\cdot|$ denotes the absolute value). Since for $F, G \in \mathbb{G}(B)$ it holds that

$$\left(FG\right)^H B \left(FG\right) = G^H \left(F^H BF\right) G = G^H BG = B$$

we have $FG \in \mathbb{G}(B)$. Thus, the automorphisms form a multiplicative subgroup of $GL_n(\mathbb{C})$ (which is sometimes called the Lie group corresponding to $[x, y] = x^H By$). The selfadjoint and skewadjoint matrices both form vector spaces over $\mathbb{C}$. Moreover, for selfadjoint matrices $J, K \in \mathbb{J}(B)$ it is easily checked that $\frac{1}{2}(JK + KJ) \in \mathbb{J}(B)$. In addition, we have $LN - LN \in \mathbb{L}(B)$ for any two skewadjoint matrices $L, N \in \mathbb{L}(B)$. With respect to these operations, selfadjoint matrices form a Jordan algebra whereas skewadjoint matrices form a Lie algebra [82, Sec. 2]. Finally, we have $M_{n \times n}(\mathbb{C}) = \mathbb{J}(B) \oplus \mathbb{L}(B)$ according to [82, Thm. 3.2]. Notice the following important facts:

- The eigenvalues of automorphisms $G \in \mathbb{G}(B)$ come in pairs $\lambda, 1/\overline{\lambda}$ since $G^\star = G^{-1}$ and $G^\star$ is similar to G^H (recall Definition 8.4 (i)). This shows that zero can never be an eigenvalue of any $G \in \mathbb{G}(B)$, that is, $\mathbb{G}(B) \subset GL_n(\mathbb{C})$.

- The eigenvalues of selfadjoint matrices $J \in \mathbb{J}(B)$ come in pairs $\lambda, \overline{\lambda}$ as $\sigma(J) = \sigma(J^\star) = \sigma(J^H) = \overline{\sigma(J)}$. Thus the spectrum of J is symmetric with respect to the real axis and the eigenspaces $E(A, \lambda)$ and $E(A^\star, \lambda)$ coincide for all eigenvalues $\lambda \in \sigma(A)$.

- The eigenvalues of skewadjoint matrices $L \in \mathbb{L}(B)$ come in pairs $\lambda, -\overline{\lambda}$ as $\sigma(L) = \sigma(-L^\star) = -\sigma(L^\star) = -\overline{\sigma(L)}$. Thus the spectrum of L is symmetric with respect to the imaginary axis and the eigenspaces $E(A, \lambda)$ and $E(A^\star, -\lambda)$ coincide for all eigenvalues $\lambda \in \sigma(A)$.

It can easily be shown that $\mathbb{J}(B)$ and $\mathbb{L}(B)$ are closed under similarity transformations with matrices from $\mathbb{G}(B)$. That is, $P^{-1} JP \in \mathbb{J}(B)$ (or $P^{-1} LP \in \mathbb{L}(B)$) holds if $J \in \mathbb{J}(B)$ (or $L \in \mathbb{L}(B)$, respectively). The following lemma states how selfadjoint and skewadjoint matrices arise. The statement of Lemma 8.2 is likely to be known although it can neither be found in [55] nor in [13] in the form below. Thus we give a short and simple proof. Notice the similarity of the results from Lemma 8.2 and the characterization of (skew)-Hamiltonian and per(skew)-Hermitian matrices from the introduction.

Lemma 8.2. *Let* $[x, y] = x^H By$ *be some nondegenerate sesquilinear form.*

(i) Let $[x,y] = x^H By$ be Hermitian. Then $A \in \mathbb{J}(B)$ $(A \in \mathbb{L}(B))$ iff $A = B^{-1}S$ for some Hermitian (skew-Hermitian) matrix $S \in \mathrm{M}_{n\times n}(\mathbb{C})$.

(ii) Let $[x,y] = x^H By$ be skew-Hermitian. Then $A \in \mathbb{L}(B)$ $(A \in \mathbb{J}(B))$ iff $A = B^{-1}S$ for some Hermitian (skew-Hermitian) matrix $S \in \mathrm{M}_{n\times n}(\mathbb{C})$.

Proof. (i) If $S \in \mathrm{M}_{n\times n}(\mathbb{C})$ is Hermitian, then $A := B^{-1}S \in \mathbb{J}(B)$ follows by direct calculation:

$$A^\star = B^{-1}A^H B = B^{-1}S^H B^{-H} B = B^{-1}SB^{-1}B = B^{-1}S = A$$

using that $B^{-H} = B^{-1}$. On the other hand, if $A^\star = B^{-1}A^H B = A$ holds, then $BA = A^H B = (B^H A)^H = (BA)^H$, so $S := BA$ is Hermitian. Moreover, $A = B^{-1}S$ holds.

Now suppose that $S \in \mathrm{M}_{n\times n}(\mathbb{C})$ is skew-Hermitian and consider $A := B^{-1}S$. Then

$$A^\star = B^{-1}S^H B^{-H} B = -B^{-1}SB^{-1}B = -B^{-1}S = -A,$$

so $A \in \mathbb{L}(B)$. In addition, if $A^\star = B^{-1}A^H B = -A$ holds, then $-BA = A^H B = (B^H A)^H = (BA)^H$, so $S := BA$ is skew-Hermitian. Again, $A = B^{-1}S$ holds and the statement is proven for $[\cdot, \cdot]$ being Hermitian. The proof for (ii) works analogously. $\square$

From Lemma 8.2 it follows that $\mathbb{J}(B)$ $(\mathbb{L}(B))$ is isomorphic to the vector space of all Hermitian (skew-Hermitian) complex matrices, if $[\cdot, \cdot]$ is Hermitian. Therefore, in this case, $\dim(\mathbb{J}(B)) = n(n+1)$ and $\dim(\mathbb{L}(B)) = n(n-1)$.

According to Definition 8.2 (ii) a sesquilinear form $[x,y] = x^H By$ is called unitary if $B = \beta U$ for some unitary $U \in \mathrm{M}_{n\times n}(\mathbb{C})$ and $\beta > 0$. Later on, those forms will be considered in more detail. For now, notice that unitary forms have the property that $\|A^\star\| = \|B^{-1}A^H B\| = \|A^H\| = \|A\|$ holds for any unitarily invariant matrix norm and any matrix $A \in \mathrm{M}_{n\times n}(\mathbb{C})$. From this we obtain the following approximation result which was first stated for Hermitian matrices in [48, Thm. 2]. Our proof closely resembles the one for Hermitian matrices in [66].

Lemma 8.3. Let $[x,y] = x^H By$ be some unitary, nondegenerate sesquilinear form.

(i) Let $A \in \mathrm{M}_n(\mathbb{C})$ and $\tilde{A} := \frac{1}{2}(A + A^\star)$. Then $\tilde{A} \in \mathbb{J}(B)$ and it holds that $\|A - \tilde{A}\| \leq \|A - C\|$ for any other matrix $C \in \mathbb{J}(B)$ and any unitarily invariant matrix norm.

(ii) Let $A \in \mathrm{M}_n(\mathbb{C})$ and $\tilde{A} := \frac{1}{2}(A - A^\star)$. Then $\tilde{A} \in \mathbb{L}(B)$ and it holds that $\|A - \tilde{A}\| \leq \|A - C\|$ for any other matrix $C \in \mathbb{L}(B)$ and any unitarily invariant matrix norm.

Proof. (i) Let $\| \cdot \|$ be some unitarily invariant matrix norm. As $(\tilde{A})^\star = \frac{1}{2}(A + A^\star)^\star = \frac{1}{2}(A^\star + A) = \tilde{A}$ we have $\tilde{A} \in \mathbb{J}(B)$. We now consider $\|A - \tilde{A}\|$ and show that $\|A - \tilde{A}\| \leqslant$

$\|A - C\|$ holds for any other matrix $C \in \mathbb{J}(B)$. Thus, let $C = B^{-1}C^H B \in \mathbb{J}(B)$ and observe that

$$
\begin{aligned}
\|A - \widetilde{A}\| = \frac{1}{2}\|A - A^\star\| &= \frac{1}{2}\|(A - C) - (A^\star - C)\| \\
&= \frac{1}{2}\|(A - C) - (A - C)^\star\| \\
&\leqslant \frac{1}{2}\big(\|A - C\| + \|(A - C)^\star\|\big) \\
&= \|A - C\|
\end{aligned}
$$

where we have used $A^\star - C = B^{-1}A^H B - B^{-1}C^H B = B^{-1}(A^H - C^H)B = (A - C)^\star$. The proof for (ii) proceeds in the same manner. □

In [55, Sec. 4] and [82] much more information on these special classes of selfadjoint and skewadjoint matrices and their properties can be found. However, the results stated in this section serve our purposes well. Before we pass on to the next section we define what we will call an automorphic diagonalization of a matrix with respect to some sesquilinear form. To this end, here and from now on $\mathbb{D}_n(\mathbb{C})$ denotes the set of all diagonal matrices in $\mathrm{M}_{n\times n}(\mathbb{C})$.

Definition 8.6. *Let $[x, y] = x^H By$ be some nondegenerate sesquilinear form on $\mathbb{C}^n$. A similarity transformation of $A \in \mathrm{M}_{n\times n}(\mathbb{C})$ such that $P^{-1}AP \in \mathbb{D}_n(\mathbb{C})$ holds with $P \in \mathbb{G}(B)$ will be called an automorphic diagonalization.*

8.3 Forms from generalized permutation matrices

Let $[x, y] = x^H By$ be some nondegenerate sesquilinear form. In the next chapter our main concern will be the automorphic diagonalization of B-normal matrices $A \in \mathrm{M}_{n\times n}(\mathbb{C})$ as explained in Definition 8.6. For this purpose, this section introduces a third subclass of forms beside orthosymmetric and unitary forms that have some desirable properties with respect to automorphic diagonalization. The introduction of this class is guided by the question what constraints on B have to be imposed to guarantee that any diagonal matrix is always B-normal.

To begin, notice that for two commuting, diagonalizable matrices $X, Y \in \mathrm{M}_{n\times n}(\mathbb{C})$ there always exists a matrix $S \in \mathrm{GL}_n(\mathbb{C})$ such that $S^{-1}XS$ and $S^{-1}YS$ are both diagonal [67, Thm. 1.3.19]. In particular, if $A \in \mathrm{M}_{n\times n}(\mathbb{C})$ is a B-normal and diagonalizable matrix, then, as $AA^\star = A^\star A$ holds, A and $A^\star$ are simultaneously diagonalizable. That means, there exists an eigenbasis $u_1, \ldots, u_n$ common to A and $A^\star$ so that $U^{-1}AU$ and $U^{-1}A^\star U$ are both diagonal for $U = [\, u_1 \, \cdots \, u_n \,]$. However, a priori U might not possess any special structure with respect to $[\cdot, \cdot]$.

To analyze this situation, we denote by $\mathbb{GP}_n(\mathbb{C})$ the group of all generalized complex permutation matrices, i.e. all matrices $P \in \mathrm{GL}_n(\mathbb{C})$ with exactly one nonzero (complex) entry per row and column. Furthermore, for $S \subseteq \mathrm{M}_{n\times n}(\mathbb{C})$ let $\mathcal{Z}(S)$ denote the set of all matrices $Z \in \mathrm{GL}_n(\mathbb{C})$ such that $Z^{-1}KZ \in S$ for all $K \in S$.

Moreover, let $\widehat{\mathbb{D}} \subset \mathbb{D}_n(\mathbb{C})$ be the subset of nonsingular diagonal matrices with pairwise distinct diagonal entries. With these definitions, we have the following result:

Lemma 8.4. *The set* $\mathcal{Z}(\mathbb{D}_n(\mathbb{C}))$ *coincides with* $\mathbb{GP}_n(\mathbb{C})$.

Proof. The inclusion $\mathbb{GP}_n(\mathbb{C}) \subseteq \mathcal{Z}(\widehat{\mathbb{D}})$ is clear and easily verified. Since $\widehat{\mathbb{D}} \subset \mathbb{D}_n(\mathbb{C})$ we obtain that any $Z \in \mathcal{Z}(\mathbb{D}_n(\mathbb{C}))$ is also an element of $\mathcal{Z}(\widehat{\mathbb{D}})$. Thus we have the inclusion $\mathcal{Z}(\mathbb{D}_n(\mathbb{C})) \subseteq \mathcal{Z}(\widehat{\mathbb{D}})$.

Now let $D \in \widehat{\mathbb{D}}$ and suppose $S = [\, s_1 \;\cdots\; s_n\,] \in \mathcal{Z}(\widehat{\mathbb{D}})$, i.e. $S^{-1}DS = F \in \widehat{\mathbb{D}}$. Certainly, D and F have the same eigenvalues, so the diagonal of F is the same as the diagonal of D up to permutation. Now multiplying $DS = SF$ with some standard unit vector $e_j, 1 \le j \le n$, yields $Ds_j = f_{jj}s_j$, i.e. s_j is an eigenvector of D with eigenvalue $f_{jj} = e_j^T F e_j$. Since the eigenvectors of D are multiples of the standard unit vectors, $s_j = \alpha e_k$ for some $\alpha \in \mathbb{C}$ and some k, $1 \le k \le n$. Since S is nonsingular, this implies that $S \in \mathbb{GP}_n(\mathbb{C})$. Thus we have $\mathcal{Z}(\widehat{\mathbb{D}}) \subseteq \mathbb{GP}_n(\mathbb{C})$ and therefore $\mathbb{GP}_n(\mathbb{C}) = \mathcal{Z}(\widehat{\mathbb{D}})$. As mentioned above, $\mathcal{Z}(\mathbb{D}_n(\mathbb{C})) \subseteq \mathcal{Z}(\widehat{\mathbb{D}})$ has to hold. Since $\mathbb{GP}_n(\mathbb{C}) \subseteq \mathcal{Z}(\mathbb{D}_n(\mathbb{C}))$ is clear, we obtain

$$\mathcal{Z}(\mathbb{D}_n(\mathbb{C})) \subseteq \mathcal{Z}(\widehat{\mathbb{D}}) = \mathbb{GP}_n(\mathbb{C}) \subseteq \mathcal{Z}(\mathbb{D}_n(\mathbb{C}))$$

which shows $\mathbb{GP}_n(\mathbb{C}) = \mathcal{Z}(\mathbb{D}_n(\mathbb{C}))$ and proves the statement. $\qquad\square$

Sesquilinear forms $[x,y] = x^H By$ with $B \in \mathbb{GP}_n(\mathbb{C})$ have some desirable properties with respect to structure-preserving diagonalization of B-normal matrices. For our purpose, the most important one is the following: let $A \in \mathrm{M}_{n \times n}(\mathbb{C})$ and assume that $P^{-1}AP = D \in \mathbb{D}_n(\mathbb{C})$ holds for some $P \in \mathbb{G}(B)$. Then, since $P^H BP = B$,

$$P^{-1}A^{\star}P = P^{-1}B^{-1}A^H BP = P^{-1}B^{-1}P^{-H}D^H P^H BP = B^{-1}D^H B = D^{\star}. \qquad (8.7)$$

Therefore, if $B \in \mathbb{GP}_n(\mathbb{C})$, we obtain the following statement on the automorphic diagonalization of nondefective matrices:

Lemma 8.5. *Let* $[x,y] = x^H By$ *be some nondegenerate and orthosymmetric sesquilinear form.*

(i) *Let* $A \in \mathrm{M}_{n \times n}(\mathbb{C})$ *be nondefective and assume that* $B \in \mathbb{GP}_n(\mathbb{C})$. *If* $P \in \mathbb{G}(B)$ *diagonalizes either* A *or* $A^{\star}$, *it simultaneously diagonalizes* A *and* $A^{\star}$.

(ii) *Assume that any* $A \in \mathrm{M}_{n \times n}(\mathbb{C})$ *and* $A^{\star}$ *are always simultaneously diagonalized whenever some* $P \in \mathbb{G}(B)$ *diagonalizes either* A *or* $A^{\star}$. *Then* $B \in \mathbb{GP}_n(\mathbb{C})$.

Proof. (i) This follows from (8.7) and the assumption $B \in \mathbb{GP}_n(\mathbb{C})$. The reverse direction starting with a diagonalization of $A^{\star}$ proceeds similarly.

(ii) It suffices to consider matrices $D \in \mathbb{D}_n(\mathbb{C})$. To see this, first observe that $I_n \in \mathbb{G}(B)$ always holds, so any $D \in \mathbb{D}_n(\mathbb{C})$ is always diagonalized as $P^{-1}DP = D \in \mathbb{D}_n(\mathbb{C})$ with $P = I_n$. Thus, according to the assumptions, $P^{-1}D^{\star}P = B^{-1}D^H B \in \mathbb{D}_n(\mathbb{C})$. As this holds for any $D \in \mathbb{D}_n(\mathbb{C})$ (in particular any $\widehat{D} \in \widehat{\mathbb{D}}$) this implies $B \in \mathbb{GP}_n(\mathbb{C})$ according to Lemma 8.4. The proof works similarly considering $A^{\star}$ instead of A first. $\qquad\square$

Notice that, whenever A has pairwise distinct eigenvalues, then $P^{-1}AP = D_1 \in \mathbb{D}_n(\mathbb{C})$ for some $P \subset \mathbb{G}(B)$ always implies $P^{-1}A^\star P = D_2 \in \mathbb{D}_n(\mathbb{C})$ regardless of the form of B. This is because

$$D_1\left(P^{-1}A^\star P\right) = P^{-1}APP^{-1}A^\star P = P^{-1}AA^\star P = P^{-1}A^\star AP = P^{-1}A^\star PP^{-1}AP$$

$$= \left(P^{-1}A^\star P\right)D_1,$$

i.e. D_1 and $P^{-1}A^\star P$ commute. Since D_1 was assumed to have pairwise distinct diagonal entries, the only matrices commuting with D_1 are diagonal themselves [67]. However, if we have multiple eigenvalues this need not be true anymore and $P^{-1}A^\star P$ need not be diagonal. In fact, if $P^{-1}AP = D_1 \in \mathbb{D}_n(\mathbb{C})$ for some $P \in \mathbb{G}(B)$, then $P^{-1}A^\star P = B^{-1}D_1^H B$ (see (8.7)) which certainly need not at all be diagonal for some arbitrary B. In addition, the simultaneous diagonalization of A and $A^\star$ via some $P \in \mathrm{GL}_n(\mathbb{C})$ does not imply $P \in \mathbb{G}(B)$. To make sure that $P^{-1}A^\star P \in \mathbb{D}_n(\mathbb{C})$ always holds whenever $P^{-1}AP \in \mathbb{D}_n(\mathbb{C})$, we have to assume that $B \in \mathbb{GP}_n(\mathbb{C})$.

The second main fact about sesquilinear forms $[x, y] = x^H By$ with $B \in \mathbb{GP}_n(\mathbb{C})$ is the following.

Lemma 8.6. *Let $[x, y] = x^H By$ be some nondegenerate and orthosymmetric sesquilinear form. Then every diagonal matrix $D \in \mathbb{D}_n(\mathbb{C})$ is B-normal iff $B \in \mathbb{GP}_n(\mathbb{C})$.*

Proof. Suppose $D \in \mathbb{D}_n(\mathbb{C})$ and $B \in \mathbb{GP}_n(\mathbb{C})$. Then

$$D^\star = B^{-1}D^H B \in \mathbb{D}_n(\mathbb{C})$$

according to Lemma 8.5 (i), and, obviously, $DD^\star = D^\star D$ as two diagonal matrices always commute. Thus any $D \in \mathbb{D}_n(\mathbb{C})$ is normal. Now suppose $DD^\star = D^\star D$ holds for every $D \in \widehat{\mathbb{D}}$. This implies $D^\star = B^{-1}D^H B$ to be diagonal for any $D \in \widehat{\mathbb{D}}$. This in turn implies that $B \in \mathcal{Z}(\widehat{\mathbb{D}}) = \mathbb{GP}_n(\mathbb{C}) = \mathcal{Z}(\mathbb{D}_n(\mathbb{C}))$ (recall Lemma 8.4 and its proof). Thus, in conclusion, B needs to be a generalized complex permutation matrix. $\square$

Using a sesquilinear form $[x, y] = x^H By$ with $B \in \mathbb{GP}_n(\mathbb{C})$ we conclude: the diagonalization $P^{-1}AP \in \mathbb{D}_n(\mathbb{C})$ of any B-normal matrix $A \in \mathrm{M}_{n \times n}(\mathbb{C})$ via some $P \in \mathbb{G}(B)$ implies that $P^{-1}A^\star P = B^{-1}D^H B$ is also diagonal. In particular, the columns of any $P \in \mathbb{G}(B)$ diagonalizing either A or $A^\star$ are a common eigenbasis to both A and $A^\star$. Moreover, whenever $B \in \mathbb{GP}_n(\mathbb{C})$, then we are assured that any matrix $D \in \mathbb{D}_n(\mathbb{C})$ is always normal with respect to $[x, y] = x^H By$. Because of these two facts, we consider the question of automorphic diagonalization in the next chapter only for sesquilinear forms $[x, y] = x^H By$ which are Hermitian or skew-Hermitian with $B \in \mathbb{GP}_n(\mathbb{C})$.

8.4   Outlook on Chapters 9 and 10

In the following two chapters our main object of investigation are matrices $A \in \mathrm{M}_{n \times n}(\mathbb{C})$ that are B-normal with respect to some indefinite sesquilinear form $[x, y] =$

$x^H By$ on $\mathbb{C}^n$. In this case, assuming the matrix at hand is nondefective, the matrix and its adjoint can always be simultaneously diagonalized. As normal matrices with respect to the standard Euclidean scalar product (and their adjoints) can always be simultaneously diagonalized by some unitary matrix, the question arises whether a similar result holds for indefinite sesquilinear forms, too. That is, we will be considering the problem of automorphic diagonalization of nondefective matrices which are B-normal with respect to some indefinite sesquilinear form $[x, y] = x^H By$ in Chapter 9. For special sesquilinear forms the similar problem of automorphic triangularization (instead of diagonalization) has already been considered, e.g. [105]. Based on the results from Section 8.3 we will confine ourselves to forms induced by generalized (complex) permutation matrices which are either Hermitian or skew-Hermitian. In Chapter 10 we will consider the situation of selfadjoint and skewadjoint matrices A with respect to some indefinite sesquilinear form that are additionally normal in the sense of the Euclidean inner product. Those matrices A are structured twice, i.e. we have $AA^H = A^H A$ (meaning normality for the Euclidean scalar product) and $A^\star = B^{-1}A^H B = \pm A$ depending on whether A is selfadjoint or skewadjoint with respect to $[x, y] = x^H By$. We will analyze these matrices and derive canonical (multiplicative) factorizations and additive decompositions of those doubly normal matrices. These results give a criterion on when a doubly normal matrix is diagonalizable by a unitary automorphism.

Section 9.1 provides some preliminary results on B-normal matrices $A \in \mathrm{M}_{n \times n}(\mathbb{R})$ for indefinite forms $[x, y] = x^H By$. Here we will derive some auxiliary statements on the eigenvectors of those matrices and their relation to the corresponding eigenvectors of the adjoint matrix $A^\star$. Based on these results, Section 9.2 presents a framework for the ordering of a common eigenbasis to A and $A^\star$ that reveals certain properties of those matrices and is the starting point for the construction of automorphisms diagonalizing both A and $A^\star$. This situation will be discussed in detail in Section 9.3 where we present necessary and sufficient conditions on when A is automorphic diagonalizable. Furthermore, we show that B-normal matrices that are additionally normal for the Euclidean scalar product (meaning that $A^H A = AA^H$ holds) are always diagonalizable by a unitary automorphism. In Section 9.4 we apply our results from Section 9.3 to sesquilinear forms occurring often in practice. First, we show that our results certainly comply with the situation for the Euclidean scalar product. Afterwards, we consider (skew-)Hamiltonian and per(skew)-Hermitian matrices as well as pseudo(skew)-Hermitian matrices (the pseudoeuclidean sesquilinear form is introduced in Section 9.4.2). For these matrices, the results from Section 9.3 can be stated in a simplified manner. In Chapter 10 we consider (skew-)Hamiltonian and per(skew)-Hermitian matrices $A \in \mathrm{M}_{n \times n}(\mathbb{R})$ which are additionally normal (i.e. $AA^H = A^H A$). We derive a canonical factorization of those matrices under unitary-automorphic transformations. We show that, as a consequence of these factorizations, these matrices always admit a certain canonical additive decomposition with some surprising properties. Chapter 10 ends with a short summary of these results in Section 10.4. The results from Chapters 8 to 10 have not been published yet, although some of these investigations are motivated by the findings from [30].

Chapter 9

Automorphic diagonalization of B-normal matrices

Irving Kaplansky (1917 – 2006), [38].

Let $[x, y] = x^H B y$ be some nondegenerate, Hermitian or skew-Hermitian sesquilinear form with $B = \pm B^H \in \mathbb{GP}_n(\mathbb{C})$. In this chapter we are concerned with the question whether a given B-normal matrix $A \in M_{n \times n}(\mathbb{C})$ can be diagonalized by a similarity transformation with some matrix $U \in \mathbb{G}(B)$. According to Lemma 8.5 we know that a diagonalization $U^{-1} A U = U^\star A U \in \mathbb{D}_n(\mathbb{C})$ of A for $U \in \mathbb{G}(B)$ also diagonalizes $A^\star$. Therefore, the columns of U are a common eigenbasis for A and $A^\star$. Based on this knowledge we will derive several relationships between the (common) eigenvectors of A and $A^\star$ and provide necessary and sufficient conditions for an automorphic diagonalization of A (and $A^\star$) to be possible. This situation is analyzed for certain sesquilinear forms in detail. For the sesquilinear form $[x, y] = x^H J_{2n} y$ with J_{2n} as given in (8.1) and (skew-)Hamiltonian matrices a lot of research has been done on their unitary/symplectic triangularization. In particular, it will turn out that our results from this chapter on symplectic diagonalization are quite similar to the ones from [50] or [105] on unitary/symplectic triangularization.

For the standard Euclidean scalar product $(x, y) = x^H y$ it is well-known that a matrix $A \in M_{n \times n}(\mathbb{C})$ that satisfies $A A^H = A^H A$ can always be brought to diagonal form by a unitary similarity transformation [67]. In particular, this applies to Hermitian and skew-Hermitian matrices as they are certainly normal. In contrast to this fact, B-normal matrices for an indefinite inner product $[x, y] = x^H B y$ need not even be diagonalizable at all. For instance consider the Hermitian sesquilinear form

$$[x, y] = x^H B y \quad \text{with} \quad B := R_n = \begin{bmatrix} & & 1 \\ & \cdot^{\cdot^{\cdot}} & \\ 1 & & \end{bmatrix} \in M_{n \times n}(\mathbb{R})$$

and an $n \times n$ Jordan block $J(\alpha)$ with real eigenvalue $\alpha \in \mathbb{R}$. Then $J(\alpha)$ is R_n-normal as it is selfadjoint, i.e. $J(\alpha) \in \mathbb{J}(R_n)$. However, $J(\alpha)$ is certainly not diagonalizable [55, Ex. 4.2.1]. Thus, for our study we generally have to assume that we are dealing with diagonalizable matrices.

In Section 9.1 we present some preliminary results on the eigenspaces of B-normal (diagonalizable) matrices $A \in M_{n \times n}(\mathbb{C})$. As the columns of any $U \in \mathbb{G}(B)$ that diagonalizes A form a common eigenbasis for A and $A^{\star}$ these results will be essential in the proofs that follow. To construct a matrix $U \in \mathbb{G}(B)$ such that $U^{-1}AU$ is diagonal from a given common eigenbasis of A and $A^{\star}$ we need to be able to transform one common eigenbasis of A and $A^{\star}$ into another one. Hereby, we have to keep track of all the individual common eigenspaces $E(A, \lambda) \cap E(A^{\star}, \mu)$ of A and $A^{\star}$ for all $\lambda \in \sigma(A), \mu \in \sigma(A^{\star})$ instead of changing the bases of $E(A, \lambda)$ and $E(A^{\star}, \mu)$ separately. To this end, in Section 9.2 we show how basis changes of all these individual common eigenspaces $E(A, \lambda) \cap E(A^{\star}, \mu)$ (with $\lambda \in \sigma(A), \mu \in \sigma(A^{\star})$) can be carried out easily and simultaneously. With this tool at hand we derive a canonical form from which we can determine whether A admits a diagonalization via some automorphism $U \in \mathbb{G}(B)$ (i.e. is automorphic diagonalizable). This is the content of Section 9.3. Our proof is constructive and shows how such an automorphism $U \in \mathbb{G}(B)$ that diagonalizes A and $A^{\star}$ can be obtained. From Section 9.4 on we consider standard examples of sesquilinear forms that arise in applications now and then. In particular, Section 9.4.1 shows how our results apply to the standard Euclidean inner product $(\cdot, \cdot)$. In Sections 9.4.2 and 9.4.3 the pseudoeuclidean inner product and the symplectic and perplectic forms $[x, y] = X^H J_{2n} y$ and $[x, y] = x^H R_{2n} y$ (where J_{2n} and R_{2n} are as introduced in Section 8) are considered and the implications of the results from Section 9.3 are analyzed. The latter two sections particularly focus on selfadjoint and skewadjoint matrices. Finally, we show in Section 9.5 that an automorphic diagonalization of a B-normal and Euclidean-normal matrix (i.e. $AA^H = A^H A$ holds) can always be constructed to be unitary as well. This chapter ends with some conclusions in Section 9.6.

9.1   Preliminary results on common eigenspaces

Consider a nondegenerate sesquilinear form $[x, y] = x^H By$ with $B = \pm B^H$ and a matrix $A \in M_{n \times n}(\mathbb{C})$ with $\lambda \in \sigma(A)$ and $\mu \in \sigma(A^{\star})$. In this section we collect several auxiliary results about the eigenspaces of A that will be needed in the sequel. To begin, it is immediate that, if $u \in E(A, \lambda)$ and $v \in E(A^{\star}, \mu)$, then

$$\overline{\lambda}[u, v] = [\lambda u, v] = [Au, v] = [u, A^{\star}v] = [u, \mu v] = \mu[u, v]. \tag{9.1}$$

Therefore, if $[u, v] \neq 0$, then $\mu = \overline{\lambda}$ has to hold. Later on, we frequently use this fact applied to a common eigenbasis $\mathbb{B} = \{u_1, \ldots, u_n\}$ of A and $A^{\star}$ which always exists if A is nondefective and B-normal (recall Section 8.3). In this particular situation, every $u_j \in \mathbb{B}$ is an eigenvector of A and $A^{\star}$ for some $\lambda \in \sigma(A)$ and $\mu \in \sigma(A^{\star})$, respectively. That is, u_j belongs to some $E(A, \lambda) \cap E(A^*, \mu)$. Lemma 9.1 states a simple consequence of (9.1) that will be needed throughout the whole section. Recall that we use the notation $E(A, \lambda, \mathbb{B})$ for $E(A, \lambda) \cap \mathbb{B}$.

Lemma 9.1. *Let $A \in \mathrm{M}_{n \times n}(\mathbb{C})$ be nondefective and B-normal with respect to some nondegenerate, orthosymmetric sesquilinear form $[x, y] = x^H B y$. Let $\mathbb{B}$ be a common eigenbasis of A and A^*. Then, if $u \in E(A, \lambda, \mathbb{B}) \cap E(A^*, \mu, \mathbb{B})$ for some $\lambda \in \sigma(A), \mu \in \sigma(A^*)$ and $v \in \mathbb{B}$ we have*

$$[u, v] \neq 0 \;\Rightarrow\; v \in E(A^*, \overline{\lambda}, \mathbb{B}) \cap E(A, \overline{\mu}, \mathbb{B}).$$

Proof. Let $u \in E(A, \lambda, \mathbb{B}) \cap E(A^*, \mu, \mathbb{B})$ and assume that $v \in \mathbb{B}$. Since v is an eigenvector of A and A^*, too, we have $v \in E(A, \nu, \mathbb{B}) \cap E(A^*, \xi, \mathbb{B})$ for two eigenvalues $\nu \in \sigma(A), \xi \in \sigma(A^*)$. First of all

$$\overline{\lambda}[u, v] = [\lambda u, v] = [Au, v] = [u, A^* v] = [u, \xi v] = \xi[u, v].$$

Therefore, if $[u, v] \neq 0$, this can only hold if $\xi = \overline{\lambda}$. On the other hand,

$$\overline{\nu}[v, u] = [\nu v, u] = [Av, u] = [v, A^* u] = [v, \mu u] = \mu[v, u].$$

According to Section 8.1, vector orthogonality is a symmetric relation for orthosymmetric sesquilinear forms. That is, $[u, v] \neq 0$ holds iff $[v, u] \neq 0$. The assumption $[u, v] \neq 0$ now implies $\nu = \overline{\mu}$. Thus, we have shown that $[u, v] \neq 0$ implies $v \in E(A, \overline{\mu}, \mathbb{B}) \cap E(A^*, \overline{\lambda}, \mathbb{B})$ and the proof is complete. $\qquad\square$

The following Corollary 9.1 reveals some important relationships between the eigenspaces of a B-normal and nondefective matrix $A \in \mathrm{M}_{n \times n}(\mathbb{C})$ and its adjoint A^*. It makes essential use of Lemma 9.1 and is the central result of this section. Keep in mind that, for any $A \in \mathrm{M}_{n \times n}(\mathbb{C})$ it holds that $\sigma(A^*) = \overline{\sigma(A)}$ as A^* and A^H are similar.

Corollary 9.1. *Let $A \in \mathrm{M}_{n \times n}(\mathbb{C})$ be nondefective and B-normal with respect to some nondegenerate orthosymmetric sesquilinear form $[x, y] = x^H B y$. Then the following statements hold:*

(i) For any $\lambda \in \sigma(A), \mu \in \sigma(A^*)$ it holds that

$$\big(E(A, \lambda) \cap E(A^*, \mu) \big) \simeq \big(E(A^*, \overline{\lambda}) \cap E(A, \overline{\mu}) \big). \tag{9.2}$$

Moreover, if $\mu \neq \overline{\lambda}$, then $E(A, \lambda) \cap E(A^*, \mu)$ and $E(A^*, \overline{\lambda}) \cap E(A, \overline{\mu})$ are disjoint and isotropic. Their direct sum is always nondegenerate whenever it is nonzero.

(ii) If $\lambda \in \sigma(A)$ and $E(A, \lambda) \cap E(A^*, \overline{\lambda})$ is nonzero, it is always a maximal nondegenerate subspace in $E(A, \lambda)$ and in $E(A^*, \overline{\lambda})$ and it holds that

$$E(A, \lambda) / \big(E(A, \lambda) \cap E(A, \lambda)^{[\perp]} \big) \simeq E(A, \lambda) \cap E(A^*, \overline{\lambda})$$

Proof. (*i*) The isomorphism (9.2) is clear if $\mu = \overline{\lambda}$. Now let $\mathbb{B}$ be a common eigenbasis of A and $A^\star$. Furthermore, let $\lambda \in \sigma(A), \mu \in \sigma(A^\star)$ with $\mu \neq \overline{\lambda}$ and consider the three vector spaces $E_\lambda := E(A, \lambda) \cap E(A^\star, \overline{\lambda}), E_{\lambda\mu} := E(A, \lambda) \cap E(A^\star, \mu)$, and $E_{\overline{\mu\lambda}} := E(A, \overline{\mu}) \cap E(A^\star, \overline{\lambda})$. Certainly,

$$E_\lambda(\mathbb{B}) := E(A, \lambda, \mathbb{B}) \cap E(A^\star, \overline{\lambda}, \mathbb{B}), \quad E_{\lambda\mu}(\mathbb{B}) := E(A, \lambda, \mathbb{B}) \cap E(A^\star, \mu, \mathbb{B})$$

$$\text{and } E_{\overline{\mu\lambda}}(\mathbb{B}) := E(A, \overline{\mu}, \mathbb{B}) \cap E(A^\star, \overline{\lambda}, \mathbb{B})$$

are bases for $E_\lambda, E_{\lambda\mu}$ and $E_{\overline{\mu\lambda}}$, respectively. Note that we allow any set $E_\lambda(\mathbb{B}), E_{\lambda\mu}(\mathbb{B})$ or $E_{\overline{\mu\lambda}}(\mathbb{B})$ (or even all three sets) to be empty. As a consequence of Lemma 9.1 $E_{\lambda\mu}$ and $E_{\overline{\mu\lambda}}$ are isotropic since, for instance, $v, w \in E_{\lambda\mu}(\mathbb{B})$ and $[v, w] \neq 0$ would imply $\mu = \overline{\lambda}$. Obviously all three spaces have pairwise empty intersections (otherwise we would again have $\mu = \overline{\lambda}$). In conclusion, $E(A, \lambda) \cap E(A^\star, \mu)$ and $E(A^\star, \overline{\lambda}) \cap E(A, \overline{\mu})$ are disjoint and isotropic.

We now collect the vectors from $E_\lambda(\mathbb{B}), E_{\lambda\mu}(\mathbb{B})$ and $E_{\overline{\mu\lambda}}(\mathbb{B})$ as columns in the matrices $U = [u_1 \cdots u_k] \in M_{n \times k}(\mathbb{C}), V \in M_{n \times \ell}(\mathbb{C})$ and $W \in M_{n \times m}(\mathbb{C})$, respectively, and all remaining elements from $\mathbb{B}$ in the matrix $X \in M_{n \times s}(\mathbb{C})$ (note that $s = n - k - \ell - m$). Then for $Z := [U \ V \ W \ X] \in GL_n(\mathbb{C})$ we obtain a matrix of the following structure:

$$Z^H B Z = \begin{bmatrix} U^H B U & 0 & 0 \\ 0 & \begin{array}{c|c} 0 & V^H B W \\ \hline W^H B V & 0 \end{array} & 0 \\ 0 & 0 & X^H B X \end{bmatrix}. \qquad (9.3)$$

The zero structure of $Z^H B Z$ follows exclusively from the fact stated in Lemma 9.1. For instance, $V^H B U = 0, W^H B U = 0$ and $X^H B U = 0$ hold since, according to Lemma 9.1, we have for $u \in E(A, \lambda, \mathbb{B}) \cap E(A^\star, \overline{\lambda}, \mathbb{B})$ and $z \in \mathbb{B}$ that $[u, z] \neq 0$ implies $z \in E(A, \overline{\lambda}, \mathbb{B}) \cap E(A^\star, \lambda, \mathbb{B})$. Since any vector $z \in E(A, \overline{\lambda}, \mathbb{B}) \cap E(A^\star, \lambda, \mathbb{B})$ already appears as a column in U, necessarily $[V \ W \ X]^H B U = 0$ follows. The remaining zero structure of $Z^H B Z$ follows by a similar argument based on Lemma 9.1.

Notice that the nonsingularity of Z (the columns of Z are the basis vectors from $\mathbb{B}$) implies the nonsingularity of $Z^H B Z$ in (9.3). We consider the matrix

$$[V \ W]^H B [V \ W] = \begin{bmatrix} 0 & V^H B W \\ \hline W^H B V & 0 \end{bmatrix}$$

which then also has to be nonsingular. Now, if either $E_{\lambda\mu}(\mathbb{B})$ or $E_{\overline{\mu\lambda}}(\mathbb{B})$ contains nontrivial elements, that is either $\ell \neq 0$ or $m \neq 0$ holds, we must have $\ell = m$ due to the nonsingularity of $[V \ W]^H B [V \ W]$. In other words, the matrices $V^H B W \in M_{\ell \times m}(\mathbb{C})$ and $W^H B V \in M_{m \times \ell}(\mathbb{C})$ are themselves square and nonsingular. On the other hand, $\ell = 0$ or $m = 0$ implies $\ell = m = 0$ and the submatrix $[V \ W]^H B [V \ W]$ in (9.3) does not exist at all. As $\ell = \dim(E_{\lambda\mu})$ and $m = \dim(E_{\overline{\mu\lambda}})$, the fact $\ell = m$ thus yields that $(E(A, \lambda) \cap E(A^\star, \mu)) \simeq (E(A^\star, \overline{\lambda}) \cap E(A, \overline{\mu}))$ since both vector spaces have the same dimension. In addition, if $\ell = m \neq 0$, the nonsingularity of $[\,V\ W\,]^H B [\,V\ W\,]$ implies that $E_{\lambda\mu} \oplus E_{\overline{\mu\lambda}} = \mathrm{span}(V) \oplus \mathrm{span}(W)$ is nondegenerate according to Definition 8.3 (ii).

(ii) Now consider again the matrix from (9.3) with the definitions from part (i). In case $E_\lambda(\mathbb{B})$ is not empty, i.e. $k \neq 0$, the matrix $U^H B U \in \mathrm{M}_{k \times k}(\mathbb{C})$ has to be nonsingular, too. That is, $E_\lambda = E(A, \lambda) \cap E(A^\star, \overline{\lambda}) = \mathrm{span}(U)$ is nondegenerate in $E(A, \lambda)$ and in $E(A^\star, \overline{\lambda})$ whenever it is nonzero. Let $M := \{s_1, \ldots, s_p\} = E(A, \lambda, \mathbb{B}) \setminus \{u_1, \ldots, u_k\}$, so M contains all $p \geq 0$ eigenvectors of A for λ that are not already eigenvectors of $A^\star$ for $\overline{\lambda}$ (i.e. we assume implicitly $\dim(E(A, \lambda)) = k + p$). Previously, the vectors in M have appeared as columns in the matrices V and X. Therefore, it is clear from (9.3) that $[u_j, v] = 0$ for all $v \in M$ and all $u_j \in E_\lambda(\mathbb{B})$. Moreover, it follows once more from Lemma 9.1 that $\mathrm{span}(M)$ is isotropic. In conclusion, we have

$$[\,U\ s_1\ \cdots\ s_p\,]^H B [\,U\ s_1\ \cdots\ s_p\,] = \left[\begin{array}{c|c} U^H B U & 0 \\ \hline 0 & 0_{p \times p} \end{array}\right]$$

According to Lemma 8.1 we obtain that E_λ is a maximal nondegenerate subspace in $E(A, \lambda)$ and that $\mathrm{span}(M) = E(A, \lambda) \cap E(A, \lambda)^{[\perp]}$ is a maximal isotropic subspace in $E(A, \lambda)$. The argumentation for $A^\star$ proceeds in a similar manner. Now statement (ii) follows. $\square$

Note that Corollary 9.1 holds in case $E(A, \lambda) \cap E(A^\star, \overline{\lambda}) = 0$. Then, in particular, $E(A, \lambda)$ is isotropic and therefore $E(A, \lambda) = E(A, \lambda) \cap E(A, \lambda)^{[\perp]}$. Since all maximal nondegenerate subspaces of $E(A, \lambda)$ must have the same dimension, it follows that $E(A, \lambda) \cap E(A^\star, \overline{\lambda}) \simeq T$ for every maximal nondegenerate subspace T in $E(A, \lambda)$. Thus, in fact a canonical choice of a maximal nondegenerate subspace in either $E(A, \lambda)$ or $E(A^\star, \overline{\lambda})$ is $E(A, \lambda) \cap E(A^\star, \overline{\lambda})$. Moreover, notice that the statement in Corollary 9.1 and the proof remains valid if A and $A^\star$ are interchanged.

9.2 An eigenbasis ordering

Let $[x, y] = x^H B y$ be some nondegenerate sesquilinear form with $B = \pm B^H \in \mathbb{GP}_n(\mathbb{C})$. Moreover, assume that $A \in \mathrm{M}_{n \times n}(\mathbb{C})$ is diagonalizable with $P = [\,p_1\ \cdots\ p_n\,] \in \mathbb{G}(B)$ such that $P^{-1}AP = D \in \mathbb{D}_n(\mathbb{C})$. According to Lemma 8.5 (i) $p_1, \ldots, p_n$ is a common eigenbasis to A and $A^\star$ (since $P^{-1}A^\star P$ is also diagonal) which we again denote by $\mathbb{B}$. Recall that $P^H B P = B$, so we have

$$B = \begin{bmatrix} b_{11} & b_{12} & \cdots & b_{1n} \\ b_{21} & b_{22} & \cdots & b_{2n} \\ \vdots & \vdots & \ddots & \vdots \\ b_{n1} & b_{n2} & \cdots & b_{nn} \end{bmatrix} = \begin{bmatrix} [p_1, p_1] & [p_1, p_2] & \cdots & [p_1, p_n] \\ [p_2, p_1] & [p_2, p_2] & \cdots & [p_2, p_n] \\ \vdots & \vdots & \ddots & \vdots \\ [p_n, p_1] & [p_n, p_2] & \cdots & [p_n, p_n] \end{bmatrix}. \tag{9.4}$$

With these assumptions, observe the following two important facts:

(i) Assume that $b_{kk} \neq 0$ for some $1 \leq k \leq n$. Then, since $P^H B P = B$, we have $[p_k, p_k] = b_{kk} \neq 0$. Thus, assuming $p_k \in E(A, \lambda, \mathbb{B}) \cap E(A^\star, \mu, \mathbb{B})$ for some $\lambda \in \sigma(A), \mu \in \sigma(A^\star)$, Lemma 9.1 yields that $p_k \in E(A, \overline{\mu}, \mathbb{B}) \cap E(A^\star, \overline{\lambda}, \mathbb{B})$. In particular, this implies $\mu = \overline{\lambda}$. Therefore, if $b_{kk} = [p_k, p_k]$ is nonzero, p_k has to be an eigenvector of A and $A^\star$ for λ and $\overline{\lambda}$, respectively.

(ii) Assume that $b_{k\ell} \neq 0$ for some $1 \leq k, \ell \leq n$. Then $P^H B P = B$ implies $b_{k\ell} = [p_k, p_\ell] \neq 0$. Now, assuming that $p_k \in E(A, \lambda, \mathbb{B}) \cap E(A^\star, \mu, \mathbb{B})$ for some $\lambda \in \sigma(A), \mu \in \sigma(A^\star)$, we obtain

$$p_\ell \in E(A, \overline{\mu}, \mathbb{B}) \cap E(A^\star, \overline{\lambda}, \mathbb{B})$$

according to Lemma 9.1. In this case $\mu \neq \overline{\lambda}$ as well as $\mu = \overline{\lambda}$ is possible.

Thus, nonzero off-diagonal entries $b_{k\ell} = [p_k, p_\ell] \neq 0$ with $k \neq \ell$ can be accomplished by eigenvectors p_k, p_ℓ of A and $A^\star$ such that the relations in (ii) above hold with either $\mu \neq \overline{\lambda}$ or $\mu = \overline{\lambda}$. On the other hand, nonzero diagonal entries $b_{kk} = [p_k, p_k] \neq 0$ can solely be produced by eigenvectors $p_k \in E(A, \lambda, \mathbb{B}) \cap E(A^\star, \overline{\lambda}, \mathbb{B})$ for some eigenvalue $\lambda \in \sigma(A)$.

Given any common eigenbasis $p_1, \ldots, p_n$ of A and $A^\star$, let $P = [\, p_1 \, \cdots \, p_n \,] \in M_{n \times n}(\mathbb{C})$. Then the matrix $P^H B P$ is always nonsingular. However, the nonzero entries in $P^H B P$ might be scattered over the whole matrix giving $P^H B P$ no structure at all. However, for our purpose it is useful that $P^H B P$ is structured somehow according to the knowledge from (i) and (ii) above. Therefore, our next goal is to order the eigenvectors $p_1, \ldots, p_n$ so that we may better determine the form of $P^H B P$. One procedure to achieve this is presented next. It is comprehensively explained in Example 9.1 below.

For this purpose, let $\lambda_1, \ldots, \lambda_{r+1}$ be the pairwise distinct eigenvalues of A and let $\Lambda = (\lambda_1, \ldots, \lambda_{r+1})$ be some ordering of them. Recall that the eigenvalues of $A^\star$ are $\overline{\lambda}_1, \ldots, \overline{\lambda}_{r+1}$ as $A^\star$ and A^H are similar. In the following we will define an ordering of the basis $\mathbb{B}$ that will be helpful in the sequel. At first, for $j = 1, \ldots, r+1$, we define the sets $U(j) := E(A, \lambda_j, \mathbb{B}) \cap E(A^\star, \overline{\lambda}_j, \mathbb{B})$,

$$V(j) := E(A, \lambda_j, \mathbb{B}) \setminus \sum_{i=1}^{j} E(A^\star, \overline{\lambda}_i, \mathbb{B}) \quad \text{and} \quad W(j) := E(A^\star, \overline{\lambda}_j, \mathbb{B}) \setminus \sum_{i=1}^{j} E(A, \lambda_i, \mathbb{B}),$$

respectively, with respect to the ordering $\Lambda = (\lambda_1, \ldots, \lambda_{r+1})$. The sets $U(j), V(j), W(j)$, $j = 1, \ldots, r+1$, form a partition of $\mathbb{B}$ as can be easily verified: in fact, for an eigenvector $v \in E(A, \lambda_j, \mathbb{B}) \cap E(A^\star, \overline{\lambda}_k, \mathbb{B})$ with $\lambda_j, \lambda_k \in \sigma(A)$ there are three possibilities:

(a) The vector v appears in $V(j)$ iff λ_j appears before λ_k in Λ.

(b) The vector v appears in $W(k)$ iff λ_k appears before λ_j in Λ.

(c) The vector v appears in $U(j)$ in case $\lambda_j = \lambda_k$, i.e. $j = k$.

Additionally we set $d_j = |U(j)|$ and $s_j := |V(j)|$ where $|\cdot|$ denotes the cardinality of the set at hand. Notice that, as a consequence of Corollary 9.1 (i), we also have $s_j = |W(j)|$. Moreover, $s_{r+1} = 0$ since it follows directly from the definition of $V(r+1)$ and $W(r+1)$ that $V(r+1) = W(r+1) = \emptyset$. Now we proceed in the following manner for ordering the sets $U(j), V(j)$ and $W(j)$:

(i) For each $j = 1, \ldots, r+1$, we arrange the vectors from $U(j), V(j)$ and $W(j)$ as the columns of the matrices $U_j \in \mathrm{M}_{n \times d_j}(\mathbb{C})$, $V_j \in \mathrm{M}_{n \times s_j}(\mathbb{C})$ and $W_j \in \mathrm{M}_{n \times s_j}(\mathbb{C})$, respectively.

(ii) In the course of step (i), the columns of V_j and W_j should again be arranged according to the ordering Λ. That is, if $v_{\ell_1}, v_{\ell_2} \in V(j)$ are eigenvectors of $A^\star$ for $\overline{\lambda}_{t_1}$ and $\overline{\lambda}_{t_2}$ $(t_1, t_2 > j)$, we apply the ordering Λ to λ_{t_1} and λ_{t_2} and the column v_{ℓ_1} should appear left to the column v_{ℓ_2} in V_j iff λ_{t_1} appears before λ_{t_2} in Λ and vice versa. The same ordering should be applied to construct W_j. I.e., if $w_{\ell_1}, w_{\ell_2} \in W(j)$ are eigenvectors of A for λ_{t_1} and λ_{t_2} $(t_1, t_2 > j)$, we apply the ordering Λ and the column w_{ℓ_1} should appear left to the column w_{ℓ_2} in W_j iff λ_{t_1} appears before λ_{t_2} in Λ and vice versa. In case two eigenvectors from $V(j)$ or $W(j)$ belong to the same eigenspace (for $A^\star$ or A, respectively), they can be ordered arbitrarily. Moreover, the columns in each U_j can be arranged in any order.

(iii) If for all $j = 1, \ldots, r+1$ the matrices U_j, V_j and W_j have been constructed as in step (ii), we set

$$Y_j := [\, U_j \; V_j \; W_j \,], j = 1, \ldots, r+1, \quad \text{and} \quad Z := [\, Y_1 \; \cdots \; Y_{r+1} \,] \in \mathrm{M}_{n \times n}(\mathbb{C}). \tag{9.5}$$

Notice that the columns of Z are still a common eigenbasis of A and $A^\star$ (the basis $\mathbb{B}$ has only been arranged in a particular order). Moreover, recall that some of the sets $U(j), V(j)$ and $W(j)$ can be empty. In this case, the corresponding matrices U_j, V_j and W_j are not existent. Before we describe the particular form of $Z^H B Z$ that we obtain, the following example should illustrate that ordering procedure.

Example 9.1. Let $[x, y] = x^H B y$ be a nondegenerate sesquilinearform on $\mathbb{C}^n$ with $B = \pm B^H$. Moreover, assume that $A \in \mathrm{M}_{10}(\mathbb{C})$ is nondefective and B-normal. Moreover, let $u_1, u_2, \ldots, u_{10}$ be a common eigenbasis to A and $A^\star$ and assume $\sigma(A) = \{\lambda, \mu, \nu\}$. This implies $\sigma(A^\star) = \{\overline{\lambda}, \overline{\mu}, \overline{\nu}\}$. Assume the following assignment of the eigenvectors $u_1, u_2, \ldots, u_{10}$ to the corresponding eigenvalues of A and $A^\star$:

	A	$A^\star$	
eigenvalue λ	u_2, u_7, u_8	u_1, u_7, u_8	eigenvalue $\overline{\lambda}$
eigenvalue μ	u_3, u_4, u_9	u_4, u_5, u_6	eigenvalue $\overline{\mu}$
eigenvalue ν	u_1, u_5, u_6, u_{10}	u_2, u_3, u_9, u_{10}	eigenvalue $\overline{\nu}$

For instance, u_3 is an eigenvector of A for μ while it is an eigenvector of $A^\star$ for $\bar{\nu}$. We consider the orderings $\Lambda_1 = (\lambda, \mu, \nu)$ and $\Lambda_2 = (\nu, \mu, \lambda)$ of $\sigma(A)$ and obtain the following sets:

	$U(1)$	$V(1)$	$W(1)$	$U(2)$	$V(2)$	$W(2)$	$U(3)$
Λ_1	u_7, u_8	u_2	u_1	u_4	u_3, u_9	u_5, u_6	u_{10}
Λ_2	u_{10}	u_1, u_5, u_6	u_2, u_3, u_9	u_4	$\emptyset$	$\emptyset$	u_7, u_8

As expected, $V(3) = W(3) = \emptyset$ for both orderings. Obviously the sets $U(1), U(2)$ and $U(3)$ for Λ_1 can be permuted according to the permutation $1 \mapsto 3, 2 \mapsto 2, 3 \mapsto 1$ to obtain the sets corresponding of Λ_2 (this is exactly the permutation giving Λ_2 from Λ_1). Notice that this is not possible with the sets $V(j), W(j), j = 1, 2$. Whenever there is more than one vector in some of the sets, the vectors have to be ordered as described. This works as follows. First consider the ordering $\Lambda_1 = (\lambda, \mu, \nu)$: The order in which u_3, u_9 should appear in V_2 depends on the eigenspaces of $A^\star$ and the ordering Λ_1. However, since both u_3 and u_9 are eigenvectors of $A^\star$ for the same eigenvalue $\bar{\nu}$ they can in fact be ordered arbitrarily. The same situation applies to W_2 so the vectors u_5, u_6 (as they are eigenvectors of A for the same eigenvalue ν) can appear in any order in W_2. This situation changes if we consider the ordering Λ_2. For $V(1)$ the vectors u_1, u_5, u_6 have to be ordered according to their eigenvalues of $A^\star$ and $\Lambda_2 = (\nu, \mu, \lambda)$. As u_5, u_6 are eigenvectors of $A^\star$ for $\bar{\mu}$ and u_1 is an eigenvector of $A^\star$ for $\bar{\lambda}$, we may set $V_1 = [\, u_5 \; u_6 \; u_1 \,]$ since μ appears before λ in Λ_2. A similar situation applies to $W(1)$: as u_3, u_9 are eigenvectors of A for μ and u_2 is an eigenvector of A for λ (and the ordering considered is (ν, μ, λ)), we may set $W_1 = [\, u_3 \; u_9 \; u_2 \,]$. If we now form $Z = [\, Y_1 \; Y_2 \; Y_3 \,]$ according to both orderings Λ_1 and Λ_2, respectively, we may use Lemma 9.1 to characterize the nonzero-structure of $Z^H B Z$. Both nonzero-structures are displayed in Figure 9.1.

In general, the following result holds:

Theorem 9.1. *Let $A \in M_{n\times n}(\mathbb{C})$ be nondefective and B-normal for some nondegenerate sesquilinear form $[x, y] = x^H B y$ with $B = \pm B^H$. Applied to a common eigenbasis $\mathbb{B}$ of A and $A^\star$ and the eigenvalue ordering $\Lambda = (\lambda_1, \ldots, \lambda_{r+1})$ for the pairwise distinct eigenvalues $\lambda_1, \ldots, \lambda_{r+1}$ of $\sigma(A)$, the ordering procedure produces a matrix $Z \in GL_n(\mathbb{C})$ in (9.5) so that $Z^{-1}AZ \in \mathbb{D}_n(\mathbb{C})$ and $Z^H B Z$ has the form*

$$
Z^H B Z = \begin{bmatrix}
\Delta_1 & 0 & & & & & \\
0 & \begin{array}{c|c} 0 & S_1 \\ \hline \widehat{S}_1 & 0 \end{array} & 0 & \cdots & & & 0 \\
& 0 & \ddots & & & & \vdots \\
& & & \Delta_r & 0 & & \\
\vdots & & & 0 & \begin{array}{c|c} 0 & S_r \\ \hline \widehat{S}_r & 0 \end{array} & 0 & \\
0 & & \cdots & & 0 & & \Delta_{r+1}
\end{bmatrix}. \tag{9.6}
$$

Moreover, for all $j = 1, \ldots, r + 1$, the following statements hold:

$$
\left[\begin{array}{cc|cc|cc|cccc}
\star & \star & & & & & & & & \\
\star & \star & & & & & & & & \\
\hline
 & & 0 & \star & & & & & & \\
 & & \star & 0 & & & & & & \\
\hline
 & & & & \star & & & & & \\
\hline
 & & & & & & 0 & 0 & \star & \star \\
 & & & & & & 0 & 0 & \star & \star \\
 & & & & & & \star & \star & 0 & 0 \\
 & & & & & & \star & \star & 0 & 0 \\
\hline
 & & & & & & & & & \star
\end{array}\right]
\qquad
\left[\begin{array}{c|cccccc|c|cc}
\star & & & & & & & & & \\
\hline
 & 0 & 0 & 0 & \star & \star & 0 & & & \\
 & 0 & 0 & 0 & \star & \star & 0 & & & \\
 & 0 & 0 & 0 & 0 & 0 & \star & & & \\
 & \star & \star & 0 & 0 & 0 & 0 & & & \\
 & \star & \star & 0 & 0 & 0 & 0 & & & \\
 & 0 & 0 & \star & 0 & 0 & 0 & & & \\
\hline
 & & & & & & & \star & & \\
\hline
 & & & & & & & & \star & \star \\
 & & & & & & & & \star & \star
\end{array}\right]
$$

FIGURE 9.1: The nonzero-structure of the matrices $Z^H BZ$ with respect to the orderings Λ_1 (left) and Λ_2 (right). Both matrices are permutation similar and can contain at most 16 nonzero entries as a consequence of Lemma 9.1.

(i) The blocks $\Delta_j = U_j^H B U_j$ have sizes $d_j \times d_j$. These matrices are either nonsingular in case $d_j > 0$ or nonexistent if $d_j = 0$. In particular, $\Delta_j = \Delta_j^H$ whenever B is Hermitian and $\Delta_j = -\Delta_j^H$ whenever B is skew-Hermitian.

(ii) The blocks $S_j = V_j^H B W_j$ and $\widehat{S}_j = W_j^H B V_j$ have sizes $s_j \times s_j$. In case B is Hermitian we have $\widehat{S}_j = S_j^H$ while we have $\widehat{S}_j = -S_j^H$ whenever B is skew-Hermitian. The matrices S_j are block-diagonal of the form

$$
S_j = \begin{bmatrix}
S_{j,j+1} & 0 & \cdots & 0 \\
0 & S_{j,j+2} & & \vdots \\
\vdots & & \ddots & 0 \\
0 & \cdots & 0 & S_{j,r+1}
\end{bmatrix}
$$

with square blocks $S_{j,k} \in \mathrm{M}_{s_{j,k} \times s_{j,k}}(\mathbb{C})$. In particular, each block $S_{j,k}$ is either nonsingular or nonexistent.

The structure in (9.6) is easily analyzed: the entries in $\Delta_j = U_j^H B U_j$ are the inner products $[x, y]$ of elements x, y from $E(A, \lambda_j, \mathbb{B}) \cap E(A^\star, \overline{\lambda_j}, \mathbb{B})$ (in particular, Δ_j is nonexistent if $E(A, \lambda_j, \mathbb{B}) \cap E(A^\star, \overline{\lambda_j}, \mathbb{B}) = \emptyset$). Moreover, the elements of $S_{j,k}$ (which is some diagonal block in $S_j = V_j^H B W_j$, $k > j$) are the inner products $[v, w]$ of elements $v \in E(A, \lambda_j, \mathbb{B}) \cap E(A^\star, \overline{\lambda_k}, \mathbb{B})$ and $w \in E(A^\star, \overline{\lambda_j}, \mathbb{B}) \cap E(A, \lambda_k, \mathbb{B})$ according to the ordering $(\lambda_1, \ldots, \lambda_{r+1})$. Therefore, Theorem 9.1 and in particular the form in (9.6) is just a direct consequence of Lemma 9.1 and the particular ordering.

Notice that the ordering used for the construction of the form (9.6) is not of particular interest. Another ordering simply effects the arrangement of the blocks in

(9.6) and thus produces nothing essentially new. To this end, recall Figure 9.1: one canonical form can be obtained from the other simply by a permutation.

Using (9.6) we can now describe the possibilities we have in changing the common eigenbasis $\mathbb{B}$ of A and $A^\star$ at hand. Certainly we are not allowed to replace the basis of some eigenspace $E(A, \lambda)$ simply by any other basis. In fact, we have to pay attention to the intersection of $E(A, \lambda)$ with the different eigenspaces of $A^\star$. As the form (9.6) groups together eigenvectors from the same eigenspace $E(A, \lambda, \mathbb{B}) \cap E(A^\star, \overline{\lambda}_k, \mathbb{B})$, $j, k = 1, \ldots, r+1$, all common eigenbases of A and $A^\star$ can be obtained from the modifications of these particular "common eigenspaces". This is the content of the following proposition.

Proposition 9.1. *Let $A \in \mathrm{M}_{n \times n}(\mathbb{C})$ be nondefective and B-normal for some nondegenerate sesquilinear form $[x, y] = x^H B y$ with $B = \pm B^H$. Moreover, let $\lambda_1, \ldots, \lambda_{r+1}$ be the distinct eigenvalues of A and assume that $Z = [\, z_1 \;\cdots\; z_n \,] \in \mathrm{GL}_n(\mathbb{C})$ in (9.5) was computed so that $Z^H B Z$ has the form in (9.6) for a common eigenbasis $z_1, \ldots, z_n$ of A and $A^\star$. Let $K = [\, k_1 \;\cdots\; k_n \,] \in \mathrm{M}_{n \times n}(\mathbb{C})$ be of the form*

$$
K := \begin{bmatrix}
\begin{matrix} Q_1 & \\ & T_1 \\ & & G_1 \end{matrix} & 0 & \cdots & 0 \\[2ex]
0 & \ddots & & \vdots \\
& & \begin{matrix} Q_r & \\ & T_r \\ & & G_r \end{matrix} & 0 \\[2ex]
0 & \cdots & 0 & Q_{r+1}
\end{bmatrix}
\tag{9.7}
$$

for matrices $Q_j \in \mathrm{M}_{d_j \times d_j}(\mathbb{C})$ and $T_j, G_j \in \mathrm{M}_{s_j \times s_j}(\mathbb{C})$ for $j = 1, \ldots, r+1$ where all T_j, G_j are assumed to have the same block-structure as the corresponding matrix S_j in (9.6). Then the following statements hold:

(i) *The vectors $Zk_1, Zk_2, \ldots, Zk_n$ are a common eigenbasis to A and $A^\star$ iff Q_j, T_j and G_j are nonsingular for all $j = 1, \ldots, r+1$ with $d_j > 0$ and $s_j > 0$. Moreover, each common eigenbasis of A and $A^\star$ can be obtained in that way.*

(ii) *Assume that K is nonsingular. If $Z^{-1} A Z = D_1 \in \mathbb{D}_n(\mathbb{C})$ and $Z^{-1} A^\star Z = D_2 \in \mathbb{D}_n(\mathbb{C})$ holds, then*

$$
(ZK)^{-1} A (ZK) = D_1 \quad \text{and} \quad (ZK)^{-1} A^\star (ZK) = D_2
$$

where $ZK = [\, Zk_1 \; Zk_2 \;\cdots\; Zk_n \,]$. That is, a similarity transformation of D_1 and D_2 with K does not change these diagonal matrices.

Proof. (i) This follows directly from the form (9.6). Indeed, to produce an eigenbasis for A and $A^\star$ different to the one given by the columns of Z, we are only allowed to change the basis for each eigenspace $E(A, \lambda_j, \mathbb{B}) \cap E(A^\star, \overline{\lambda}_k, \mathbb{B})$ independently. This is accomplished only by a matrix of the form of K iff K is nonsingular.

(ii) The sizes and position of the blocks in K are (by construction) adapted to the form of D_1 and D_2. In particular, for $K^{-1}D_1K$ each block from K is multiplied by a (diagonal) block with the same eigenvalue in D_1. The same holds for D_2. That is, D_1 and D_2 do not change at all. □

9.3 Towards an automorphic diagonalization

Recall that a diagonal matrix $D \in \mathbb{D}_n(\mathbb{C})$ is always normal with respect to $[x, y] = x^H By$ as long as $B \in \mathbb{GP}_n(\mathbb{C})$. In this section we investigate the question whether all B-normal and diagonalizable matrices $A \in M_{n\times n}(\mathbb{C})$ can be diagonalized by some $U \in \mathbb{G}(B)$. In particular, we assume $B \in \mathbb{GP}_n(\mathbb{C})$ is either Hermitian or skew-Hermitian.

For any Hermitian or skew-Hermitian matrix $M = [m_{ij}]_{ij} \in \mathbb{GP}_n(\mathbb{C})$ two numbers play an important role in the subsequent analysis. These are the number of nonzero diagonal entries $m_{jj} \neq 0$ of M, $j = 1, \ldots, n$, which we denote by $i(M)$, and the number of nonzero entries $0 \neq m_{jk} = \pm\overline{m}_{kj}$ $(j, k = 1, \ldots, n)$ of M which is denoted by $t(M)$. Nonzero entries $m_{jk} = \pm\overline{m}_{kj}$ will simply be called nonzero off-diagonal pairs for short. Thus, for any Hermitian or skew-Hermitian (complex) generalized permutation matrix M we obviously have the relation $t(M) = \frac{1}{2}(n - i(M))$. For instance, if

$$
M = \begin{bmatrix} 1 & & & \\ & & 1 & \\ & 1 & & \\ & & & -1 \end{bmatrix},
$$

then $i(M) = 2$ and $t(M) = \frac{1}{2}(4 - 2) = 1$. Note that, in case $M = M^H$ all diagonal entries of M are real while they are purely imaginary in case $M = -M^H$. Moreover, the spectrum of any Hermitian or skew-Hermitian matrix $M \in \mathbb{GP}_n(\mathbb{C})$ is easily determined. For both cases, nonzero diagonal entries $m_{jj} \neq 0$ are certainly eigenvalues of M. In addition, in the Hermitian case, any nonzero off-diagonal pair $m_{jk} = \overline{m}_{kj}$ contributes a pair of real eigenvalues $+|m_{jk}|, -|m_{jk}|$ to $\sigma(M)$ (with $|\cdot|$ denoting the absolute value of a complex number) while in the skew-Hermitian case each nonzero off-diagonal pair $m_{jk} = -\overline{m}_{kj}$ generates two purely imaginary eigenvalues $+|m_{jk}|i, -|m_{jk}|i$. We begin with a simple but crucial observation:

Proposition 9.2. *Let $A \in M_{n\times n}(\mathbb{C})$ be nondefective and B-normal for some non-degenerate sesquilinear form $[x, y] = x^H By$ with $B = \pm B^H \in \mathbb{GP}_n(\mathbb{C})$. Then A is automorphic diagonalizable iff there exists a nonsingular matrix $Z \in M_{n\times n}(\mathbb{C})$ with $Z^{-1}AZ \in \mathbb{D}_n(\mathbb{C})$ and $Z^H BZ \in \mathbb{GP}_n(\mathbb{C})$ so that*

$$
i(Z^H BZ) = i(B) \quad \text{and} \quad t(Z^H BZ) = t(B).
$$

Proof. $\Rightarrow$: Assume that A is diagonalizable via $U \in \mathbb{G}(B)$. Then take $Z = U$.
$\Leftarrow$: Now assume $Z \in M_{n\times n}(\mathbb{C})$ satisfies all criteria stated in the proposition, in

particular $Z^H B Z \in \mathbb{GP}_n(\mathbb{C})$ with $i(Z^H B Z) = i(B)$ and $t(Z^H B Z) = t(B)$. Then there exists a (real) permutation matrix $P \in \mathrm{M}_{n \times n}(\mathbb{R})$ so that $P^H(Z^H B Z)P$ has the same nonzero pattern as B. In fact, P only permutes nonzero diagonal entries and nonzero off-diagonal pairs of $Z^H B Z$ into the proper positions according to the nonzero structure of B. Certainly, still $P^H(Z^{-1}AZ)P \in \mathbb{D}_n(\mathbb{C})$ holds. Now notice that $\widetilde{B} := (ZP)^H B(ZP)$ and B are congruent, cf. Proposition 8.2.

(i) If B is Hermitian so is $\widetilde{B}$. Recall that each nonzero off-diagonal pair in B constributes a pair of real eigenvalues $+\lambda, -\lambda \in \mathbb{R}$ to its spectrum. The same is true for $\widetilde{B}$. Therefore, as $t(\widetilde{B}) = t(B) =: t$ both matrices have t positive and t negative real eigenvalues (counted with multiplicities) that rise from their nonzero off-diagonal pairs. As B and $\widetilde{B}$ are congruent (i.e. they have the same inertia) and $i(\widetilde{B}) = i(B)$ holds, both matrices must therefore have the same number of positive and negative (real) entries along their diagonals as well. Therefore, w.l.o.g. we can take P so that the positions of the positive and negative real diagonal entries in $\widetilde{B}$ and B match.

(ii) If B is skew-Hermitian so is $\widetilde{B}$. Now each nonzero off-diagonal pair in B constributes a pair of purely imaginary eigenvalues $+\lambda, -\lambda \in i\mathbb{R}$ to its spectrum. The same is true for $\widetilde{B}$. Therefore, as $t(\widetilde{B}) = t(B) =: t$ both matrices have t positive and t negative purely imaginary eigenvalues (counted with multiplicities) that rise from their nonzero off-diagonal pairs. As B and $\widetilde{B}$ are congruent (i.e. they have the same inertia) and $i(\widetilde{B}) = i(B)$ holds, both matrices must therefore have the same number of positive and negative (purely imaginary) entries along their diagonals as well. Therefore, w.l.o.g. we can take P so that the positions of the positive and negative purely imaginary diagonal entries in $\widetilde{B}$ and B match.

Finally, in both cases there exists a nonsingular diagonal matrix $D \in \mathbb{D}_n(\mathbb{C})$ that scales the entries of $\widetilde{B}$ appropriately so that $B = (ZPD)^H B(ZPD)$. Of course $(ZPD)^{-1}A(ZPD)$ remains to be diagonal.   □

Notice that in the proof of Proposition 9.2 we constructed an automorphism $U := ZPD \in \mathbb{G}(B)$ (that diagonalizes A) from Z simply by permutation and scaling of the columns of Z. Taking Theorem 8.5 (i) into account this shows that the columns of Z (under the conditions from Propositions 9.2) must have been a common eigenbasis to A and $A^\star$ before these changes have taken place. In view of Proposition 9.2 we must therefore consider the question whether it is possible to achieve that the number of nonzero diagonal entries and nonzero off-diagonal pairs in $Z^H B Z$ of the form (9.6) becomes equal to $i(B)$ and $t(B)$ by a transformation K as given in Proposition 9.1. We will now derive the necessary and sufficient conditions for this to be possible.

To this end, let $A \in \mathrm{M}_{n \times n}(\mathbb{C})$ be B-normal with respect to $[x, y] = x^H B y$ with $B = \pm B^H \in \mathbb{GP}_n(\mathbb{C})$ and assume $Z^{-1}AZ \in \mathbb{D}_n(\mathbb{C})$ so that $Z^H B Z$ is as in (9.6). First of all we are interested in turning (9.6) into a slightly more condensed form using Proposition 9.1.

First assume that B is Hermitian. Then, the matrices $Q_j \in \mathrm{GL}_{d_j}(\mathbb{C}), j = 1, \ldots, r+1$, in Proposition 9.1 can be chosen so that $Q_j^H \Delta_j Q_j$ is a diagonal signature matrix $\widehat{I}_{d_j}$, where $\widehat{I}_{d_j} = \mathrm{diag}[\pm 1, \ldots, \pm 1] \in \mathbb{D}_{d_j}(\mathbb{R})$ for a certain number of $+1$'s and -1's (recall that each Δ_j is Hermitian and nonsingular). Of course, the number of $+1$'s and -1's coincides with the inertia of Δ_j. In a second step, we choose $T_j = (S_j^H)^{-1}$ and $G_j = I_{s_j}$ for each $j = 1, \ldots, r$. Notice that T_j always has the same block structure as S_j (as required by Proposition 9.1) and that not all blocks Δ_j, T_j need to exist (in which case no modification is required).

If B is skew-Hermitian, we may proceed analogously: the matrices $Q_j \in \mathrm{Gl}_{d_j}(\mathbb{C})$, $j = 1, \ldots, r+1$, are chosen according to Proposition 9.1 so that $Q_j^H \Delta_j Q_j$ is a diagonal complex signature matrix $\widehat{I}_{d_j} = \mathrm{diag}[\pm i, \ldots, \pm i] \in \mathbb{D}_{d_j}(\mathbb{R})$ for some combination of $+i$'s and $-i$'s. For each $j = 1, \ldots, r$ we choose $T_j = (S_j^H)^{-1}$ and $G_j = I_{s_j}$ as before. With these choices, setting up K, the transformation $K^H(Z^H BZ)K$ yields a matrix of the form

$$
K^H Z^H BZK =
\begin{bmatrix}
\widehat{I}_{d_1} & 0 & & & & \\
0 & \begin{matrix} 0 & I_{s_1} \\ L_{s_1} & 0 \end{matrix} & 0 & \cdots & & 0 \\
& 0 & \ddots & & & \vdots \\
\vdots & & & \begin{matrix} \widehat{I}_{d_r} & 0 \\ 0 & \begin{matrix} 0 & I_{s_r} \\ L_{s_r} & 0 \end{matrix} \end{matrix} & 0 & \\
0 & & \cdots & & 0 & \widehat{I}_{d_{r+1}}
\end{bmatrix}
\tag{9.8}
$$

for certain (real/complex) signature matrices $\widehat{I}_{d_1}, \ldots, \widehat{I}_{d_{r+1}}$ with $L_{s_j} = I_{s_j}$ for all $j = 1, \ldots, r$ in case B is Hermitian or $L_{s_j} = -I_{s_j}$ otherwise. Notice that the matrix $(ZK)^H BZK$ in (9.8) is a generalized permutation matrix with $i((ZK)^H BZK) = \sum_{j=1}^{r+1} d_j$ and $t((ZK)^H BZK) = \sum_{j=1}^{r} s_j$. To simplify the notation, we use the abbreviation δ_A for $\sum_{j=1}^{r+1} d_j$ (recall that d_j denoted the size of $\Delta_j, j = 1, \ldots, r+1$). Certainly δ_A is a characteristic number for the particular matrix A (as is each $d_j, j = 1, \ldots, r+1$) and not dependent on the decomposition (9.6) or any modifications from K. With this form at hand, we now distinguish between three possible cases.

1. Case $\delta_A < i(B)$. In this case, the automorphic diagonalization of A is not possible. This is because there are not enough eigenvectors from nonisotropic eigenspaces $E(A, \lambda_j) \cap E(A^\star, \overline{\lambda}_j)$, $\lambda_j \in \sigma(A)$, that would be necessary to produce the $i(B)$ nonzero diagonal entries in B. Recall also the discussion at the beginning of Section 9.2 where we showed that $[v, v] \neq 0$ is only possible for some common eigenvector v of A and $A^\star$ if $v \in E(A, \lambda) \cap E(A^\star, \overline{\lambda})$ for some $\lambda \in \sigma(A)$. If there exist less than $i(B)$ common eigenvectors of A and $A^\star$ with this property, an automorphic diagonalization cannot exist.

2. Case $\delta_A = i(B)$. For this case, consider the matrix $\widetilde{Z} := ZK$ and the form $\widetilde{Z}^H B \widetilde{Z}$ in (9.8). In particular, notice that $\delta_A = i(\widetilde{Z}^H B \widetilde{Z})$. As $\delta_A = i(B)$ we have $i(B) = i(\widetilde{Z}^H B \widetilde{Z})$ and (as a consequence) $t(B) = t(\widetilde{Z}^H B \widetilde{Z})$. Moreover, as $\widetilde{Z}$ is nonsingular, all conditions from Proposition 9.2 are satisfied. This means that A is automorphic diagonalizable and a diagonalization can be found as described in the proof of Proposition 9.2.

3. Case $\delta_A > i(B)$. This is the most complicated case as the form (9.8) does not have enough nonzero off-diagonal pairs compared to B. Thus, additional nonzero off-diagonal pairs have to be created from the (real/complex) signature matrices $\widehat{I}_{d_1}, \ldots, \widehat{I}_{d_{r+1}}$ [1]. Notice that the number of missing off-diagonal pairs is $\frac{1}{2}(\delta_A - i(B))$ as there are $\frac{1}{2}(n - i(B))$ and $\frac{1}{2}(n - \delta_A)$ off-diagonal pairs in B and (9.8), respectively. We will now show how nonzero off-diagonal pairs can be created from such a signature matrix, say $\widehat{I}_{d_j}$.

A nonzero off-diagonal pair contributes either a pair of real eigenvalues $+1, -1$ or imaginary eigenvalues $+i, -i$ to the matrix in (9.8) dependent on B being Hermitian or skew-Hermitian. Thus, in fact a pair of $+1, -1$ or $+i, -i$ entries in $\widehat{I}_{d_j}$ can be turned into a Hermitian or skew-Hermitian off-diagonal pair $+1, +1$ or $+1, -1$. In the Hermitian case, the necessary transformation is given by

$$V^H \begin{bmatrix} -1 & 0 \\ 0 & 1 \end{bmatrix} V = \begin{bmatrix} 0 & 1 \\ 1 & 0 \end{bmatrix} \text{ for } V = \frac{1}{\sqrt{2}} \begin{bmatrix} -1 & 1 \\ 1 & 1 \end{bmatrix} \tag{9.9}$$

while it is given as

$$U^H \begin{bmatrix} -i & 0 \\ 0 & +i \end{bmatrix} U = \begin{bmatrix} 0 & 1 \\ -1 & 0 \end{bmatrix} \text{ for } U = \frac{1}{\sqrt{2}} \begin{bmatrix} -i & 1 \\ i & 1 \end{bmatrix} \tag{9.10}$$

in the skew-Hermitian case. Thus, whenever there are two diagonal entries $-1, +1$ or $+i, -i$ side by side in $\widehat{I}_{d_j}$, we can once more apply Proposition 9.1 to the matrix from (9.8). Thereby, we choose Q_j as an $d_j \times d_j$ identity matrix with V or U incorporated in the appropriate position along the main diagonal to generate one nonzero off-diagonal pair, i.e.

$$Q_j^H \widehat{I}_{d_j} Q_j = \begin{bmatrix} \widehat{I}_p & & \\ & \begin{matrix} 0 & 1 \\ \pm 1 & 0 \end{matrix} & \\ & & \widehat{I}_{d_j - p - 2} \end{bmatrix} \tag{9.11}$$

where $\widehat{I}_p$ and $\widehat{I}_{d_j - p - 2}$ are still real/complex signature matrices. As long as there are more pairs of diagonal entries $-1, +1$ or $+i, -i$ in $Q_j^H \widehat{I}_{d_j} Q_j$ in (9.11)

---

[1] This is the main difference between case 1 and case 3: under some circumstances, an off-diagonal pair can be created from two diagonal entries. However, a diagonal entry can never be created from any nonzero off-diagonal pair.

this process can be continued: by first choosing Q_j to be some particular permutation matrix the diagonal entries $-1, +1$ or $+i, -i$ can be permuted to appear side by side and additional nonzero off-diagonal pairs can be created by yet another suitable choice of Q_j as described above. In conclusion, whenever $\widehat{I}_{d_j}$ has $i_+(\widehat{I}_{d_j})$ $+1$'s/$+i$'s and $i_-(\widehat{I}_{d_j})$ -1's/$-i$'s along its main diagonal, we can construct a matrix Q_j according to Proposition 9.1 such that there are $\min\{i_+(\widehat{I}_{d_j}), i_-(\widehat{I}_{d_j})\}$ nonzero off-diagonal pairs in $Q_j^H \widehat{I}_{d_j} Q_j$ while all remaining entries of $Q_j^H \widehat{I}_{d_j} Q_j$ are still ± 1 or $\pm i$ distributed along the diagonal[2].

This procedure can now be applied to all the matrices $\widehat{I}_{d_1}, \ldots, \widehat{I}_{d_{r+1}}$. Thus, if the $+1$ ($+i$) and -1 ($-i$) entries are well distributed among those matrices, the creation of the missing off-diagonal pairs might succeed. However, notice that only $\min\{i_+(\widehat{I}_{d_j}), i_-(\widehat{I}_{d_j})\}$ off-diagonal pairs can be created from each $\widehat{I}_{d_j}$, so the overall creation of at most $\sum_{j=1}^{r+1} \min\{i_+(\widehat{I}_{d_j}), i_-(\widehat{I}_{d_j})\}$ off-diagonal pairs is possible. If the number of missing off-diagonal pairs $\frac{1}{2}(\delta_A - i(B))$ is less than this, the creation of those pairs is possible. Moreover, if $\frac{1}{2}(\delta_A - i(B))$ is strictly less than $\sum_{j=1}^{r+1} \min\{i_+(\widehat{I}_{d_j}), i_-(\widehat{I}_{d_j})\}$, we can choose from which signature matrices $\widehat{I}_{d_j}, j = 1, \ldots, r+1$, the missing pairs should be created. On the other hand, we are certainly not allowed to apply the transformation (9.9) or (9.10) to pairs from different matrices $\widehat{I}_{d_j}, \widehat{I}_{d_k}$ even if the two $\pm 1/\pm i$ entries appear in (9.8) side by side (for instance in the situation when we consider $\widehat{I}_{d_j}$ and $\widehat{I}_{d_{j+1}}$ and $s_j = 0$). This would mix different eigenspaces and therefore destroy the diagonal form of either $Z^{-1}AZ$ or $Z^{-1}A^\star Z$ or both. However, as the form of the transformation matrices as in (9.7) we apply is adapted to the forms (9.6) and (9.8), the previously described situation of mixing eigenspaces can in fact not occur in the described procedure.

If we succeed in creating the missing nonzero off-diagonal pairs from the matrices $\widehat{I}_{d_1}, \ldots, \widehat{I}_{d_{r+1}}$, we again are in the situation of Proposition 9.2, i.e. the new matrix has the same number of nonzero diagonal entries and nonzero off-diagonal pairs as B. Then Proposition 8.1 guarantees that the signs of the remaining diagonal entries of the resulting matrix matches those of B (modulo permutation). Then A is automorphic diagonalizable according to Proposition 9.2. Before we state our main theorem concerning automorphic diagonalization notice once more that it is possible that the diagonal entries with the same sign cluster too much in certain matrices among $\widehat{I}_{d_1}, \ldots, \widehat{I}_{d_{r+1}}$ so that there are not enough $\pm$ pairs in the single matrices $\widehat{I}_{d_j}$ to construct the necessary off-diagonal pairs. In this case we can not construct an automorphic diagonalization of A. We summarize our findings in the next theorem which is the main result of this section.

Theorem 9.2. Let $A \in \mathrm{M}_{n \times n}(\mathbb{C})$ with pairwise distinct eigenvalues $\lambda_1, \ldots, \lambda_{r+1}$ be nondefective and B-normal with respect to $[x, y] = x^H By$. Moreover, assume that

---

[2]For simplicity, we described that procedure in several steps. However, certainly all transformations on $\widehat{I}_{d_j}$ can be accomplished by the choice of one particular Q_j.

$B \in \mathbb{GP}_n(\mathbb{C})$ *is either Hermitian or skew-Hermitian and* $i(B)$ *denotes the number of nonzero diagonal entries of* B. *For each eigenspace* $E(A, \lambda_j) \cap E(A^*, \overline{\lambda}_j)$, $\lambda_j \in \sigma(A)$, *let* d_j *denote its dimension and let* $i_+(\lambda_j)$ *and* $i_-(\lambda_j)$ *be either the numbers of positive and negative eigenvalues of* $\Delta_j := U_j^H B U_j = \Delta_j^H$ *in case* $B = B^H$ *or the numbers of positive and negative imaginary eigenvalues of* $\Delta_j = -\Delta_j^H$ *in case* $B = -B^H$ *for some matrix* $U_j = [\, u_1 \; \cdots \; u_{d_j} \,] \in \mathrm{M}_{n \times d_j}(\mathbb{C})$ *whose columns are a basis of* $E(A, \lambda_j) \cap E(A^*, \overline{\lambda}_j)$. *Then* $i_+(\lambda_j)$ *and* $i_-(\lambda_j)$ *are independent of the particular choice of* U_j *and* A *is automorphic diagonalizable iff*

$$2 \cdot \sum_{j=1}^{r+1} \min \left\{ i_+(\lambda_j), i_-(\lambda_j) \right\} \geq \sum_{j=1}^{r+1} d_j - i(B) \geq 0 \qquad (9.12)$$

Proof. First notice that $\delta_A - i(B) = \sum_{j=1}^{r+1} d_j - i(B)$ is always even. In fact, if $i(B)$ is the number of nonzero diagonal entries in B, then $\frac{1}{2}(n - i(B))$ coincides with the number of nonzero off-diagonal pairs in B. Thus $n - i(B)$ is even. For A consider the form obtained in (9.8). Then $\delta_A = \sum_{j=1}^{r+1} d_j$ is the number of nonzero diagonal entries and, respectively, $\frac{1}{2}(n - \delta_A)$ is the number of nonzero off-diagonal pairs in (9.8). Thus, $n - \delta_A$ is even, too. Now notice that $(n - i(B)) - (n - \delta_A) = \delta_A - i(B)$ must be even. Thus $\frac{1}{2}(\sum_{j=1}^{r+1} d_j - i(B)) \in \mathbb{N}_0$. Comparing the form (9.8) and the number $\frac{1}{2}(n - i(B))$ off nonzero off-diagonal pairs in B, $\frac{1}{2}(\delta_A - i(B))$ is exactly the number of nonzero off-diagonal pairs that are missing in the form (9.8) compared to B. Thus, these are the pairs that have to be generated from the signature matrices $\hat{I}_{d_1}, \ldots, \hat{I}_{d_{r+1}}$ in (9.6) as described previously. From each ± 1 or $\pm i$ pair in any $\hat{I}_{d_j}$ exactly one such pair can be generated. Moreover, in $\hat{I}_{d_j}$ at most $\min\{i_+(\lambda_j), i_-(\lambda_j)\}$ off-diagonal pairs can be accomplished, where $i_+(\lambda_j)$ and $i_-(\lambda_j)$ denote the number of $+1/+i$'s and $-1/-i$'s along the main diagonal of $\hat{I}_{d_j}$, respectively. Therefore, if (9.12) is satisfied, the needed nonzero off-diagonal pairs can be created from $\hat{I}_{d_1}, \ldots, \hat{I}_{d_{r+1}}$. The fact that $i_+(\lambda_j)$ and $i_-(\lambda_j)$ are independent of the particular choice of U_j is clear from Proposition 8.2. In case A is automorphic diagonalizable, the relation (9.12) follows immediately from this diagonalization. $\square$

Notice that all three distinct cases discussed earlier are incorporated in (9.12). In fact, if (as in case 1) $\delta_A < i(B)$, then $\frac{1}{2}(\sum_{j=1}^{r+1} d_j - i(B)) < 0$ so the condition (9.12) does not hold. On the other hand if (as in case 2) $\delta_A = i(B)$, then $\frac{1}{2}(\sum_{j=1}^{r+1} d_j - i(B)) = 0$. As $\sum_{j=1}^{r+1} \min\{i_+(\lambda_j), i_-(\lambda_j)) \geq 0$ always holds, (9.12) is always satisfied. Thus, in conclusion, the complicated appearance of (9.12) is essentially due to case 3: $\delta_A > i(B)$.

9.4   Frequently arising indefinite forms: examples

In this section we discuss the implications from Theorem 9.2 for several standard sesquilinear forms. As in applications selfadjoint and skewadjoint matrices occur frequently, we focus on these classes of transformations. Moreover, since the standard

Euclidean scalar product $(x, y) = x^H y$ on $\mathbb{C}^n$ is a special (positive definite) sesquilinear form, our result from (9.12) certainly has to hold for Hermitian and skew-Hermitian matrices $A \in M_{n \times n}(\mathbb{C})$ as well. This will briefly be discussed in Section 9.4.1. In Section 9.4.2 we investigate the consequences of Theorem 9.2 for the pseudoeuclidean sesquilinear form before we analyze the symplectic and perplectic forms $[x, y] = x^H J_{2n} y$ and $[x, y] = x^H R_{2n} y$ in Section 9.4.3.

Before we start, Remark 9.1 recalls some facts about selfadjoint and skewadjoint matrices $A \in M_{n \times n}(\mathbb{C})$ that already appeared in Section 8.2. In the following subsections these will be frequently used.

Remark 9.1. *Let* $[x, y] = x^H B y$ *with* $B = \pm B^H \in \mathbb{GP}_n(\mathbb{C})$ *be some nondegenerate sesquilinear form on* $\mathbb{C}^n$. *Moreover, let* $A \in M_{n \times n}(\mathbb{C})$.

(i) *Assume* $A \in \mathbb{J}(B)$, *i.e.* $A = A^\star$. *Then the eigenvalues of* A *appear in pairs* $(\lambda, \overline{\lambda})$ *and all eigenspaces of* A *and* $A^\star$ *coincide, that is it holds that* $E(A, \lambda) = E(A^\star, \lambda)$ *for all* $\lambda \in \sigma(A)$. *Assume that* $v \in E(A, \lambda_j)$ *for some* $\lambda_j \in \sigma(A)$. *Then* $v \in E(A^\star, \overline{\lambda}_j)$ *holds iff* $\lambda_j = \overline{\lambda}_j$, *i.e. iff* $\lambda_j \in \mathbb{R}$. *In other words we have* $E(A, \lambda_j) \cap E(A^\star, \overline{\lambda}_j) \neq \{0\}$ *iff* $\lambda_j \in \sigma(A)$ *is real.*

(ii) *Now assume that* $A \in \mathbb{L}(B)$, *that is* $A = -A^\star$. *The eigenvalues now appear in pairs* $(\lambda, -\overline{\lambda})$ *and we have the relation* $E(A, \lambda) = E(A^\star, -\lambda)$ *for the eigenspaces of* A *and* $A^\star$ *and all eigenvalues* $\lambda \in \sigma(A)$. *Furthermore, if* $v \in E(A, \lambda_j)$, *then* $v \in E(A^\star, \overline{\lambda}_j)$ *holds iff* $-\lambda_j = \overline{\lambda}_j$, *i.e.* $\lambda_j = -\overline{\lambda}_j$. *This means that* $E(A, \lambda_j) \cap E(A^\star, \overline{\lambda}_j) \neq \{0\}$ *iff* $\lambda_j \in \sigma(A)$ *is purely imaginary.*

For each selfadjoint or skewadjoint matrix $A \in M_{n \times n}(\mathbb{C})$ considered in the sequel, let always d_j denote the dimension of $E(A, \lambda_j) \cap E(A^\star, \overline{\lambda}_j)$, $\lambda_j \in \sigma(A)$, as a subspace of $\mathbb{C}^n$. Moreover, for the sesquilinear form $[x, y] = x^H B y$ with $B = \pm B^H \in \mathbb{GP}_n(\mathbb{C})$ we continue denoting the number of nonzero diagonal entries of B by $i(B)$ and the number of nonzero off-diagonal pairs by $t(B) = \frac{1}{2}(n - i(B))$.

9.4.1 The standard Euclidean scalar product

We start our discussion of selfadjoint and skewadjoint matrices $A \in M_{n \times n}(\mathbb{C})$ for $[x, y] = x^H B y$ with the standard Euclidean scalar product. As this is the special case $B = I_n$, Theorem 9.2 certainly has to hold. Indeed, first assume that $A \in \mathbb{J}(I_n)$, i.e. $A = A^H$. Then $\sigma(A) = \{\lambda_1, \ldots, \lambda_{r+1}\} \subset \mathbb{R}$, i.e. all eigenvalues of A are real. With the observation from Remark 9.1 (i) this implies that $E(A, \lambda_j) \cap E(A^\star, \overline{\lambda}_j) = E(A, \lambda_j) = E(A^\star, \lambda_j)$ for all eigenspaces of A and $A^\star$. In particular, we have $\sum_{j=1}^{r+1} d_j = n$. Moreover, as $i(I_n) = n$ we obtain $\sum_{j=1}^{r+1} d_j - i(I_n) = 0$. Moreover, the left hand side of (9.12) is always zero, too, as $Z^H I_n Z = Z^H Z$ has entirely positive eigenvalues, i.e. $\min\{i_+(\lambda_j), i_-(\lambda_j)\} = 0$ for all $\lambda_j \in \sigma(A)$. Thus the relation (9.12) always holds. With the observation from Remark 9.1 (ii) the result follows in a similar manner for skew-Hermitian matrices $A \in M_{n \times n}(\mathbb{C})$, i.e. $A = -A^H$, noting that those matrices have entirely purely imaginary eigenvalues. Finally, notice that these derivations only show that a diagonalizable Hermitian or skew-Hermitian matrix is always unitarily diagonalizable. Our discussion does not show that such matrices are in fact always diagonalizable.

9.4.2   The pseudoeuclidean inner product

Given two natural numbers $p, q \in \mathbb{N}_0$, $p + q = n$, the pseudoeuclidean sesquilinear form $[x, y] = x^H \Sigma_{p,q} y$ on $\mathbb{C}^n$ is induced by the matrix

$$B = \Sigma_{p,q} = \left[\begin{array}{c|c} -I_p & \\ \hline & I_q \end{array} \right] \in \mathrm{M}_{n \times n}(\mathbb{R}).$$

Matrices $A \in \mathrm{M}_{n \times n}(\mathbb{C})$ from $\mathbb{G}(\Sigma_{p,q}), \mathbb{J}(\Sigma_{p,q})$ or $\mathbb{L}(\Sigma_{p,q})$ are usually referred to as pseudounitary, pseudohermitian and pseudoskew-Hermitian, respectively [82]. For instance, pseudohermitian matrices play a key role in solving Hermitian generalized eigenproblems $Ax = \lambda Bx$ with $A = A^H, B = B^H$, see [15, 124]. Their properties have already been studied in, e.g., [93, 98] (see also the references therein). Here we have $i(\Sigma_{p,q}) = n$ and $t(\Sigma_{p,q}) = 0$ regardless of the particular choice of p and q.

Now, let $A \in \mathrm{M}_{n \times n}(\mathbb{C})$ be nondefective and $\Sigma_{p,q}$-normal with pairwise distinct eigenvalues $\lambda_1, \ldots, \lambda_{r+1}$. Then, according to Theorem 9.2, as $i(\Sigma_{p,q}) = n$ the term $\sum_{j=1}^{r+1} d_j - i(\Sigma_{p,q})$ in (9.12) will always be smaller than zero unless $\sum_{j=1}^{r+1} d_j = i(\Sigma_{p,q}) = n$. Recall that the observations (i) and (ii) from Remark 9.1 yield that $d_j > 0$ holds iff $\lambda_j \in \sigma(A)$ is real/purely imaginary for $A \in \mathbb{J}(\Sigma_{p,q})$ or $A \in \mathbb{L}(\Sigma_{p,q})$, respectively. Now Theorem 9.2 gives the following results:

Theorem 9.3. *Let $A \in \mathrm{M}_{n \times n}(\mathbb{C})$ be nondefective. If $A \in \mathbb{J}(\Sigma_{p,q})$, then A is pseudounitary diagonalizable iff all eigenvalues of A are real. Furthermore, if $A \in \mathbb{L}(\Sigma_{p,q})$, then A is pseudounitary diagonalizable iff all eigenvalues of A are purely imaginary.*

9.4.3   The perplectic and symplectic forms

The perplectic and symplectic sesquilinear forms $[x, y] = x^H By$ on $\mathbb{C}^{2n}$ are induced by the matrices

$$B = R_{2n} := \left[\begin{array}{ccc} & & 1 \\ & \cdot^{\cdot^{\cdot}} & \\ 1 & & \end{array} \right] \quad \text{and} \quad B = J_{2n} := \left[\begin{array}{cc} & I_n \\ -I_n & \end{array} \right].$$

For the sesquilinear form $[x, y] = x^H R_{2n} y$, matrices $A \in \mathrm{M}_{2n \times 2n}(\mathbb{C})$ from $\mathbb{G}(R_{2n})$ are called perplectic whereas matrices from $\mathbb{J}(R_{2n})$ and $\mathbb{L}(R_{2n})$ are called perhermitian and perskew-Hermitian. Furthermore, for $[x, y] = y^H J_{2n} y$ those matrices are called symplectic, skew-Hamiltonian and Hamiltonian, respectively. Notice that J_m (in the sense as above) can only be defined for m being even (which we assumed to be $2n$). However, R_m is well-defined for m being odd, too. Nevertheless, we will only discuss both cases for $2n$ here.

Hamiltonian and skew-Hamiltonian matrices are well-studied, e.g. [30, 41, 47], and arise frequently in the study of algebraic matrix equations [75]. A concise numerical treatment of symplectic matrices can be found in [40]. Further (numerical) aspects have been discussed in, e.g., [39] or [24] (see also the references therein). Structured matrices according to $[x, y] = x^H R_{2n} y$ occur in the control of mechanical and electrical vibrations, see [110]. They have already been analyzed in [81]. For R_{2n} and J_{2n} we have $i(R_n) = i(J_{2n}) = 0$ and $t(R_n) = t(J_{2n}) = n$. Therefore, according to Theorem 9.2 both indefinite forms must behave quite similar with respect to automorphic diagonalization although one is Hermitian while the other is skew-Hermitian. Consider once more the formula from Theorem 9.2 for the case $i(R_{2n}) = i(J_{2n}) = 0$, i.e.

$$2 \cdot \sum_{j=1}^{r+1} \min \left\{ i_+(\lambda_j), i_-(\lambda_j) \right\} \geq \sum_{j=1}^{r+1} d_j \geq 0 \qquad (9.13)$$

As $d_j \geq 0$ for all $j = 1, \ldots, r + 1$, the right inequality is always satisfied. To derive a suitable condition from the left inequality, recall that it always holds that $\min\{i_+(\lambda_j), i_-(\lambda_j)\} \leq \frac{1}{2} d_j$ for each eigenvalue $\lambda_j \in \sigma(A)$. In particular, in this situation, we can at most achieve equality for the left condition in (9.13). Equality, in turn, is achieved iff each (nonzero) d_j is even and $\min\{i_+(\lambda_j), i_-(\lambda_j)\} = \frac{1}{2} d_j$. Taking (i) and (ii) from Remark 9.1 into account, we obtain the following result for the perplectic form $[x, y] = x^H R_{2n} y$:

Theorem 9.4. Let $A \in \mathrm{M}_{n \times n}(\mathbb{C})$ be nondefective.

(i) Assume that $A \in \mathbb{J}(R_{2n})$. Then A is perplectic diagonalizable iff each real eigenvalue $\lambda_j \in \sigma(A)$ has even algebraic (= geometric) multiplicity m_j and, given any basis $v_1, \ldots, v_{m_j}$ of $E(A, \lambda_j)$, the matrix $V^H R_{2n} V$ for $V = [v_1 \cdots v_{m_j}]$ has equally many positive and negative (real) eigenvalues.

(ii) Let $A \in \mathbb{L}(R_n)$. Then A is perplectic diagonalizable iff each purely imaginary eigenvalue $\lambda_j \in \sigma(A)$ has even algebraic (= geometric) multiplicity m_j and, given any basis $v_1, \ldots, v_{m_j}$ of $E(A, \lambda_j)$, the matrix $V^H R_n V$ for $V = [v_1 \cdots v_{m_j}]$ has equally many positive and negative (real) eigenvalues.

Recall that, considering matrices $A \in \mathbb{J}(J_{2n}), \mathbb{L}(J_{2n})$ with $\sigma(A) = \{\lambda_1, \ldots, \lambda_{r+1}\}$, the products $[v, w] = v^H J_{2n} w$ for vectors $v, w \in E(A, \lambda_j) \cap E(A^\star, \overline{\lambda}_j)$ constitute the blocks Δ_j in (9.6). In particular, each matrix Δ_j is skew-Hermitian. Moreover, as each matrix Δ_j is nonsingular, too, their dimensions must all be even. This follows immediately since any skew-Hermitian matrix always has even rank [67]. Thus we may reformulate Theorem 9.4 for the symplectic form $[x, y] = x^H J_{2n} y$ without the assumption on even multiplicities. Regarding the reduction of a Hamiltonian matrix to upper (block-)triangular form, similar results can be found in, e.g., [50].

Theorem 9.5. Let $A \in \mathrm{M}_{n \times n}(\mathbb{C})$ be nondefective.

(i) *Assume that $A \in \mathbb{J}(J_{2n})$. Then A is symplectic diagonalizable iff, given any basis $v_1, \ldots, v_{m_j}$ of $E(A, \lambda_j)$ for any real eigenvalue $\lambda_j \in \sigma(A)$, the matrix $V^H J_{2n} V$ for $V = [\, v_1 \cdots v_{m_j} \,]$ has equally many positive and negative imaginary eigenvalues.*

(ii) *Let $A \in \mathbb{L}(J_{2n})$. Then A is symplectic diagonalizable iff, given any basis $v_1, \ldots, v_{m_j}$ of $E(A, \lambda_j)$ for any purely imaginary eigenvalue $\lambda_j \in \sigma(A)$, the matrix $V^H J_{2n} V$ for $V = [\, v_1 \cdots v_{m_j} \,]$ has equally many positive and negative imaginary eigenvalues.*

Theorem 9.4 and Theorem 9.5 admit a simple and immediate Corollary. Notice that the same criterion as stated in Corollary 9.2 about the existence of a symplectic diagonalization of a Hamiltonian matrix showed up in [105, Thm. 3.1] where the Hamiltonian Schur form was developed.

Corollary 9.2. *Let $A \in M_{n \times n}(\mathbb{C})$ be nondefective. Then $A \in \mathbb{J}(R_{2n})$ $(A \in \mathbb{J}(J_{2n}))$ is always perplectic (symplectic) diagonalizable if A has no purely real eigenvalues. Moreover, $A \in \mathbb{L}(R_{2n})$ $(A \in \mathbb{L}(J_{2n}))$ is always perplectic (symplectic) diagonalizable if A has no purely imaginary eigenvalues.*

9.5 Unitary structure-preserving diagonalization

Let $[x, y] = x^H B y$ be either the perplectic form $[x, y] = x^H R_{2n} y$ or the symplectic form $[x, y] = x^H J_{2n} y$ on $\mathbb{C}^{2n}$. Before we pass on to the next chapter where we exclusively consider matrices from $\mathbb{J}(R_{2n}), \mathbb{L}(R_{2n}), \mathbb{J}(J_{2n})$ and $\mathbb{L}(J_{2n})$ that are additionally normal with respect to the Euclidean inner product $(x, y) = x^H y$, we briefly discuss the implications of Theorem 9.2 on B-normal and (Euclidean)-normal matrices. That is, these matrices $A \in M_{n \times n}(\mathbb{C})$ satisfy $AA^H = A^H A$ as well as $A^\star A = AA^\star$ where $A^\star$ denotes the adjoint of A, $A^\star = B^{-1} A^H B$. The next theorem shows that, if such a "doubly normal" matrix is automorphic diagonalizable, then the unitary diagonalization comes "for free".

Theorem 9.6. *Let $A \in M_{2n \times 2n}(\mathbb{C})$ be normal with respect to the Euclidean scalar product (i.e. $A^H A = AA^H$ holds) and normal with respect to either the perplectic form $[x, y] = x^H R_{2n} y$ or the symplectic form $[x, y] = x^H J_{2n} y$. Then, if A is nondefective and the conditions stated in Theorem 9.2 hold, A is always diagonalizable by a unitary automorphism.*

In other words Theorem 9.6 states that, whenever a (Euclidean)-normal and R_{2n}-normal (J_{2n}-normal) matrix $A \in M_{2n \times 2n}(\mathbb{C})$ is automorphic diagonalizable, then the perplectic (symplectic) matrix that transforms A to diagonal form can always be chosen to be unitary as well. Thus, a unitary-perplectic (unitary-symplectic) diagonalization of a normal and R_{2n}-normal (J_{2n}-normal) matrix A does not require further conditions beside those stated in Theorem 9.2. The proof is outlined below for the perplectic form and works in a similar manner for the symplectic inner product.

It essentially uses one important fact: as the normality of A implies the normality of $A^\star$ (this can be verified by a direct calculation), the R_{2n}-normality $A^\star A = AA^\star$ yields that there exists a common orthogonal eigenbasis to A and $A^\star$ [103, Thm. 2]. Then the process of constructing an automorphism diagonalizing A described in Section 9.3 can be carried out on a unitary level (i.e. all transformation matrices can be chosen to be unitary).

Proof. Assume that $A \in M_{2n \times 2n}(\mathbb{C})$ is normal with respect to the Euclidean scalar product $(x, y) = x^H y$ and the perplectic form $[x, y] = x^H R_{2n} y$ and that the conditions stated in Theorem 9.2 hold. Let the columns of $Z \in M_{2n \times 2n}(\mathbb{C})$ be an orthogonal eigenbasis of A and $A^\star$ such that $Z^H R_{2n} Z$ has the canonical form (9.6). Notice that $Z^H R_{2n} Z$ is a congruence and similarity transformation of R_{2n} and, as R_{2n} is unitary, $Z^H R_{2n} Z$ is unitary as well. This implies that each single block in the form (9.6) is unitary for itself. Thus, the transformations with Q_j and T_j, $j = 1, \ldots, r + 1$, that led to the form (9.8) can be chosen to be unitary, too. This implies that $Q_j^H \Delta_j Q_j$ is a similarity transformation and the eigenvalues of each Δ_j are only $+1$ and -1 (as these are the only eigenvalues of R_{2n}). Moreover, the transformation of diagonal entries $+1, -1$ in (9.8) into nonzero off-diagonal pairs $+1, +1$ (if necessary) by means of the matrix V in (9.9) is unitary. Finally, a permutation matrix that permutes the nonzero off-diagonal $+1, +1$ pairs into the right positions (and thus generating R_{2n}) is unitary. In conclusion, if the diagonalization condition (9.12) is satisfied, the perplectic transformation can be constructed to be unitary. $\square$

9.6 Conclusions

In this section we analyzed the structure-preserving diagonalizability of matrices that are normal with respect to some nondegenerate sesquilinear form $[x, y] = x^H By$. As those matrices need not be diagonalizable at all, we confined ourselves to nondefective matrices. Moreover, we focused on forms that are induced by a Hermitian or skew-Hermitian generalized (complex) permutation matrix $B \in M_{n \times n}(\mathbb{C})$. In this case, an automorphic diagonalization $P^{-1}AP$ of $A \in M_{n \times n}(\mathbb{C})$ always implies that $P^{-1}A^\star P$ is also diagonal. We presented some basic facts concerning the common eigenspaces of any B-normal, nondefective matrix $A \in M_{n \times n}(\mathbb{C})$ and its adjoint $A^\star$. These facts have been necessary to be able to order a common eigenbasis of A and $A^\star$ in a way that revealed sufficient and necessary conditions for an automorphic diagonalization. In particular, we derived a canonical form from which we were able to construct an automorphic diagonalization if this is possible. We discussed several indefinite sesquilinear forms that show up frequently in applications. Among them, we focused on selfadjoint and skewadjoint matrices from the pseudoeuclidean inner product as well as from the symplectic and perplectic indefinite forms. Finally, we showed that an automorphic diagonalization of a B-normal and (Euclidean)-normal matrix can always be chosen to be unitary, too.

Chapter 10

Structured normal matrices

Emil Artin (1898 – 1962), [4].

In this section we consider matrices in $\mathbb{J}(B)$ and $\mathbb{L}(B)$ in more detail, where the indefinite sesquilinear form $[x, y] = x^H B y$ on $\mathbb{C}^{2n}$ under consideration is either the perplectic form, $B = R_{2n}$, or the symplectic form, $B = J_{2n}$. In particular, we are interested in the characterization of matrices $A \in \mathrm{M}_{2n \times 2n}(\mathbb{C})$ that belong to either $\mathbb{J}(B)$ or $\mathbb{L}(B)$ and are, additionally, normal with respect to the standard Euclidean scalar product $(x, y) = x^H y$. Therefore, these matrices are B-normal and normal. Those "doubly normal" matrices include, for instance, bihermitian matrices, which are Hermitian (i.e. normal) and perhermitian (i.e. R_{2n}-normal). These have already been considered in [81]. To avoid any misconception, we strictly distinguish between normality (meaning $A^H A = A A^H$) and B-normality, meaning $A A^\star = A^\star A$ with $A^\star = B^{-1} A^H B$ denoting the adjoint of A, from now on.

In Section 10.1 we present some facts about isotropic subspaces for the perplectic and symplectic forms on $\mathbb{C}^{2n}$. These results will be useful in Section 10.2 where we derive a canonical (multiplicative) factorization of normal and perhermitian matrices through a similarity transformation. The corresponding Theorem 10.1 is stated here with complete proof although some aspects follow a similar argumentation as in Section 9.3. This factorization and the characterization of the set of unitary and perplectic matrices leads us to an additive decomposition of normal perhermitian matrices. In particular, those matrices can always be expressed as a special sum of three normal matrices. We will show that this decomposition provides additional information of the matrix at hand, for instance with respect to the existence of a maximal invariant isotropic subspace. Moreover, it characterizes matrices which are unitary and perplectic diagonalizable in a different fashion compared to the discussion in Section 9. In Section 10.2.1 we discuss the same decomposition for normal perskew-Hermitian matrices. These results can actually be easily derived from Section 10.2. Matrices that are normal and (skew-)Hamiltonian are discussed in Section 10.3. This chapter ends with some conclusions in Section 10.4.

10.1   Lagrangian subspaces

Let $[x, y] = x^H B y$ be either the perplectic form with $B = R_{2n}$ or the symplectic form $B = J_{2n}$ on $\mathbb{C}^{2n}$. In this section we briefly collect some information about isotropic subspaces with respect to $[\cdot, \cdot]$. These facts, in particular Corollary 10.1, will be of advantage for our discussion in the sequel. The results from this section (in particular Corollary 10.1) are likely to be known although they are not readily found in the literature.

At first, it is obvious that the set of all isotropic subspaces in $\mathbb{C}^{2n}$ constitutes a partial order under the relation of set-inclusion [53, Def. O-1.6][1]. Moreover, for any chain of isotropic subspaces $F_1 \subseteq F_2 \subseteq \cdots \subseteq F_k$ the space F_k contains all other spaces from this chain [53, Def. O-1.6]. In other words, each chain of subspaces has an isotropic subspace as an upper bound. According to the lemma of Zorn [131], these facts lead to the observation that the (partially ordered) set of isotropic subspaces has maximal elements. In accordance with the discussion in Section 8.1 we call an isotropic subspace maximal (in $\mathbb{C}^{2n}$) if it is not properly contained in any isotropic subspace of larger dimension. The next proposition presents an upper bound for the dimensions of isotropic subspaces.

Proposition 10.1. *For the symplectic form* $[x, y] = x^H J_{2n} y$ *and the perplectic form* $[x, y] = x^H R_{2n} y$ *on* $\mathbb{C}^{2n}$ *the maximal possible dimension of an isotropic subspace is* n.

Proof. For the Hermitian form $[x, y] = x^H R_{2n} y$ on $\mathbb{C}^{2n}$ the statement is proven in [55, Thm. 2.3.4] noting that R_{2n} has only the eigenvalues $+1$ and -1 with multiplicity n. The statement for the symplectic form follows from the same theorem taking into account that the skew-Hermitian form $[x, y] = x^H J_{2n} y$ and the Hermitian form $[x, y] = x^H (i J_{2n}) y$ have the same isotropic subspaces and $i J_{2n}$ has eigenvalues $+1$ and -1 again with multiplicities n.   □

For both indefinite forms it is easily seen that $\mathrm{span}(e_1, e_2, \ldots, e_n) \subseteq \mathbb{C}^{2n}$ constitutes an isotropic subspace of dimension n. Thus, the bound given in Proposition 10.1 is in fact sharp. Furthermore, it is clear that $\mathrm{span}(e_1, e_2, \ldots, e_n)$ has to be a maximal isotropic subspace. The following proposition makes a statement on the dimensions of all other maximal isotropic subspaces.

Proposition 10.2. *For the symplectic form* $[x, y] = x^H J_{2n} y$ *and the perplectic form* $[x, y] = x^H R_{2n} y$ *on* $\mathbb{C}^{2n}$ *all maximal isotropic subspaces have the same dimension. In particular, an isotropic subspace is maximal iff it has dimension* n.

Proof. The statement for the Hermitian form $[x, y] = x^H R_{2n} y$ is proven in [14, § 4.2]. The statement on the symplectic form $[x, y] = x^H J_{2n} y$ follows again from the fact that the Hermitian form $[x, y] = x^H (i J_{2n}) y$ has the same isotropic subspaces as $[x, y] = x^H J_{2n} y$.   □

---

[1]That is, for any isotropic subspaces $F, G, H \subseteq \mathbb{C}^{2n}$ we have reflexivity ($F \subseteq F$), transitivity ($F \subseteq G, G \subseteq H$ yields $F \subseteq H$) and anti-symmetry ($F \subseteq G, G \subseteq F$ yields $F = G$)

As we have already seen that span$(e_1, e_2, \ldots, e_n)$ is a maximal isotropic subspace, it follows from Proposition 10.2 that all maximal isotropic subspaces have dimension n. It is common to call maximal isotropic subspaces whose dimension is exactly half the dimension of the whole vector space *Lagrangian*, cf. [50, Def. 1.2]. The statement of the following corollary will be important in the upcoming sections.

Corollary 10.1. *For the symplectic form* $[x, y] = x^H J_{2n} y$ *and the perplectic form* $[x, y] = x^H R_{2n} y$ *on* $\mathbb{C}^{2n}$, *each isotropic subspace of* $\mathbb{C}^{2n}$ *is contained in a Lagrangian subspace.*

Proof. Let $F \subseteq \mathbb{C}^{2n}$ be any isotropic subspace. As the set of all isotropic subspaces of $\mathbb{C}^{2n}$ is partially ordered and has maximal elements, there is always a maximal isotropic subspace $G \subseteq \mathbb{C}^{2n}$ that contains S. As all maximal isotropic subspaces have dimension n (i.e. are Lagrangian) according to Proposition 10.2 the statement follows. □

10.2 Normal perhermitian matrices

We begin this section on normal perhermitian matrices with a canonical factorization under unitary and perplectic similarity transformations. According to Theorem 9.6 normal perhermitian matrices are unitary-perplectic diagonalizable if the conditions posed in Theorem 9.2 hold. However, as this will certainly not always be the case, Theorem 10.1 presents a canonical form that always exists no matter whether (9.12) is satisfied or not. The proof heavily relies on the results from Section 9.1, 9.2 and Section 9.4.3 on the eigenspaces of perhermitian matrices taking into account that a full eigenbasis with orthonormal eigenvectors is now available. Notice that Theorem 10.1 and its proof appeared in [8] the first time.

Theorem 10.1. *Let* $A \in M_{2n \times 2n}(\mathbb{C})$ *be normal and perhermitian. Then there exists a unitary and perplectic matrix* $Z \in M_{2n \times 2n}(\mathbb{C})$ *so that*

$$Z^H A Z = \begin{bmatrix} \Lambda & & \\ & X & \\ & & \Lambda^\star \end{bmatrix} \tag{10.1}$$

where $\Lambda \in M_{d \times d}(\mathbb{C})$ *is a diagonal matrix with* $d \in \mathbb{N}_0$ *being the number of eigenvalues with nonzero imaginary parts of* A *(counted with multiplicities) and* $\sigma(\Lambda) \in \mathbb{C} \setminus \mathbb{R}$. *In addition, the matrix* $X \in M_{2(n-d) \times 2(n-d)}(\mathbb{R})$ *is real Hermitian and perhermitian (thus* $\sigma(X) \subset \mathbb{R}$*) and has only entries along its diagonal and its antidiagonal.*

The matrix $X \in M_{2(n-d) \times 2(n-d)}(\mathbb{R})$ in Theorem 10.1 can structurally be depicted as

$$X = \boxed{\diagbox} \in M_{2(n-d) \times 2(n-d)}(\mathbb{R}).$$

We give a complete proof of Theorem 10.1 although it is to some extend similar to the procedure described for automorphic diagonalization in Section 9.3 and some arguments closely resemble the ones from the proof of Theorem 9.6. However, here our goal is not a diagonalization and several different aspects have to be taken into account. Keep in mind, that for any perhermitian matrix $A \in M_{n \times n}(\mathbb{C})$ we have $E(A, \lambda_j) \cap E(A^\star, \overline{\lambda}_j) \neq \{0\}$ iff $\lambda_j \in \sigma(A)$ is purely real, that is, has zero imaginary part.

Proof. Let $\lambda_1, \ldots, \lambda_s$ be the (pairwise distinct) eigenvalues of A with nonzero, positive imaginary parts each λ_j having algebraic (= geometric) multiplicity d_j, $j = 1, \ldots, s$. Moreover, set $d := \sum_{j=1}^s d_j$. As $A^\star$ and A^H are similar, note that $\overline{\lambda}_1, \ldots, \overline{\lambda}_s \in \sigma(A)$ and that d_j also equals the multiplicity of $\overline{\lambda}_j$. Furthermore, let $\mu_1, \ldots, \mu_r$ be the (pairwise distinct) purely real eigenvalues of A with multiplicities $s_j, j = 1, \ldots, r$.

Now let $Q \in M_{2n \times 2n}(\mathbb{C})$ be a unitary matrix so that

$$Q^H A Q = \begin{bmatrix} D & & \\ & B & \\ & & R_d D^H R_d \end{bmatrix} \in \mathbb{D}_{2n}(\mathbb{C})$$

with $D = \mathrm{diag}[D(\lambda_1), \ldots, D(\lambda_s)] \in M_{d \times d}(\mathbb{C})$, $D(\lambda_j) = \lambda_j I_{d_j}$ for $j = 1, \ldots, s$ and $B = \mathrm{diag}[\mu_1, \ldots, \mu_r] \in M_{2(n-d) \times 2(n-d)}(\mathbb{R})$ holds. Due to the eigenvalue pairing $\lambda_j, \overline{\lambda}_j \in \sigma(A)$ and the normality of A such a unitary matrix Q always exists. Taking Proposition 9.1 and the facts from Remark 9.1 into account we may easily determine the form of $Q^H R_{2n} Q$ similarly as this was done before. In particular, we obtain

$$Q^H R_{2n} Q = \begin{bmatrix} 0 & 0 & S \\ & \begin{matrix} \Delta_1 & & \\ & \ddots & \\ & & \Delta_r \end{matrix} & \\ 0 & & 0 \\ S^H & 0 & 0 \end{bmatrix}, \text{ with } S = \begin{bmatrix} & & S_1 \\ & \cdot^{\,\cdot} & \\ S_s & & \end{bmatrix}. \quad (10.2)$$

Hereby, the blocks S_j are of sizes $d_j \times d_j$, $j = 1, \ldots, s$, and the blocks Δ_j are of size $s_j \times s_j$, $j = 1, \ldots, r$, respectively. It is important to note that the matrix $Q^H R_{2n} Q$ remains to be unitary as R_{2n} and Q are both unitary. Notice further that the form (10.2) can be explained in a similar manner as this was done for the form (9.6) in Section 9.2. Now each matrix Δ_j, S_j in (10.2) is nonsingular as Q and R_{2n} are nonsingular. Moreover, it can easily be seen that each matrix Δ_j, S_j is necessarily unitary for itself. In addition, the blocks $\Delta_j, j = 1, \ldots, r$, and therefore $\Delta := \mathrm{diag}[\Delta_1, \ldots, \Delta_r]$ are Hermitian.

Now we define

$$T_1 = \mathrm{diag}\big[S_1 R_{d_1}, \ldots, S_s R_{d_s}\big] \in M_{d \times d}(\mathbb{C})$$

which is unitary. As $\sigma(\Delta) \subset \sigma(R_{2n})$ ($R_{2n} \mapsto Q^H R_{2n} Q$ is a similarity transformation) and $\sigma(R_{2n}) = \{+1, -1\}$ we additionally find a unitary $W \in M_{2(n-d) \times 2(n-d)}(\mathbb{C})$ with the same block-structure as B so that $W^H \Delta W$ is diagonal and a real signature matrix, i.e. $W^H \Delta W = \text{diag}[\pm 1, \ldots, \pm 1]$. Then, with $T_2 = \text{diag}[T_1, W, I_d]$ (which is unitary as well) we obtain

$$
T_2^H(Q^H RQ)T_2 = \begin{bmatrix} 0 & 0 & R_d \\ \hline & \pm 1 & & \\ 0 & & \ddots & & 0 \\ & & & \pm 1 \\ \hline R_d & 0 & 0 \end{bmatrix} , \text{ with } R_d = \begin{bmatrix} & & R_{d_1} \\ & \ddots & \\ R_{d_s} & & \end{bmatrix} .
$$

Do not overlook that $T_2^H(Q^H AQ)T_2 = \text{diag}[D, B, R_d D^H R_d]$ still holds. Moreover, as $\sigma(T_2^H(Q^H R_{2n}Q)T_2) = \sigma(R_{2n})$, Proposition 8.1 immediately implies that the $2(n - d) \times 2(n - d)$ block $\text{diag}[\pm 1, \ldots, \pm 1]$ must have the same number of -1's and $+1$'s. Thus, we assume w.l.o.g. that W has already been chosen so that -1's and $+1$'s occur alternatingly. Now we turn each $+1/-1$ pair of diagonal entries into a pair of nonzero off-diagonal entries $+1/+1$. This is again achieved by the matrix $V \in M_{2 \times 2}(\mathbb{C})$ from (9.9), i.e. we have

$$
V^H \begin{bmatrix} -1 & 0 \\ 0 & 1 \end{bmatrix} V = \begin{bmatrix} 0 & 1 \\ 1 & 0 \end{bmatrix} \text{ for } V = \frac{1}{\sqrt{2}} \begin{bmatrix} -1 & 1 \\ 1 & 1 \end{bmatrix} .
$$

Recall that V is unitary as well. Thus, forming $T_3 = \text{diag}[I_d, V, \ldots, V, I_d] \in M_{2n \times 2n}(\mathbb{R})$ which is orthogonal (i.e. unitary) we obtain

$$
T_3^H(T_2^H(Q^H R_{2n}Q)T_2)T_3 = \begin{bmatrix} 0 & 0 & R_d \\ \hline & 0 \; 1 & & \\ & 1 \; 0 & & \\ 0 & & \ddots & & 0 \\ & & & 0 \; 1 \\ & & & 1 \; 0 \\ \hline R_d & 0 & 0 \end{bmatrix} .
$$

For $T_3^T(T_2^H(Q^H AQ)T_2)T_3$ we now obtain

$$
T_3^T(T_2^H(Q^H AQ)T_2)T_3 = \text{diag}[D, \widehat{B}, R_d D^H R_d]
$$

where $\widehat{B} = \text{diag}[\widehat{B}_1, \ldots, \widehat{B}_{n-d}] \in M_{2(n-d) \times 2(n-d)}(\mathbb{R})$ with 2×2 blocks $\widehat{B}_j, j = 1, \ldots, n - d$, which are real, Hermitian and perhermitian. In fact, $\widehat{B}_j = V^H \text{diag}[c_1, c_2]V$ yields a diagonal matrix in case $c_1 = c_2$ but a real, Hermitian and perhermitian matrix for any $c_1 \neq c_2 \in \mathbb{R}$. Finally, there exists a real permutation matrix $P \in M_{2(n-d) \times 2(n-d)}(\mathbb{R})$ so that $P^T \text{diag}[R_2, R_2, \ldots, R_2]P = R_{2(n-d)}$. The final transformation yields

$$
P^T(T_3^T(T_2^H(Q^H R_{2n}Q)T_2)T_3P = R_{2n},
$$

so $Z := QT_2T_3P$ is unitary (as it is a product of unitary matrices) and perplectic. Notice that the last transformation P applied to $T_3^T(T_2^H(Q^H AQ)T_2)T_3$ transforms the block $\widehat{B}$ into X-form, so $Z^H AZ$ has the desired form. This completes the proof. $\square$

Notice that the antidiagonal part of X in (10.1) need not always be completely equipped with nonzero elements. In fact, if multiple real eigenvalues of A are suitably arranged (in pairs of two) side by side in B and the entries in Δ in the same positions are of different signs, then the transformation with V does not change the form of B. In fact, if the conditions from Theorem 9.4 (i) are satisfied, all transformations with V in the previous proof can be arranged so that B remains diagonal (i.e. there is no contribution to the antidiagonal part of X at all). This had to be the case as such a matrix is unitary-perplectic diagonalizable according to Theorem 9.6.

However, if A is not unitary-perplectic diagonalizable, we need to apply the transformation with V to two real diagonal entries $c_1 \neq c_2$ in B at some time (in this case, this cannot be avoided). We obtain for $V^H \mathrm{diag}[c_1, c_2] V$ a full matrix of the form

$$V^H \begin{bmatrix} c_1 & 0 \\ 0 & c_2 \end{bmatrix} V = \begin{bmatrix} \alpha & \beta \\ \beta & \alpha \end{bmatrix}, \quad \alpha, \beta \in \mathbb{R}. \tag{10.3}$$

The case $c_1 \neq c_2$ corresponds to a mixing of eigenspaces, the situation we did not allow in Section 9.3 and which creates the nonzero off-diagonal entries. Thus, after a suitable permutation of the diagonal of B, it might be possible to reduce the number of nonzero entries on the antidiagonal in X. Finally, to eliminate the whole antidiagonal part of X, it is sufficient and necessary that the conditions from Theorem 9.2 hold.

The next proposition characterizes matrices $Z \in M_{2n \times 2n}(\mathbb{C})$ which are both unitary and perplectic in detail. From now on, we use the short-hand-notation $\mathcal{UP}(2n)$ for this class of matrices.

Proposition 10.3. *The set $\mathcal{UP}(2n)$ of unitary and perplectic matrices in $M_{2n \times 2n}(\mathbb{C})$ can be characterized as*

$$\mathcal{UP}(2n) = \left\{ \begin{bmatrix} V & R_{2n} V R_n \end{bmatrix} \in M_{2n \times 2n}(\mathbb{C}) \mid V \in M_{2n \times n}(\mathbb{C}), V^H V = I_n, V^H R_{2n} V = 0 \right\}.$$

Proof. Let $Q = [Q_1 \ Q_2] \in \mathcal{UP}(2n)$ with $Q_1, Q_2 \in M_{2n \times n}(\mathbb{C})$. As Q is unitary we have $Q^H Q = I_{2n}$ and as it is perplectic $Q^H R_{2n} Q = R_{2n}$ holds. Multiplying the latter with Q from the left gives $R_{2n} Q = Q R_{2n}$, so Q commutes with R_{2n}. Matrices satisfying this condition are known as centrosymmetric [110]. It is easy to see that any centrosymmetric matrix $C \in M_{2n \times 2n}(\mathbb{C})$ is symmetric with respect to the "center" of it and thus can be expressed as $C = [W \ R_{2n} W R_n]$ for some $W \in M_{2n \times n}(\mathbb{C})$. Moreover, any matrix of the form of C is centrosymmetric for any W. As Q is perplectic, i.e.

$$\begin{bmatrix} Q_1^H \\ Q_2^H \end{bmatrix} R_{2n} \begin{bmatrix} Q_1 & Q_2 \end{bmatrix} = \begin{bmatrix} Q_1^H R_{2n} Q_1 & Q_1^H R_{2n} Q_2 \\ Q_2^H R_{2n} Q_1 & Q_2^H R_{2n} Q_2 \end{bmatrix} = \begin{bmatrix} & R_n \\ R_n & \end{bmatrix},$$

we have that $Q_1^H R_{2n} Q_1 = 0$, so $\mathrm{im}(Q_1)$ is a Lagrangian subspace. Finally, as Q is unitary,

$$\begin{bmatrix} Q_1^H \\ Q_2^H \end{bmatrix} \begin{bmatrix} Q_1 & Q_2 \end{bmatrix} = \begin{bmatrix} Q_1^H Q_1 & Q_1^H Q_2 \\ Q_2^H Q_1 & Q_2^H Q_2 \end{bmatrix} = \begin{bmatrix} I_n & \\ & I_n \end{bmatrix},$$

it holds that $Q_1^H Q_1 = I_n$. This completes the proof. □

Beside Theorem 10.1 the following Theorem 10.2 is the main result of this section. It states how a canonical additive decomposition of normal perhermitian matrices can be obtained from the one in (10.1).

Theorem 10.2. *Let $A \in M_{2n \times 2n}(\mathbb{C})$ be normal and perhermitian. Then there exists a normal matrix $N \in M_{2n \times 2n}(\mathbb{C})$ with $NN^\star = N^\star N = 0$ and a Hermitian and perhermitian matrix $Z \in M_{2n \times 2n}(\mathbb{C})$ so that $A = N + N^\star + Z$. Moreover, it holds that $\mathrm{rank}(N) \le n$ and $\mathrm{im}(N)$ is an isotropic subspace for R_{2n}. If $\mathrm{rank}(N) = n$ the subspace $\mathrm{im}(N)$ is Lagrangian.*

Proof. Assume that $A \in M_{2n \times 2n}(\mathbb{C})$ is normal and perhermitian. Then, according to Theorem 10.1, there exists a $U \in \mathcal{UP}(2n)$ so that $U^H A U = \Delta$ is in canonical form (10.1). In particular, according to Proposition 10.3, $U = [\, V \ R_{2n} V R_n \,]$ for some $V \in M_{2n \times n}(\mathbb{C})$ with $V^H V = I_n$ and $V^H R_{2n} V = 0$ (i.e. the columns of V span a Lagrangian subspace). We write $\Delta = D + \widehat{R}$ where $D = \mathrm{diag}[\Delta]$ is the diagonal of Δ. Notice that $\widehat{R}$ has only entries along its antidiagonal and is real, Hermitian and perhermitian. Moreover, we decompose D as $D = \mathrm{diag}[\widetilde{D}, R_n \widetilde{D}^H R_n]$ with $\widetilde{D} \in \mathbb{D}_n(\mathbb{C})$. Then, $A = U \Delta U^H = U D U^H + U \widehat{R} U^H$ and, due to the form of U, the matrix A can be written as

$$A = V \widetilde{D} V^H + R_{2n} V R_n \left(R_n \widetilde{D}^H R_n \right) R_n V^H R_{2n} + U \widehat{R} U^H$$
$$= V \widetilde{D} V^H + R_{2n} V \widetilde{D}^H V^H R_{2n} + U \widehat{R} U^H = N + N^\star + Z$$

with $N = V \widetilde{D} V^H$ and $Z = U \widehat{R} U^H$. The matrix $N \in M_{2n \times 2n}(\mathbb{C})$ has rank $\le n$ and is normal. In fact we have $NN^H = V \widetilde{D} V^H (V \widetilde{D} V^H)^H = V \widetilde{D} V^H V \widetilde{D}^H V^H = V \widetilde{D} \widetilde{D}^H V^H$ and $N^H N = V \widetilde{D}^H \widetilde{D} V^H$, so both expressions coincide (as $\widetilde{D}$ is diagonal). It follows immediately that $N^\star$ is normal as well. Moreover, we obtain

$$NN^\star = (V \widetilde{D} V^H)(R_{2n} V \widetilde{D}^H V^H R_{2n}) = V \widetilde{D} V^H R_{2n} V \widetilde{D}^H V^H R_{2n}$$
$$= V \widetilde{D} (V^H R_{2n} V) \widetilde{D}^H V^H R_{2n}$$

so $NN^\star = 0$ since $V^H R_{2n} V = 0$. Similarly $N^\star N = 0$ follows, so in particular N is R_{2n}-normal. Furthermore, as $\mathrm{im}(N) \subseteq \mathrm{im}(V)$, $\mathrm{im}(N)$ is always isotropic and Lagrangian in case $\dim(\mathrm{im}(N)) = \mathrm{rank}(N) = n$. Finally, as $\widehat{R}$ is Hermitian and perhermitian, Z is Hermitian and perhermitian as U is unitary and perplectic. $\square$

It is immediate that the matrix $Z = U \widehat{R} U^H$ in the proof of Theorem 10.2 vanishes if $\widehat{R} = 0$. This in turn can be achieved iff the matrix at hand is unitary-perplectic diagonalizable according to Theorem 9.6. Therefore, we obtain the statement in Corollary 10.2 *(i)* below as a special case of Theorem 10.2. Moreover, in order to show that even the reverse statement of Corollary 10.2 *(i)* holds, we take the result from Corollary 10.1 into account.

Corollary 10.2. *Let $A \in M_{2n \times 2n}(\mathbb{C})$.*

(i) *Assume that A is normal and perhermitian and unitary-perplectic diagonalizable. Then there exists a normal matrix $N \in \mathrm{M}_{2n \times 2n}(\mathbb{C})$ with $NN^\star = N^\star N = 0$ so that $A = N + N^\star$ holds.*

(ii) *Assume that A can be expressed as $A = N + N^\star$ for some normal matrix $N \in \mathrm{M}_{2n \times 2n}(\mathbb{C})$ with $NN^\star = N^\star N = 0$. Then A is normal and perhermitian and can be diagonalized by a unitary-perplectic similarity transformation.*

In both cases $\mathrm{rank}(N) \leq n$. *Furthermore,* $\sigma(A) = (\sigma(N) \cup \sigma(N^\star)) \setminus \{0\}$ *holds in case A is nonsingular and* $\sigma(A) = \sigma(N) \cup \sigma(N^\star)$ *otherwise. In addition,* $\mathrm{im}(N)$ *and* $\mathrm{im}(N^\star)$ *are both invariant subspaces for A which are always isotropic. They are both Lagrangian if* $\mathrm{rank}(N) = n$.

Proof. (i) The statement follows in accordance with the proof of Theorem 10.2 noting that $\widehat{R} = 0$ and that the columns of V are all eigenvectors of A.

(ii) Assume that the conditions stated above hold. Then $A = N + N^\star$ is perhermitian as $A^\star = (N + N^\star)^\star = N^\star + N = A$. Moreover, we have

$$
\begin{aligned}
A^H A &= (N + N^\star)^H (N + N^\star) = N^H N + (N^\star)^H N + N^H N^\star + (N^\star)^H N^\star \\
AA^H &= (N + N^\star)(N + N^\star)^H = NN^H + N^\star N^H + N(N^\star)^H + N^\star(N^\star)^H.
\end{aligned}
\tag{10.4}
$$

Keeping in mind that the normality of $N^\star$ follows from the normality of N we find that both expressions coincide. This is since the assumption $NN^\star = N^\star N$ and the normality of $N^\star$ imply $N(N^\star)^H = (N^\star)^H N$ according to [60, (6)][2]. Similarly, we obtain $N^\star N^H = N^H N^\star$. With these observations, we find that the expressions in (10.4) coincide and A is normal.

Now notice that $NN^\star = N^\star N = 0$ implies $\mathrm{im}(N^\star) \subseteq \mathrm{null}(N)$ and $\mathrm{im}(N) \subseteq \mathrm{null}(N^\star)$. Therefore, since $\mathrm{rank}(N) = \mathrm{rank}(N^\star)$, this is only possible if $\mathrm{rank}(N) \leq n$. The normality of N implies that there exists a diagonal matrix $\widetilde{D} \in \mathrm{M}_{n \times n}(\mathbb{C})$, $\mathrm{rank}(\widetilde{D}) = \mathrm{rank}(N)$, and a matrix $V \in \mathrm{M}_{2n \times n}(\mathbb{C})$ with orthonormal columns (i.e. $V^H V = I_n$) so that $N = V\widetilde{D}V^H$. If $\mathrm{rank}(N) = k < n$, then $\widetilde{D}$ has $n - k$ eigenvalues equal to zero and, w.l.o.g., we assume that these zeros appear in the trailing $n - k$ diagonal positions in $\widetilde{D}$. Notice that the expression of N implies $N^\star = R_{2n} N^H R_{2n} = R_{2n} V \widetilde{D}^H V^H R_{2n}$. Therefore, A can be expressed as

$$
\begin{aligned}
A = N + N^\star &= V\widetilde{D}V^H + R_{2n} V \widetilde{D}^H V^H R_{2n} \\
&= V\widetilde{D}V^H + R_{2n} V R_n (R_n \widetilde{D}^H R_n) R_n V^H R_{2n}.
\end{aligned}
\tag{10.5}
$$

We now define $D \in \mathrm{M}_{2n \times 2n}(\mathbb{C})$ as $D := \mathrm{diag}[\widetilde{D}, R_n \widetilde{D}^H R_n]$ and observe in accordance with (10.5) that

$$
\begin{bmatrix} V & R_{2n} V R_n \end{bmatrix}
\begin{bmatrix} \widetilde{D} & \\ & R_n \widetilde{D}^H R_n \end{bmatrix}
\begin{bmatrix} V^H \\ R_n V^H R_{2n} \end{bmatrix}
= V\widetilde{D}V^H + R_{2n} V \widetilde{D}^H V^H R_{2n} = A.
\tag{10.6}
$$

---

[2]In operator theory this result is known as Fuglede's Theorem and holds true for bounded operators over complex Hilbert spaces, [52].

Let $U := [\, V \ R_{2n} V R_n \,]$. Then, obviously, $U^H A U = D$ is diagonal. Finally, in order to prove the unitary-perplectic diagonalizability of A we show that $U \in \mathcal{UP}(2n)$ by applying Proposition 10.3. That is, we need to show that $\mathrm{im}(V)$ is a Lagrangian subspace. However, in the expressions (10.5) and (10.6) this need a priori not be the case but can always be achieved as follows.

First, notice that $\mathrm{im}(N) \subseteq \mathrm{im}(V)$ always holds (since $N = V\widetilde{D}V^H$). Moreover, using the condition $N^\star N = 0$, i.e. $R_{2n}N^H R_{2n}N = 0$, the multiplication with R_{2n} (which is nonsingular) from the left yields $N^H R_{2n}N = 0$ so $\mathrm{im}(N)$ is an isotropic subspace. As $AN = (N + N^\star)N = N^2$, $\mathrm{im}(N)$ is also invariant for A (the same holds for $\mathrm{im}(N^\star)$). Now we distinguish between two cases:

(*i*) Assume that $\mathrm{rank}(N) = n$. Then, as $\dim(\mathrm{im}(V)) = n$ we have $\mathrm{im}(N) = \mathrm{im}(V)$. Therefore, $\mathrm{im}(V)$ is an isotropic subspace of dimension n, i.e. Lagrangian, and Proposition 10.3 yields that U is unitary-perplectic. Obviously, as $U^H A U = D$, $\mathrm{im}(V)$ is invariant for A as well. The nonzero eigenvalues of A are are exactly the nonzero eigenvalues of N and $N^\star$, cf. (10.6). In particular, if A is nonsingular we have $\sigma(A) = (\sigma(N) \cup \sigma(N^\star)) \setminus \{0\}$ whereas zero need not be excluded whenever it is already an eigenvalue of A.

(*ii*) Now assume $\mathrm{rank}(N) = k < n$. Recall that we assumed the $n - k$ eigenvalues of $\widetilde{D}$ which are equal to zero to appear in its trailing $n - k$ diagonal positions. Then, it is immediate that $\mathrm{im}(N)$ coincides with the span of the first k columns $v_1, \ldots, v_k$ of V. In other words, the last $n - k$ columns $v_{k+1}, \ldots, v_n$ of V have no contribution to the matrices N, $N^\star$ or A at all. Therefore, as long as the orthogonality constraint is met, $v_{k+1}, \ldots, v_n$ can be replaced by any other columns without changing the expression of A in (10.6).

Now we take Corollary 10.1 into account. As $\mathrm{span}(v_1, \ldots, v_k) = \mathrm{im}(N)$ is an isotropic subspace (of dimension k), it is properly contained in a Lagrangian subspace. Therefore, there exist $n - k$ vectors $\widetilde{v}_{k+1}, \ldots, \widetilde{v}_n \in \mathbb{C}^{2n}$ such that $\mathrm{span}(v_1, \ldots, v_k, \widetilde{v}_{k+1}, \ldots, \widetilde{v}_n)$ is a Lagrangian subspace. If $\widetilde{v}_{k+1}, \ldots, \widetilde{v}_n$ are chosen so that

$$\widetilde{V} := \begin{bmatrix} v_1 & \cdots & v_k & \widetilde{v}_{k+1} & \cdots & \widetilde{v}_n \end{bmatrix} \in M_{2n \times n}(\mathbb{C})$$

has orthonormal columns, i.e. $\widetilde{V}^H \widetilde{V} = I_n$, we obtain

$$\begin{bmatrix} \widetilde{V} & R_{2n}\widetilde{V}R_n \end{bmatrix} \begin{bmatrix} \widetilde{D} & \\ & R_n\widetilde{D}^H R_n \end{bmatrix} \begin{bmatrix} \widetilde{V}^H \\ R_n\widetilde{V}^H R_{2n} \end{bmatrix} = \widetilde{V}\widetilde{D}\widetilde{V}^H + R_{2n}\widetilde{V}\widetilde{D}^H\widetilde{V}^H R_{2n} = A. \tag{10.7}$$

Now the matrix $\widetilde{U} := [\, \widetilde{V} \ R_{2n}\widetilde{V}R_n \,] \in M_{2n \times 2n}(\mathbb{C})$ is unitary and perplectic according to Proposition 10.3. Therefore, $\widetilde{U}^H A \widetilde{U}$ is a similarity transformation of A with a unitary-perplectic matrix transforming A to diagonal form. The statements on the spectrum of A, N and $N^\star$ follow immediately from (10.7) similarly as in (*i*). $\qquad\square$

In other words, Theorem 10.2 states that a normal perhermitian matrix can be diagonalized by a unitary-perplectic similarity transformation iff A can be expressed as $A = N + N^\star$ for some normal matrix $N \in \mathrm{M}_{2n \times 2n}(\mathbb{C})$ satisfying $NN^\star = N^\star N = 0$. Recall that, according to Corollary 9.2, the results from Corollary 10.2 always hold if A has no purely real eigenvalues. On the other hand, if $A \in \mathrm{M}_{2n \times 2n}(\mathbb{C})$ is normal and perhermitian with only real eigenvalues, it must be Hermitian as well. In this context, Theorem 10.1 states that for any Hermitian and perhermitian matrix A there exists some $U \in \mathcal{UP}(2n)$ so that

$$U^H A U = X, \text{ where } X = \left[\times \right] \in \mathrm{M}_{2n \times 2n}(\mathbb{R}).$$

Of course X is real, Hermitian and perhermitian. This decomposition has already been noticed in [81].

Theorem 10.1 can also be used to determine whether a normal and perhermitian matrix $A \in \mathrm{M}_{2n \times 2n}(\mathbb{C})$ has purely real eigenvalues or not. Notice that, for normal matrices (without additional structure) Corollary 9.2 applies in the same way.

Corollary 10.3. Let $A \in \mathrm{M}_{2n \times 2n}(\mathbb{C})$ be normal and perhermitian. Then A has purely real eigenvalues iff $A - A^H$ is singular. Moreover, $2n - \mathrm{rank}(A - A^H)$ is the number of purely real eigenvalues of A (counted with multiplicities).

Proof. Let $\Delta = \mathrm{diag}[\Lambda, X, \Lambda^\star] = U^H A U$ be the canonical decomposition from Theorem 10.1. Then $A - A^H = U\Delta U^H - U\Delta^H U^H = U\left(\Delta - \Delta^H\right)U^H$ and in particular

$$\Delta - \Delta^H = \begin{bmatrix} \Lambda - \Lambda^H & & \\ & 0 & \\ & & \Lambda^\star - (\Lambda^\star)^H \end{bmatrix}.$$

As $\Lambda - \Lambda^H$ and $\Lambda^\star - (\Lambda^\star)^H$ will (by construction) never have zero diagonal entries, $\Delta - \Delta^H$ will be singular if and only if the X block in decomposition (10.1) of A exists, i.e. if A has purely real eigenvalues. The dimension of X determines the rank-deficiency of $A - A^H$ and is equal to the number of purely real eigenvalues of A (counted with multiplicities). $\square$

From Corollary 10.3 we may directly derive the following sufficiency result about the existence of a invariant Lagrangian subspace for $A \in \mathrm{M}_{2n \times 2n}(\mathbb{C})$.

Proposition 10.4. Let $A \in \mathrm{M}_{2n \times 2n}(\mathbb{C})$ be normal and perhermitian. Then A has an invariant Lagrangian subspace if $A - A^H$ is nonsingular.

Proof. As noticed in Corollary 10.3, A has purely real eigenvalues if and only if $A - A^H$ is singular. Thus, if $A - A^H$ is nonsingular, Corollary 10.2 (i) applies and $\mathrm{im}(N)$ with $\dim(\mathrm{im}(N)) = n$ is an invariant Lagrangian subspace for A. $\square$

10.2.1   Normal perskew-Hermitian matrices

The results obtained so far apply in a similar manner to normal perskew-Hermitian matrices. For these matrices, the canonical decomposition analogous to Theorem 10.1 is stated in Theorem 10.3 below. It essentially uses the fact that a matrix $A \in M_{2n \times 2n}(\mathbb{C})$ is perhermitian iff $B := iA$ is perskew-Hermitian (while the normality if maintained).

Theorem 10.3. *Let* $A \in M_{2n \times 2n}(\mathbb{C})$ *be a normal and perskew-Hermitian matrix. Then there exists a unitary and perplectic matrix* $Z \in M_{2n \times 2n}(\mathbb{C})$ *so that*

$$Z^H A Z = \begin{bmatrix} \Lambda & & \\ & X & \\ & & -\Lambda^\star \end{bmatrix} \qquad (10.8)$$

where $\Lambda \in M_{d \times d}(\mathbb{C})$ *is a diagonal matrix with* $d \in \mathbb{N}_0$ *being the number of eigenvalues with nonzero real parts of* A *(counted with multiplicities) and* $\sigma(\Lambda) \in \mathbb{C} \setminus i\mathbb{R}$. *In addition, the matrix* $X \in M_{2(n-d) \times 2(n-d)}(\mathbb{R})$ *is purely imaginary skew-Hermitian and perskew-Hermitian (thus* $\sigma(X) \subset i\mathbb{R}$) *and has only entries along its diagonal and its antidiagonal.*

Proof. If $A \in M_{2n \times 2n}(\mathbb{C})$ is normal and perskew-Hermitian, then $B = iA$ is perhermitian. Moreover, B is still normal, so Theorem 10.1 applies to B. Let $U^H B U = \Delta = \text{diag}[\Lambda, X, \Lambda^\star]$ be its canonical decomposition. As $B = U \Delta U^H$ and $B = iA$ we have $A = U(-i\Delta)U^H$. In particular, the eigenvalues of $\widetilde{\Lambda} := -i\Lambda$ all have nonzero real parts and $-i\Lambda^\star = -i R_d \Lambda^H R_d = -R_d i \Lambda^H R_d = -R_d(-i\Lambda)^H R_d = -\widetilde{\Lambda}^\star$. Moreover, $\widetilde{X} := -iX$ is skew-Hermitian and perskew-Hermitian (as X was Hermitian and perhermitian). The eigenvalues of $\widetilde{X}$ are thus purely imaginary. This completes the proof.   □

The following Theorem 10.4 states the analogous result to Theorem 10.2 for normal and perskew-Hermitian matrices. Compared to the previous result, notice the change of sign.

Theorem 10.4. *Let* $A \in M_{2n \times 2n}(\mathbb{C})$ *be normal and perskew-Hermitian. Then there exists a normal matrix* $N \in M_{2n \times 2n}(\mathbb{C})$ *with* $NN^\star = N^\star N = 0$ *and a skew-Hermitian and perskew-Hermitian matrix* $Z \in M_{2n \times 2n}(\mathbb{C})$ *so that* $A = N - N^\star + Z$. *Moreover, it holds that* $\text{rank}(N) \leq n$ *and* $\text{im}(N)$ *is an isotropic subspace for* R_{2n}. *If* $\text{rank}(N) = n$ *the subspace* $\text{im}(N)$ *is Lagrangian.*

The proof of Theorem 10.4 follows along the same lines as the proof of Theorem 10.2. For that reason it is omitted. In case $A \in M_{2n \times 2n}(\mathbb{C})$ is unitary-perplectic diagonalizable, it can be achieved that the antidiagonal part of X in (10.8) disappears (i.e. all its entries are zero). This implies the matrix Z in Theorem 10.4 to be zero and leads to the following result analogous to Corollary 10.2. The proof proceeds in a similar manner as before. Do not overlook the change of sign which comes from the perskew-symmetry.

Corollary 10.4. *Let* $A \in \mathrm{M}_{2n \times 2n}(\mathbb{C})$.

(*i*) *Assume that* A *is normal and perskew-Hermitian and unitary-perplectic diagonalizable. Then there exists a normal matrix* $N \in \mathrm{M}_{2n \times 2n}(\mathbb{C})$ *with* $NN^* = N^*N = 0$ *so that* $A = N - N^*$ *holds.*

(*ii*) *Assume that* A *can be expressed as* $A = N - N^*$ *for some normal matrix* $N \in \mathrm{M}_{2n \times 2n}(\mathbb{C})$ *with* $NN^* = N^*N = 0$. *Then* A *is normal and perskew-Hermitian and can be diagonalized by a unitary-perplectic similarity transformation.*

In both cases $\mathrm{rank}(N) \leq n$. *Furthermore,* $\sigma(A) = (\sigma(N) \cup \sigma(-N^*)) \setminus \{0\}$ *holds in case* A *is nonsingular and* $\sigma(A) = \sigma(N) \cup \sigma(-N^*)$ *otherwise. In addition,* $\mathrm{im}(N)$ *and* $\mathrm{im}(N^*)$ *are both invariant subspaces for* A *which are always isotropic. They are both Lagrangian if* $\mathrm{rank}(N) = n$.

As before, if $\sigma(A) \subset \mathbb{C} \setminus i\mathbb{R}$, i.e. A has no purely imaginary eigenvalues, Corollary 10.4 applies. Similarly to the preceding discussion, a normal and perkew-Hermitian matrix $A \in \mathrm{M}_{2n \times 2n}(\mathbb{C})$ has purely imaginary eigenvalues iff $A + A^H$ is singular. On the other hand, if $A + A^H$ is nonsingular, it is guaranteed that A has an invariant Lagrangian subspace. These facts are seen easily according to Corollary 10.3 and Proposition 10.4 from the previous section.

10.3   Normal Hamiltonian matrices

In this section we consider normal Hamiltonian and skew-Hamiltonian matrices $A \in \mathrm{M}_{2n \times 2n}(\mathbb{C})$. It is not immediate that the results from Theorem 10.2 and Theorem 10.4 (and, as a consequence, the results from Corollary 10.2 and Corollary 10.4) actually hold in the same manner (adapted to the new structures) to those matrices. In fact, the canonical factorization presented in Theorem 10.5 of normal Hamiltonian matrices used for the upcoming proof appears quite differently to the ones from Theorem 10.1 and Theorem 10.3. However its consequence for the additive decomposition discussed in the previous two sections is very much the same.

Theorem 10.5 ([8, Thm. 2.2]). *Let* $A \in \mathrm{M}_{2n \times 2n}(\mathbb{C})$ *be a normal and Hamiltonian matrix. Then there exists a unitary and symplectic matrix* $Z \in \mathrm{M}_{2n \times 2n}(\mathbb{C})$ *so that*

$$Z^H A Z = \begin{bmatrix} D_1 & 0 & 0 & 0 \\ 0 & D_2 & 0 & D_3 \\ 0 & 0 & -D_1^H & 0 \\ 0 & -D_3 & 0 & D_2 \end{bmatrix} \tag{10.9}$$

where $D_j, j = 1, 2, 3$ *are diagonal matrices,* $D_1 \in \mathrm{M}_{n_1 \times n_1}(\mathbb{C}), D_2 \in \mathrm{M}_{n_2 \times n_2}(i\mathbb{R}), D_3 \in \mathrm{M}_{n_2 \times n_2}(\mathbb{R}), n_1 + n_2 = n$, *and* $\begin{bmatrix} D_2 & D_3 \\ -D_3 & D_2 \end{bmatrix}$ *is a Hamiltonian matrix with purely imaginary eigenvalues.*

Before we derive similar decompositions as in the previous sections, the next proposition characterizes matrices $Z \in \mathrm{M}_{2n \times 2n}(\mathbb{C})$ which are both unitary and symplectic. For this class of matrices we use the abbreviation $\mathcal{US}(2n)$. Notice that this result can also be found in [105, Sec. 2] or can easily be derived from the characterization of symplectic matrices in [50, Prop. 1.4].

Proposition 10.5. *The set* $\mathcal{US}(2n)$ *of unitary and symplectic matrices in* $\mathrm{M}_{2n \times 2n}(\mathbb{C})$ *can be characterized as*

$$\mathcal{US}(2n) = \left\{ \left[\, V \; J_{2n}^{H} V \, \right] \in \mathrm{M}_{2n \times 2n}(\mathbb{C}) \, \middle| \, V \in \mathrm{M}_{2n \times n}(\mathbb{C}), V^{H} V = I_n, V^{H} J_{2n} V = 0 \right\}.$$

Proof. The proof proceeds analogous to that of Proposition 10.3. Let $Q = [\, Q_1 \, Q_2 \,] \in \mathcal{US}(2n)$ with $Q_1, Q_2 \in \mathrm{M}_{2n \times n}(\mathbb{C})$. As Q is unitary we have $Q^{H} Q = I_{2n}$ and as it is symplectic $Q^{H} J_{2n} Q = J_{2n}$ holds. Multiplying the latter with Q from the left gives $J_{2n} Q = Q J_{2n}$, so Q commutes with J_{2n}. From this relation it follows that $J_{2n} Q_1 = -Q_2$, i.e. $Q_2 = J_{2n}^{H} Q_1$. As Q is symplectic, i.e.

$$\begin{bmatrix} Q_1^{H} \\ Q_2^{H} \end{bmatrix} J_{2n} \begin{bmatrix} Q_1 & Q_2 \end{bmatrix} = \begin{bmatrix} Q_1^{H} J_{2n} Q_1 & Q_1^{H} J_{2n} Q_2 \\ Q_2^{H} J_{2n} Q_1 & Q_2^{H} J_{2n} Q_2 \end{bmatrix} = \begin{bmatrix} & I_n \\ -I_n & \end{bmatrix},$$

we have that $Q_1^{H} J_{2n} Q_1 = 0$, so $\mathrm{im}(Q_1)$ is a Lagrangian subspace. Finally, as Q is unitary,

$$\begin{bmatrix} Q_1^{H} \\ Q_2^{H} \end{bmatrix} \begin{bmatrix} Q_1 & Q_2 \end{bmatrix} = \begin{bmatrix} Q_1^{H} Q_1 & Q_1^{H} Q_2 \\ Q_2^{H} Q_1 & Q_2^{H} Q_2 \end{bmatrix} = \begin{bmatrix} I_n & \\ & I_n \end{bmatrix},$$

so $Q_1^{H} Q_1 = I_n$. This completes the proof. $\qquad\square$

The additive decomposition stated in Theorem 10.4 for normal and perskew-Hermitian matrices $A \in \mathrm{M}_{2n \times 2n}(\mathbb{C})$ takes the following form for normal Hamiltonian ones:

Theorem 10.6. *Let* $A \in \mathrm{M}_{2n \times 2n}(\mathbb{C})$ *be normal and Hamiltonian. Then there exists a normal matrix* $N \in \mathrm{M}_{2n \times 2n}(\mathbb{C})$ *with* $N N^{\star} = N^{\star} N = 0$ *and a skew-Hermitian and Hamiltonian matrix* $Z \in \mathrm{M}_{2n \times 2n}(\mathbb{C})$ *so that* $A = N - N^{\star} + Z$. *Moreover, it holds that* $\mathrm{rank}(N) \leq n$ *and* $\mathrm{im}(N)$ *is an isotropic subspace for* J_{2n}. *If* $\mathrm{rank}(N) = n$ *the subspace* $\mathrm{im}(N)$ *is Lagrangian.*

The proof of Theorem 10.6 is similar to the one from Theorem 10.2 and is stated here for completeness.

Proof. According to Theorem 10.5 there exists a $U \in \mathcal{US}(2n)$ so that $U^{H} A U = \Delta$ is in canonical form (10.9). In particular, according to Proposition 10.5, $U = [\, V \; J_{2n}^{H} V \,]$ for some $V \in \mathrm{M}_{2n \times n}(\mathbb{C})$ with $V^{H} V = I_n$ and $V^{H} J_{2n} V = 0$. We write $\Delta = D + \widehat{R}$ where $D = \mathrm{diag}[\Delta]$. Notice that $\widehat{R}$ is real, skew-Hermitian and Hamiltonian. Moreover, we decompose D as $D = \mathrm{diag}[\widetilde{D}, -\widetilde{D}^{H}]$ with $\widetilde{D} \in \mathrm{M}_n(\mathbb{C})$. Then, $A = U \Delta U^{H} = U D U^{H} + U \widehat{R} U^{H}$. Now the matrix A can be written as

$$A = V \widetilde{D} V^{H} - J_{2n}^{H} V \widetilde{D}^{H} V^{H} J_{2n} + U \widehat{R} U^{H} = N - N^{\star} + Z$$

with $N = V\widetilde{D}V^H$ and $Z = U\widehat{R}U^H$. The matrix $N \in \mathrm{M}_n(\mathbb{C})$ has rank $\leq n$ and is normal as we have $NN^H = V\widetilde{D}V^H(V\widetilde{D}V^H)^H = V\widetilde{D}V^H V\widetilde{D}^H V^H = V\widetilde{D}\widetilde{D}^H V^H$ and $N^H N = V\widetilde{D}^H \widetilde{D}V^H$. Additionally, we obtain

$$NN^\star = (V\widetilde{D}V^H)(J_{2n}^H V\widetilde{D}^H V^H J_{2n}) = V\widetilde{D}V^H(-J_{2n})V\widetilde{D}^H V^H J_{2n}$$
$$= -V\widetilde{D}(V^H J_{2n} V)\widetilde{D}^H V^H J_{2n}$$

so $NN^\star = 0$ since $V^H J_{2n} V = 0$. Similarly $N^\star N = 0$ follows. Furthermore, as $\mathrm{im}(N) \subseteq \mathrm{im}(V)$, $\mathrm{im}(N)$ is always isotropic and Lagrangian in case $\dim(\mathrm{im}(N)) = \mathrm{rank}(N) = n$. Moreover, as $\widehat{R}$ is Hamiltonian and skew-Hermitian, Z is Hamiltonian and skew-Hermitian as $U \in \mathcal{US}(2n)$. □

As in the previous section the matrix $Z = U\widehat{R}U^H$ in the proof of Theorem 10.2 vanishes if $\widehat{R} = 0$. This can be achieved if A is unitary-symplectic diagonalizable. Thus, we obtain the analogous result to Corollary 10.4 for Hamiltonian matrices.

Corollary 10.5. *Let* $A \in \mathrm{M}_{2n \times 2n}(\mathbb{C})$.

(i) *Assume that A is normal and Hamiltonian and unitary-symplectic diagonalizable. Then there exists a normal matrix* $N \in \mathrm{M}_{2n \times 2n}(\mathbb{C})$ *with* $NN^\star = N^\star N = 0$ *so that $A = N - N^\star$ holds.*

(ii) *Assume that A can be expressed as $A = N - N^\star$ for some normal matrix* $N \in \mathrm{M}_{2n \times 2n}(\mathbb{C})$ *with* $NN^\star = N^\star N = 0$. *Then A is normal and Hamiltonian and can be diagonalized by a unitary-symplectic similarity transformation.*

In both cases $\mathrm{rank}(N) \leq n$. *Furthermore,* $\sigma(A) = (\sigma(N) \cup \sigma(-N^\star)) \setminus \{0\}$ *holds in case A is nonsingular and* $\sigma(A) = \sigma(N) \cup \sigma(-N^\star)$ *otherwise. In addition,* $\mathrm{im}(N)$ *and* $\mathrm{im}(N^\star)$ *are both invariant subspaces for A which are always isotropic. They are both Lagrangian if* $\mathrm{rank}(N) = n$.

The statement from Corollary 10.5 (i) follows from the proof of Theorem 10.6 noting that $\widehat{R} = 0$ and that the columns of V are all eigenvectors of A. The proof of part (ii) proceeds in a similar fashion as the one for Corollary 10.2 (ii) taking Corollary 10.1 into account. Finally, we may again determine the number of purely imaginary eigenvalues of a normal Hamiltonian matrix $A \in \mathrm{M}_{2n \times 2n}(\mathbb{C})$ by considering the rank of $A + A^H$.

Corollary 10.6. *Let* $A \in \mathrm{M}_{2n \times 2n}(\mathbb{C})$ *be normal and Hamiltonian. Then A has purely imaginary eigenvalues iff $A + A^H$ is singular. Moreover,* $2n - \mathrm{rank}(A + A^H)$ *is the number of purely imaginary eigenvalues of A (counted with multiplicities).*

Corollary 10.6 follows in a similar manner as Corollary 10.3. In particular, it is guaranteed that A has an invariant Lagrangian subspace if $A + A^H$ is nonsingular (as, in this case, A does not have any purely imaginary eigenvalues and the situation from Corollary 10.5 (i) applies).

Applying the previous discussion to skew-Hamiltonian matrices, we certainly obtain similar results to those of Theorem 10.5, Theorem 10.6 and Corollary 10.5. Moreover, analogous statements to Corollary 10.3 and Proposition 10.4 can easily be derived. For completeness, we only state the analogous result to Corollary 10.5 for skew-Hamiltonian matrices and omit a further discussion of those matrices as no substantially new results occur.

Corollary 10.7. *Let $A \in \mathrm{M}_{2n \times 2n}(\mathbb{C})$.*

(i) *Assume that A is normal and skew-Hamiltonian and unitary-symplectic diagonalizable. Then there exists a normal matrix $N \in \mathrm{M}_{2n \times 2n}(\mathbb{C})$ with $NN^\star = N^\star N = 0$ so that $A = N + N^\star$ holds.*

(ii) *Assume that A can be expressed as $A = N + N^\star$ for some normal matrix $N \in \mathrm{M}_{2n \times 2n}(\mathbb{C})$ with $NN^\star = N^\star N = 0$. Then A is normal and skew-Hamiltonian and can be diagonalized by a unitary-symplectic similarity transformation.*

In both cases $\mathrm{rank}(N) \leq n$. *Furthermore,* $\sigma(A) = (\sigma(N) \cup \sigma(N^\star)) \setminus \{0\}$ *holds in case A is nonsingular and* $\sigma(A) = \sigma(N) \cup \sigma(N^\star)$ *otherwise. In addition,* $\mathrm{im}(N)$ *and* $\mathrm{im}(N^\star)$ *are both invariant subspaces for A which are always isotropic. They are both Lagrangian if* $\mathrm{rank}(N) = n$.

10.4 Conclusions

In this chapter we considered matrices $A \in \mathrm{M}_{2n \times 2n}(\mathbb{C})$ that are normal (i.e. $AA^H = A^H A$) and belong to the classes $\mathbb{J}(R_{2n}), \mathbb{L}(R_{2n}), \mathbb{J}(J_{2n})$ or $\mathbb{L}(J_{2n})$. We provided several auxiliary facts on maximal isotropic subspaces that led to the result that any isotropic subspace is always contained in a maximal isotropic subspace. Those maximal isotropic subspaces are of dimension n and have been called Lagrangian. We derived canonical factorizations for normal matrices from $\mathbb{J}(R_{2n}), \mathbb{L}(R_{2n}), \mathbb{J}(J_{2n})$ or $\mathbb{L}(J_{2n})$ under unitary-perplectic/unitary-symplectic similarity transformations and showed how these factorizations lead to special additive decompositions of those matrices. These decompositions are, in both cases, of the form $A = N \pm N^\star + Z$ where $N \in \mathrm{M}_{2n \times 2n}(\mathbb{C})$ is a normal matrix with $NN^\star = N^\star N = 0$ and $Z \in \mathrm{M}_{2n \times 2n}(\mathbb{C})$ is a Hermitian/skew-Hermitian matrix in the same structure class as A (so Z is indeed structured twice). Moreover, the eigenvalues of Z are always either purely real or purely imaginary. We showed that if A is unitary-perplectic/unitary-symplectic diagonalizable, the matrix Z can be chosen to be identically zero so that A can be expressed as $A = N \pm N^\star$ for some normal matrix N. Furthermore, we proved that a decomposition $A = N \pm N^\star$ for some normal matrix N satisfying $NN^\star = N^\star N = 0$ indeed implies that A is structured according to the particular structure-class, normal and diagonalizable by a unitary automorphism. Moreover, we motivated how this decomposition can be used to derive statement on the number of purely real/imaginary eigenvalues of A and the existence of an invariant Lagrangian subspace. In particular, if $A = N \pm N^*$, then the image of N is an invariant isotropic subspace for A. If, in addition, $\mathrm{rank}(N) = n$ holds, then $\mathrm{im}(N)$ is Lagrangian.

Bibliography

[1] Amir Amiraslani, Robert M. Corless, and Peter Lancaster. "Linearization of Matrix Polynomials expressed in polynomial Bases". In: *IMA Journal of Numerical Analysis* 29 (2009), pp. 141–157.

[2] Miguel F. Anjos, Sven Hammarling, and Christopher C. Paige. *Solving the Generalized Symmetric Eigenvalue Problem.* MIMS EPrint 2008.67 (1992). URL: http://eprints.maths.manchester.ac.uk/1120/1/AHP.pdf.

[3] Efstathios N. Antoniou and Stavros Vologiannidis. "A new family of Companion Forms of Polynomial Matrices". In: *Electronic Journal of Linear Algebra* 11 (2004), pp. 78–87.

[4] Emil Artin. *Geometric Algebra.* Dover Books on Mathematics. Mineola, New York: Dover Publications, 2016.

[5] Zhaojun Bai, James Demmel, and Ming Gu. "An inverse free parallel spectral Divide and Conquer Algorithm for Nonsymmetric Eigenproblems". In: *Numerische Mathematik* 76 (1997), pp. 279–308.

[6] Zhaojun Bai et al. *Templates for the Solution of Algebraic Eigenvalue Problems.* Philadelphia: Society for Industrial and Applied Mathematics, 2000.

[7] Stephen Barnett. "A Companion Matrix analogue for orthogonal Polynomials". In: *Linear Algebra and its Applications* 12 (1975), pp. 197–202.

[8] Erna Begovic, Heike Faßbender, and Philip Saltenberger. *On normal and structured Matrices under unitary structure preserving Transformations.* ArXiv Preprint 1810.03369 (2018). URL: https://arxiv.org/pdf/1810.03369.pdf.

[9] Peter Benner, Volker Mehrmann, and Hongguo Xu. "A numerically stable, structure preserving Method for computing the Eigenvalues of real Hamiltonian or Symplectic Pencils". In: *Numerische Mathematik* 78 (1998), pp. 329–358.

[10] Peter Benner et al. "Numerical Computation of Deflating Subspaces of Skew-Hamiltonian/Hamiltonian Pencils". In: *SIAM Journal on Matrix Analysis and Applications* 24 (2002), pp. 165–190.

[11] Timo Betcke et al. "NLEVP: A Collection of Nonlinear Eigenvalue Problems". In: *ACM Transactions on Mathematical Software* 39 (2013), pp. 1–28.

[12] Dario A. Bini, Luca Gemignani, and Françoise Tisseur. "The Ehrlich–Aberth Method for the Nonsymmetric Tridiagonal Eigenvalue Problem". In: *SIAM Journal on Matrix Analysis and Applications* 27 (2005), pp. 153–175.

[13]   János Bognár. *Indefinite Inner Product Spaces.* Berlin, Heidelberg: Springer-Verlag, 1974.

[14]   Nicolas Bourbaki. *Algèbre (Chapitre 9: Formes sesquilinéaires et formes quadratique).* Berlin, Heidelberg: Springer-Verlag, 2007.

[15]   Michael A. Brebner and Janez Grad. "Eigenvalues of Ax = λBx for real symmetric Matrices A and B computed by Reduction to a pseudosymmetric Form and the HR process". In: *Linear Algebra and its Applications* 43 (1982), pp. 99–118.

[16]   William C. Brown. *Matrices over Commutative Rings.* Monographs and Textbooks in pure and applied Mathematics, Vol. 169. New York: Dekker, 1993.

[17]   Maribel I. Bueno, Kyle Curlett, and Susana Furtado. "Structured Strong Linearizations from Fiedler Pencils with Repetition I". In: *Linear Algebra and its Applications* 460 (2014), pp. 51–80.

[18]   Maribel I. Bueno and Susana Furtado. "Palindromic Linearizations of a Matrix Polynomial of odd Degreee obtained from Fiedler Pencils with Repetition". In: *Electronic Journal of Linear Algebra* 23 (2012), pp. 562–577.

[19]   Maribel I. Bueno and Susana Furtado. "Structured Strong Linearizations from Fiedler Pencils with Repetition II". In: *Linear Algebra and its Applications* 463 (2014), pp. 282–321.

[20]   Maribel I. Bueno, Fernando de Terán, and Froilán M. Dopico. "Recovery of Eigenvectors and Minimal Bases of Matrix Polynomials from Generalized Fiedler Linearizations". In: *SIAM Journal on Matrix Analysis and Applications* 32 (2011), pp. 463–483.

[21]   Maribel I. Bueno et al. *A unified Approach to Fiedler-like Pencils via Strong Block Minimal Bases Pencils.* ArXiv Preprint 1611.07170v1 (2016). URL: https://arxiv.org/pdf/1611.07170.pdf.

[22]   Maribel I. Bueno et al. "Large Vector Spaces of block-symmetric Strong Linearizations of Matrix Polynomials". In: *Linear Algebra and its Applications* 477 (2015), pp. 165–210.

[23]   Angelika Bunse-Gerstner. "An Analysis of the HR Algorithm for computing the Eigenvalues of a Matrix". In: *Linear Algebra and its Applications* 35 (1981), pp. 155–173.

[24]   Angelika Bunse-Gerstner and Heike Faßbender. "Breaking Van Loan's Curse: A Quest for Structure-Preserving Algorithms for Dense Structured Eigenvalue Problems". In: *Numerical Algebra, Matrix Theory, Differential-Algebraic, Equations and Control Theory.* Heidelberg, New York, London: Springer International Publishing (2015).

[25]   Angelika Bunse-Gerstner and Volker Mehrmann. "The HHDR Algorithm and its Application to optimal Control Problems". In: *RAIRO. Automatique-Productique, Informatique Industrielle* 23 (1989), pp. 309–329.

[26]   Carmen Campos and Jose E. Roman. "Restarted Q-Arnoldi-type Methods exploiting Symmetry in quadratic Eigenvalue Problems". In: *BIT Numerical Mathematics* 56 (2016), pp. 1213–1236.

[27]   Robert M. Corless and Gurjeet Litt. *Generalized Companion Matrices for Polynomials not expressed in Monomial Bases.* Ontario Research Centre for Computer Algebra, Technical Report (2001). URL: http://www.apmaths.uwo.ca/~rcorless/frames/PAPERS/PABV/CMP.pdf.

[28]   Biswajit Das and Shreemayee Bora. *Vector Spaces of Generalized Linearizations for rectangular Matrix Polynomials.* ArXiv Preprint 1808.00517 (2018). URL: https://arxiv.org/pdf/1808.00517.pdf.

[29]   Philip I. Davies, Nicholas J. Higham, and Françoise Tisseur. "Analysis of the Cholesky Method with Iterative Refinement for Solving the Symmetric Definite Generalized Eigenproblem". In: *SIAM Journal on Matrix Analysis and Applications* 23 (2001), pp. 472–493.

[30]   Ralph J. de la Cruz and Heike Faßbender. "On the Diagonalizability of a Matrix by a symplectic Equivalence, Similarity or Congruence Transformation". In: *Linear Algebra and its Applications* (2016), pp. 288–306.

[31]   Fernando De Terán, Froilán M. Dopico, and D. Steven Mackey. "Linearizations of singular Matrix Polynomials and the Recovery of Minimal Indices". In: *Electronic Journal of Linear Algebra* 18 (2009), pp. 371–402.

[32]   Fernando De Terán, Froilán M. Dopico, and D. Steven Mackey. "Spectral equivalence of matrix polynomials and the Index Sum Theorem". In: *Linear Algebra and its Applications* 459 (2014), pp. 264–333.

[33]   Lokenath Debnath and Piotr Mikusiński. *Hilbert Spaces with Applications.* 3rd Edition. Amsterdam and Boston: Elsevier Academic Press, 2005.

[34]   Froilán M. Dopico et al. *Block Kronecker Linearizations of Matrix Polynomials and their Backward Errors.* MIMS Eprint 2016.34 (2016). URL: http://eprints.maths.manchester.ac.uk/2481/1/kron-pencils-backerror-fro.pdf.

[35]   Froilán M. Dopico et al. "Block Kronecker Linearizations of Matrix Polynomials and their Backward Errors". In: *Numerische Mathematik* 140 (2018), pp. 373–426.

[36]   Cedric Effenberger and Daniel Kressner. "Chebyshev Interpolation for nonlinear Eigenvalue Problems". In: *BIT Numerical Mathematics* 52 (2012), pp. 933–951.

[37]   Louis W. Ehrlich. "A modified Newton Method for Polynomials". In: *Communications of the ACM* 10 (1967), pp. 107–108.

[38]   John H. Ewing and Frederick W. Gehring. *PAUL HALMOS Celebrating 50 Years of Mathematics.* New York: Springer-Verlag, 1991.

[39] Heike Faßbender. "Structured Eigenvalue Problems - Structure-Preserving Algorithms, Structured Error Analysis". In: *Handbook of Linear Algebra, 2nd Edition*. Boca Raton, FL: CRC Press, Taylor and Francis Group (2014).

[40] Heike Faßbender. *Symplectic Methods for the Symplectic Eigenproblem*. Boston: Kluwer Academic Publishers, 2002.

[41] Heike Faßbender and Kh.D. Ikramov. "Several Observations on Symplectic, Hamiltonian, and skew-Hamiltonian matrices". In: *Linear Algebra and its Applications* 400 (2005), pp. 15–29.

[42] Heike Faßbender and Philip Saltenberger. "Block Kronecker Ansatz Spaces for Matrix Polynomials". In: *Linear Algebra and its Applications* 542 (2018), pp. 118–148.

[43] Heike Faßbender and Philip Saltenberger. "On a Modification of the EVEN-IRA Algorithm for the solution of T-even polynomial Eigenvalue Problems". In: *Proceedings in Applied Mathematics and Mechanics (PAMM). Special issue: 89th Annual Meeting of the International Association of Applied Mathematics and Mechanics (GAMM)* 18 (2018).

[44] Heike Faßbender and Philip Saltenberger. *On a new kind of Ansatz Spaces for Matrix Polynomials*. ArXiv Preprint 1610.05988 (2016). URL: https://arxiv.org/pdf/1610.05988.pdf.

[45] Heike Faßbender and Philip Saltenberger. "On Vector Spaces of Linearizations for Matrix Polynomials in orthogonal Bases". In: *Linear Algebra and its Applications* 525 (2017), pp. 59–83.

[46] Heike Faßbender and Philip Saltenberger. *Some Notes on the Linearization of Matrix Polynomials in Standard and Tschebyscheff Basis*. Technical Report (2016). URL: https://www.tu-braunschweig.de/Medien-DB/numerik/preprints/linpoltscheb.pdf.

[47] Heike Faßbender et al. "Hamiltonian square Roots of skew-Hamiltonian Matrices". In: *Linear Algebra and its Applications* 287.1-3 (1999), pp. 125–159.

[48] Ky Fan and A. J. Hoffman. "Some metric Inequalities in the Space of Matrices". In: *Proceedings of the American Mathematical Society* 6 (1955).

[49] Miroslav Fiedler. "A note on Companion Matrices". In: *Linear Algebra and its Applications* 372 (2003), pp. 325–331.

[50] Gerhard Freiling, Volker Mehrmann, and Hongguo Xu. "Existence, Uniqueness, and Parametrization of Lagrangian Invariant Subspaces". In: *SIAM Journal on Matrix Analysis and Applications* 23 (2002), pp. 1045–1069.

[51] Ferdinand Georg Frobenius. "Theorie der linearen Formen mit ganzen Coefficienten". In: *Journal für die reine und angewandte Mathematik* 86 (1879), pp. 147–208.

[52] Bent Fuglede. "A Commutativity Theorem for Normal Operators". In: *Proceedings of the National Academy of Sciences* 36 (1950), pp. 35–40.

[53] Gerhard Gierz et al. *Continuous Lattices and Domains*. Encyclopedia of mathematics and its applications, Vol. 93. Cambridge: Cambridge University Press, 2003.

[54] Johann Wolfgang von Goethe and Adolf Muschg. *Wilhelm Meisters Wanderjahre oder die Entsagenden*. 10. Auflage. Insel-Taschenbuch, Vol. 575. Frankfurt am Main: Insel-Verlag, 2006.

[55] Israel Gohberg, Peter Lancaster, and Leiba Rodman. *Indefinite Linear Algebra and Applications*. Basel: Birkhäuser Verlag, 2005.

[56] Yisrael Z. Gohberg, Peter Lancaster, and Leiba Rodman. *Matrix Polynomials*. Classics in Applied Mathematics, Vol. 58. Philadelphia: Society for Industrial and Applied Mathematics, 2009.

[57] Gene H. Golub and Charles van Loan. *Matrix Computations*. 4th Edition. Johns Hopkins Studies in Mathematical Sciences. Baltimore: Johns Hopkins Univ. Press, 2013.

[58] Irving J. Good. "The Colleague Matrix, a Chebyshev Analogue of the Companion Matrix". In: *The Quarterly Journal of Mathematics* 12 (1961), pp. 61–68.

[59] Roger G. Grimes, John G. Lewis, and Horst D. Simon. "A Shifted Block Lanczos Algorithm for Solving Sparse Symmetric Generalized Eigenproblems". In: *SIAM Journal on Matrix Analysis and Applications* 15 (1994), pp. 228–272.

[60] Robert Grone et al. "Normal Matrices". In: *Linear Algebra and its Applications* 87 (1987), pp. 213–225.

[61] Nicholas J. Higham. *Accuracy and Stability of Numerical Algorithms*. Philadelphia: Society for Industrial and Applied Mathematics, 1996.

[62] Nicholas J. Higham, Ren-Cang Li, and Françoise Tisseur. "Backward Error of Polynomial Eigenproblems Solved by Linearization". In: *SIAM Journal on Matrix Analysis and Applications* 29 (2008), pp. 1218–1241.

[63] Nicholas J. Higham, D. Steven Mackey, and Françoise Tisseur. "Definite Matrix Polynomials and their Linearization by Definite Pencils". In: *SIAM Journal on Matrix Analysis and Applications* 31 (2009), pp. 478–502.

[64] Nicholas J. Higham, D. Steven Mackey, and Françoise Tisseur. "The Conditioning of Linearizations of Matrix Polynomials". In: *SIAM Journal on Matrix Analysis and Applications* 28 (2006), pp. 1005–1028.

[65] Nicholas J. Higham et al. "Symmetric Linearizations for Matrix Polynomials". In: *SIAM Journal on Matrix Analysis and Applications* 29 (2007), pp. 143–159.

[66] Nicolas J. Higham. *Matrix Nearness Problems and Applications*. URL: http://citeseerx.ist.psu.edu/viewdoc/download?doi=10.1.1.35.2899&rep=rep1&type=pdf.

[67]   Roger A. Horn and Charles R. Johnson. *Matrix Analysis*. Cambridge: Cambridge Univ. Press, 1999.

[68]   Tsung-Min Hwang, Wen-Wei Lin, and Volker Mehrmann. "Numerical Solution of Quadratic Eigenvalue Problems with structure-preserving Methods". In: *SIAM Journal on Scientific Computing* 24 (2003), pp. 1283–1302.

[69]   James Kestyn, Eric Polizzi, and Ping Tak Peter Tang. *FEAST Eigensolver for Non-Hermitian Problems*. ArXiv Preprint 1506.04463 (2015). URL: https://arxiv.org/pdf/1506.04463.pdf.

[70]   Lukas Krämer et al. "Dissecting the FEAST Algorithm for Generalized Eigenproblems". In: *Journal of Computational and Applied Mathematics* 244 (2013), pp. 1–9.

[71]   Daniel Kressner. *Numerical Methods for General and Structured Eigenvalue Problems*. Lecture Notes in Computational Science and Engineering, Vol. 46. Berlin, Heidelberg: Springer-Verlag, 2005.

[72]   Daniel Kressner and Jose E. Roman. "Memory-efficient Arnoldi Algorithms for Linearizations of Matrix Polynomials in Chebyshev Basis". In: *Numerical Linear Algebra with Applications* 21 (2014), pp. 569–588.

[73]   Daniel Kressner, Christian Schröder, and David S. Watkins. "Implicit QR algorithms for palindromic and even eigenvalue problems". In: *Numerical Algorithms* 51 (2009), pp. 209–238.

[74]   Peter Lancaster. "Symmetric Transformation of the Companion Matrix". In: *NABLA: Bulletin of the Malayan Math. Soc.* 8 (1961), pp. 146–148.

[75]   Peter Lancaster and Leiba Rodman. *Algebraic Riccati equations*. Oxford Science Publications. Oxford: Clarendon Press, 2002.

[76]   Peter Lancaster et al. *Lambda-Matrices and Vibrating Systems*. International Series in Pure and Applied Mathematics, Vol. 94. Burlington: Elsevier Science, 2014.

[77]   Leonid Lerer and Miron Tismenetsky. "The Bezoutian and the eigenvalue-separation Problem for Matrix Polynomials". In: *Integral Equations and Operator Theory* 5 (1982), pp. 386–445.

[78]   Kurt Lewin. "Field Theory in Social Science: Selected Theoretical Papers." In: *American Journal of Sociology* 57 (1951), pp. 86–87.

[79]   Wen-Wei Lin, Volker Mehrmann, and Hongguo Xu. "Canonical Forms for Hamiltonian and Symplectic Matrices and Pencils". In: *Linear Algebra and its Applications* 302-303 (1999), pp. 469–533.

[80]   D. Steven Mackey. *Structured Linearizations for Matrix Polynomials*. PhD Thesis, MIMS Eprint 2006.68 (2006). URL: http://eprints.maths.manchester.ac.uk/226/1/mackey06.pdf.

[81] D. Steven Mackey, Niloufer Mackey, and Daniel M. Dunlavy. "Structure Preserving Algorithms for Perplectic Eigenproblems". In: *The Electronic Journal of Linear Algebra* 13 (2005), pp. 10–39.

[82] D. Steven Mackey, Niloufer Mackey, and Francoise Tisseur. *On the Definition of two natural Classes of Scalar Product*. MIMS Eprint 2007.64 (2007). URL: http://eprints.ma.man.ac.uk/1236/1/covered/MIMS_ep2007_64.pdf.

[83] D. Steven Mackey, Niloufer Mackey, and Françoise Tisseur. "Structured Factorizations in Scalar Product Spaces". In: *SIAM Journal on Matrix Analysis and Applications* 27 (2005), pp. 821–850.

[84] D. Steven Mackey and Vasilije Perović. "Linearizations of Matrix Polynomials in Bernstein Bases". In: *Linear Algebra and its Applications* 501 (2016), pp. 162–197.

[85] D. Steven Mackey and Vasilije Perović. "Linearizations of Matrix Polynomials in Newton Bases". In: *Linear Algebra and its Applications* 556 (2018), pp. 1–45.

[86] D. Steven Mackey et al. "Möbius Transformations of Matrix Polynomials". In: *Linear Algebra and its Applications* 470 (2015), pp. 120–184.

[87] D. Steven Mackey et al. "Structured Polynomial Eigenvalue Problems: Good Vibrations from Good Linearizations". In: *SIAM Journal on Matrix Analysis and Applications* 28 (2006), pp. 1029–1051.

[88] D. Steven Mackey et al. *Vector Spaces of Linearizations for Matrix Polynomials*. Preprint (2005). URL: http://page.math.tu-berlin.de/~mehl/papers/genlin.pdf.

[89] D. Steven Mackey et al. "Vector Spaces of Linearizations for Matrix Polynomials". In: *SIAM Journal on Matrix Analysis and Applications* 28 (2006), pp. 971–1004.

[90] John Maroulas and Stephen Barnett. "Polynomials with respect to a general Basis. I. Theory". In: *Journal of Mathematical Analysis and Applications* 72 (1979), pp. 177–194.

[91] MATLAB. *Version 9.4.0.813654 (R2018a)*. Natick, Massachusetts: The MathWorks Inc., 2010.

[92] Christian Mehl. URL: http://page.math.tu-berlin.de/~mehl/talks/siamala2009.pdf.

[93] Christian Mehl, Volker Mehrmann, and Hongguo Xu. "Canonical forms for doubly structured Matrices and Pencils". In: *Electronic Journal of Linear Algebra* 7 (2000), pp. 112–151.

[94] Volker Mehrmann, Christian Schröder, and Valeria Simoncini. "An implicitly-restarted Krylov Subspace Method for real symmetric/skew-symmetric Eigenproblems". In: *Linear Algebra and its Applications* 436 (2012), pp. 4070–4087.

[95]   Volker Mehrmann and Heinrich Voss. "Nonlinear Eigenvalue Problems: A Challenge for modern Eigenvalue Methods". In: *GAMM-Mitteilungen* 27 (2004), pp. 121–152.

[96]   Volker Mehrmann and David S. Watkins. "Polynomial Eigenvalue Problems with Hamiltonian Structure". In: *Electronic Transactions on Numerical Analysis* 13 (2002), pp. 106–118.

[97]   Volker Mehrmann and David S. Watkins. "Structure-Preserving Methods for Computing Eigenpairs of Large Sparse Skew-Hamiltonian/Hamiltonian Pencils". In: *SIAM Journal on Scientific Computing* 22 (2001), pp. 1905–1925.

[98]   Volker Mehrmann and Hongguo Xu. "Structured Jordan Canonical Forms for structured Matrices that are Hermitian, skew Hermitian or unitary with respect to indefinite Inner Products". In: *The Electronic Journal of Linear Algebra* 5 (1999), pp. 67–103.

[99]   Cleve Moler and Gilbert W. Stewart. "An Algorithm for the Generalized Matrix Eigenvalue Problem $Ax = \lambda Bx$". In: *SIAM Journal on Numerical Analysis* 10 (1973), pp. 241–256.

[100] Yuji Nakatsukasa and Nicholas J. Higham. "Stable and Efficient Spectral Divide and Conquer Algorithms for the Symmetric Eigenvalue Decomposition and the SVD". In: *SIAM Journal on Scientific Computing* 35 (2013), A1325–A1349.

[101] Yuji Nakatsukasa, Vanni Noferini, and Alex Townsend. "Vector Spaces of Linearizations for Matrix Polynomials: a bivariate polynomial Approach". In: *SIAM Journal on Matrix Analysis and Applications* 38 (2017), pp. 1–29.

[102] Yuji Nakatsukasa et al. *Reduction of Matrix Polynomials to simpler Forms*. MIMS Eprint 2017.13 (2017). URL: http://eprints.maths.manchester.ac.uk/2539/1/nttz17.pdf.

[103] Morris Newman. "Two classical Theorems on commuting Matrices". In: *Journal of Research of the National Bureau of Standards* (1967), pp. 69 –71.

[104] Vanni Noferini and Javier Pérez. "Fiedler–Comrade and Fiedler–Chebyshev Pencils". In: *SIAM Journal on Matrix Analysis and Applications* 37 (2016), pp. 1600–1624.

[105] Chris Paige and Charles van Loan. "A Schur Decomposition for Hamiltonian Matrices". In: *Linear Algebra and its Applications* 41 (1981), pp. 11–32.

[106] Beresford N. Parlett. *The Symmetric Eigenvalue Problem*. Philadelphia: Society for Industrial and Applied Mathematics, 1998.

[107] Eric Polizzi. "Density-matrix-based Algorithm for solving Eigenvalue Problems". In: *Physical Review B* 79 (2009).

[108] Cornelis Praagman. "Invariants of Polynomial Matrices". In: *Proceedings of the First European Control Conference* 2 (1991), pp. 1274–1277.

[109]   Constance Reid. *Hilbert*. Berlin, Heidelberg: Springer-Verlag, 1970.

[110]   Russell M. Reid. "Classroom Note: Some Eigenvalue Properties of Persymmetric Matrices". In: *SIAM Review* 39.2 (1997), pp. 313–316.

[111]   Axel Ruhe. "Rational Krylov: A Practical Algorithm for Large Sparse Nonsymmetric Matrix Pencils". In: *SIAM Journal on Scientific Computing* 19 (1998), pp. 1535–1551.

[112]   Yousef Saad. *Numerical Methods for large Eigenvalue Problems*. Classics in applied mathematics. Philadelphia, PA.: Society for Industrial and Applied Mathematics, 2011.

[113]   Tetsuya Sakurai and Hiroto Tadano. "CIRR: A Rayleigh-Ritz type Method with Contour Integral for Generalized Eigenvalue Problems". In: *Hokkaido Mathematical Journal* 36 (2007), pp. 745–757.

[114]   Danny C. Sorensen. "Implicit Application of Polynomial Filters in a k-Step Arnoldi Method". In: *SIAM Journal on Matrix Analysis and Applications* 13 (1992), pp. 357–385.

[115]   Danny C. Sorensen. "Implicitly Restarted Arnoldi/Lanczos Methods for Large Scale Eigenvalue Calculations". In: *Parallel Numerical Algorithms. ICASE/LaRC Interdisciplinary Series in Science and Engineering, Vol 4*. Dordrecht: Springer Netherlands (1997).

[116]   Gilbert W. Stewart. "A Krylov–Schur Algorithm for Large Eigenproblems". In: *SIAM Journal on Matrix Analysis and Applications* 23 (2002), pp. 601–614.

[117]   Gilbert W. Stewart. *Matrix Algorithms, Volume II: Eigensystems*. Philadelphia: Society for Industrial and Applied Mathematics, 2001.

[118]   Gábor Szegö. *Orthogonal Polynomials*. Colloquium Publications, Vol. 23. Providence, RI: American Mathematical Society, 2003.

[119]   Fernando de Terán and Froilán M. Dopico. "First order Spectral Perturbation Theory of square singular Matrix Polynomials". In: *Linear Algebra and its Applications* 432 (2010), pp. 892–910.

[120]   Fernando de Terán, Froilán M. Dopico, and D. Steven Mackey. "Fiedler Companion Linearizations and the Recovery of Minimal Indices". In: *SIAM Journal on Matrix Analysis and Applications* 31 (2010), pp. 2181–2204.

[121]   Fernando de Terán, Froilán M. Dopico, and D. Steven Mackey. "Fiedler Companion Linearizations for rectangular Matrix Polynomials". In: *Linear Algebra and its Applications* 437 (2012), pp. 957–991.

[122]   Fernando de Terán, Froilán M. Dopico, and Paul van Dooren. "Matrix Polynomials with Completely Prescribed Eigenstructure". In: *SIAM Journal on Matrix Analysis and Applications* 36 (2015), pp. 302–328.

[123]   Françoise Tisseur. URL: http://www.ma.man.ac.uk/~ftisseur/talks/iwasep.pdf.

[124] Françoise Tisseur. "Tridiagonal-Diagonal Reduction of Symmetric Indefinite Pairs". In: *SIAM Journal on Matrix Analysis and Applications* 26 (2004), pp. 215–232.

[125] Françoise Tisseur and Karl Meerbergen. "The Quadratic Eigenvalue Problem". In: *SIAM Review* 43 (2001), pp. 235–286.

[126] Frank Uhlig. "The DQR algorithm, basic Theory, Convergence, and conditional Stability". In: *Numerische Mathematik* 76 (1997), pp. 515–553.

[127] Paul van Dooren. "The computation of Kronecker's Canonical Form of a singular Pencil". In: *Linear Algebra and its Applications* 27 (1979), pp. 103–140.

[128] David S. Watkins. "On Hamiltonian and Symplectic Lanczos Processes". In: *Linear Algebra and its Applications* 385 (2004), pp. 23–45.

[129] T. Wiedemann. *Das HDR Verfahren zur Lösung des allgemeinen Eigenwertproblems Ax = λBx*. Diplomarbeit, Universität Bielefeld. (1989).

[130] Frank Wilczek. "The Dirac Equation". In: *International Journal of Modern Physics A* 19 (2004), pp. 45–74.

[131] Max Zorn. "A Remark on Method in transfinite Algebra". In: *Bulletin of the American Mathematical Society* 41 (1935), pp. 667–670.

Index